Metallic Glasses and Their Composites

D.V. Louzguine (Luzgin)

The formation of metallic glasses and dual-phase composite/hybrid materials is reviewed, as well as the glass transition process and the resulting structural phenomena. These materials exhibit high strength, extreme hardness, good wear resistance and large elastic deformation.

Due to their excellent structural, functional, magnetic, chemical and biological properties metallic glasses are suitable for a great number of applications, including in such areas as microelectromechanical devices, pressure sensors, biological implants and precision surgical instruments.

This book is suitable for materials scientists and students as well as for readers working in other disciplines requiring a higher level of understanding of the present subject.

Metallic Glasses and Their Composites

D.V. Louzguine (Luzgin)

WPI Advanced Institute for Materials Research,

Tohoku University, Japan

Published by **Materials Research Forum LLC**
Millersville, PA 17551, USA

Published as part of the book series
Materials Research Foundations
Volume 19 (2018)
ISSN 2471-8890 (Print)
ISSN 2471-8904 (Online)

Print ISBN 978-1-945291-42-5
ePDF ISBN 978-1-945291-43-2

Distributed worldwide by

Materials Research Forum LLC
105 Springdale Lane
Millersville, PA 17551
USA
http://www.mrforum.com

Manufactured in the United States of America
10 9 8 7 6 5 4 3 2 1

Table of Contents

Preface

The formation process of both conventional metallic glasses and bulk metallic glasses through melt solidification using the glass-transition phenomenon (either by kinetic freezing or by a kind of phase transition) is to this day still not fully understood. Two general mechanisms of glass formation related to stability of the supercooled liquid are presented here: the limiting of crystal nucleation and the limiting of their growth in relation to the formation of crystals and quasicrystals. Since the properties of metallic glasses, especially mechanical, are defined by their atomic structure, the research conducted on the structure of glassy alloys is emphasized as a particularly important topic in modern physical metallurgy. Due to the absence of a crystalline lattice and dislocations, a unique deformation mechanism can be observed in bulk glassy alloys, which allows them to exhibit high strength, extreme hardness, good wear resistance and large elastic deformation. While metastable, but long-lived at room temperature, glassy alloys devitrify/crystallize, upon heating. Devitrification term includes both formation of the supercooled liquid and crystallization upon heating. In some cases it is used to produce crystal/glassy composites with enhanced mechanical properties. Magnetic materials make up another extremely practically important and developing application field of metallic glassy alloys. Due to the excellent structural, functional, magnetic, chemical and biological properties metallic glasses are suitable for different applications, including sportive goods, cases, wire cords, elastic hinges, springs, diaphragms, membranes, knife blades, electromagnetic wave shields, optical mirrors, power inductors, Coriolis flow meters, biological implants, etc. Especially, metallic glasses are well applicable for high added value components where their relatively high material cost is not the limiting factor, such as, ferromagnetic cores of mini transformers, micro-geared motor parts in microelectromechanical devices, pressure sensors, orthopaedic screws, precision surgical instruments and so on. Future development of metallic glasses is likely going to be connected with dual-phase composite/hybrid materials utilizing attractive properties of both: metallic glassy phase and an additional one.

Chapter 1

Formation of metallic glasses and their general physical properties

Metallic glasses and bulk metallic glasses (BMGs), in particular, are usually produced by solidification of a melt while amorphous alloys can be produced by various other techniques including physical and chemical vapor deposition, electrodeposition, mechanical attrition, ion implantation, sintering and others. The formation mechanisms of bulk metallic glasses and their general physical properties are discussed in the present chapter.

Contents

1.1 Formation of metallic glasses

The first metallic glassy alloy (or metallic glass) [1], namely Au-Si, was produced by rapid solidification of the melt in 1960 [2]. Ternary Au-Si-Ge [3] and Au-Pb-Sb [4] alloys followed by other system alloys [5,6] developed later demonstrated a better glass-forming ability (GFA). In general, metallic glasses can be produced using various processing methods. Amorphous metallic materials were produced since the middle of the past Century by thin film deposition [7]. Recently this method has been extended and nanostructured metallic glasses were produced [8,9]. However, they and the alloys produced by a solid state reaction using mechanical attrition [10], such as ball milling [11] or by severe plastic deformation [12,13] and later compacted using thermal (hot pressing in a WC die [14]), microwave [15] or spark plasma sintering (SPS) (Fig. 1.1) [16] are usually called amorphous. Nevertheless, Ni-based bulk metallic glassy samples [17,18] with a size of 20 mm and nearly 100% relative density (Fig. 1.2) were fabricated by spark plasma sintering of the gas-atomized $Ni_{52.5}Nb_{10}Zr_{15}Ti_{15}Pt_{7.5}$ glassy powders (here and elsewhere the chemical composition is given in nominal at.%). As in some cases sintering is performed within the supercooled liquid their properties are similar to those formed by casting.

Fig. 1.1. Schematic representation of the spark plasma sintering technique.

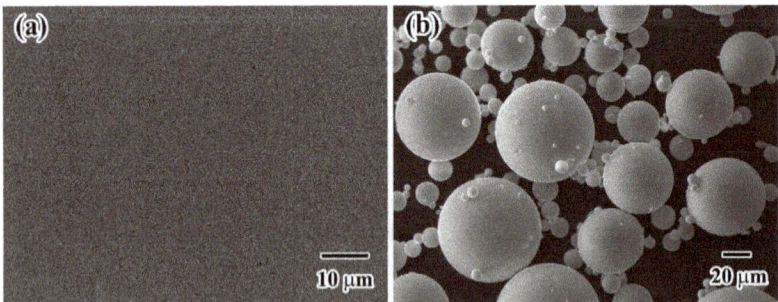

Fig. 1.2. SEM micrograph of the cross section of the sintered $Ni_{52.5}Nb_{10}Zr_{15}Ti_{15}Pt_{7.5}$ specimen obtained at a sintering temperature of 773 K (a). The micrograph of the original powder (b) is also shown for comparison. Reproduced from Ref. [17] with permission of American Institute of Physics.

The application of pulsed voltage induces various phenomena caused by electrical and thermal effects, providing advantages which could not be realized using conventional sintering processes. In the SPS process, pulse electric current directly flows through the powder material being sintered, and a high heating efficiency is achieved. These phenomena can be summarized as follows [17]: electrical breakdown of a surface oxide film and removal of the contaminated layer on the particle surface by spark generation and sputtering effect; destruction of the surface oxide film and necks formation between the powder particles; Joule heating at the neck, and enhanced migration of an atom by the temperature difference between the neck and the particle core. Therefore, the sintering can be carried out in a shorter time than during conventional processes. Thus, the SPS process can be applied to materials that require suppressing crystallization and grain growth. Al-based bulk glassy samples of high relative density were obtained by warm extrusion of the atomized amorphous powders [19] because they have rather low glass-forming ability owing to a high density of so-called quenched-in nuclei in some glasses. Electrodeposition from a solution can also be used [20,21].

Much more productive is solidification from a liquid phase by Cu-mold casting, for example, using a quartz crucible through a nozzle (Fig. 1.3 (a)) or water cooling technique (Fig. 1.4). The initial ingots of metallic glasses are usually prepared either by arc-melting or by induction melting mixtures of high purity elemental metals having more than 99.9 mass.% purity in an argon atmosphere (Fig. 1.3 (b)). Also, large scale 3D metallic glassy samples of complex geometry were recently produced by the Laser Foil Printing technology either by foil-welding by a continuous wave fiber laser or by foil-cutting by a pulsed laser [22].

After general saturation of the research activities in the field of conventional crystalline alloys since the end of the past century [23] bulk metallic glasses and glassy-crystal dual phase materials [24,25], owing to their good mechanical, chemical, magnetic and other properties, are at the research frontier among metallic materials [26]. These materials having a non-crystalline structure [27] exhibit high strength [28], wear [29] and corrosion resistance [30]. Bulk glassy alloys having an extraordinary high GFA [4-6,31] among metallic alloys have been widely produced after the breakthrough achieved in the end of 1980s and beginning of 1990s [32,33]. They can be produced at the cooling rates of the order of 100, 10, 1 K/s and even less, which is much lower than those of 10^4-10^6 K/s required for vitrification of marginal glass-formers by using a rapid solidification technique.

Fig. 1.3. Bulk glassy sample casting technique, scheme (a) and a photograph of an ingot prepared in an arc-melter (b).

Fig. 1.4. Water cooling process of a BMG glass-forming melt encapsulated in a quartz tube under an Ar atmosphere.

The photographs of the cast BMG alloy samples produced by using Ar gas pressure casting (samples up to 10 mm size) and gravity casting (from 5 to 25 mm size) are shown in Fig. 1.5. Larger size samples are usually produced by water cooling in quartz or steel tubes.

(a)

(b)

Fig. 1.5. Two photographs of the cast BMG alloy samples produced by the Ar gas pressure casting (a) and the gravity casting (b) techniques.

Bulk metallic glasses of 1-2 mm diameter were first produced in the noble metals-based metallic systems [34]. A large number of bulk metallic glasses also called bulk metallic glassy alloys defined as 3-dimentional massive glassy articles with a size of not less than 1 mm (by another definition not less than 10 mm) in every spatial dimension have been produced during the last 20 years in the thickness range of 10^0-10^2 mm by using various casting processes [35,36,37,38,39].

Bulk metallic glasses can be roughly classified into the metal-metal group glasses (the majority of BMGs like Zr-Cu-Al, for example) and metal-metalloid glasses (Fe-B-C-P with other elements, Pd-Ni-Cu-P, etc…). Although well-known noble-metals- [34] and lanthanide-metals- [40] based bulk metallic glasses (BMGs) are not particularly attractive for engineering applications due to their high cost and low oxidation resistance, respectively, Ti [41], Zr [42,43] and Mg-based [44] bulk metallic glasses discovered later on are quite promising engineering materials. BMG alloys with exceptionally large critical size were obtained in Ca- [45,46], Cu-[47], Co-[48,49], Fe-[50], La-[51], Mg-[52], Nd-[53], Ni-[54], Pd- [55], Pt- [56], Ti-Zr- [57] and Zr-based [36] alloy systems. Large size ingots of Ti-Zr-Ni-Be and Zr-Ti-Ni-Be BMGs are produced because of Be addition [58]. The addition of Be suppresses crystallization of Zr and Ti alloys because it forms deep eutectic and there are no stable intermetallic compounds in the Zr, Ti-rich area. Moreover, Be-rich intermetallic compounds are mostly incongruently melting, and thus, difficult to form from the melt as well as in case of Mg-Y system [59].

Although Al-based metallic glasses are rather marginal glass-formers the addition of Co [60] improved the glass-forming ability of the Al-RE (RE-rare earth metal) [61] and ternary Al-RE-TM (TM-transition metals) [62] alloys (critical ingot size reached 0.5 mm and even larger) and increased their mechanical strength. Mischmetal(Mm) can also be added to substitute RE metals in Al-[63], Mg-[64,65] and Ge-based [66] alloys thus reducing the total cost. Quaternary $Al_{85}Y_8Ni_5Co_2$ alloy shows one of the widest supercooled liquid region on heating among Al-based metallic glasses (all alloy compositions are given in nominal atomic percents (at.%)) [60]. The GFA of this alloy was further improved by the addition of Ca [67]. Based on these principles Al-based BMGs with a critical size of 1 mm were also produced recently [68,69].

Bulk metallic glass -metallic crystal [70] -ceramics [71,72], -polymer [73,74] composites were also produced by various methods including casting (both in-situ and ex-situ composites) and sintering (ex-situ composites). A secondary phase in the glassy matrix, for example the crystalline one acts as active barriers for shear band propagation, enabling multiple shear deformation and preventing sudden brittle rupture. For example, in-situ $Cu_{50}Zr_{30}Ti_{10}Nb_{10}$ bulk composite alloy with micrometer-size crystals of $Cu_{10}Zr_7$ phase embedded in the glassy matrix formed by copper mold casting showed a high ultimate strength of 2000 MPa, a yield strength of 1770 MPa and enhanced ductility with 2.3 % plastic deformation [75].

The ex-situ $Mg_{58}Cu_{28.5}Gd_{11}Ag_{2.5}$ bulk metallic glass composites dispersion strengthened by spherical Ti particles pre-mixed in the melt with different volume fractions and particle sizes were synthesized by injection casting. The Ti powder particles absorb the deformation energy and branch the primary shear bands into multiple shear bands. They decrease the stress concentration in front of a shear band and enhance plasticity. A high-quality interface was found between Ti particles and the amorphous matrix. For a given volume fraction of Ti of 40% it is found that smaller particles improve the compression plasticity from 12 % plastic strain (at the particle size of 90 μm) up to 25 % strain (at the size of 50 μm). Higher volume fraction of Ti leads to even larger compression plasticity [76]. This subject will also be discussed in Chapter 5.

1.2 General physical properties

From Fig. 1.4 one can admit good surface quality (shiny surfaces) of bulk metallic glasses suitable for near-net-shape processing. Their mass density is somewhat lower than that in the crystalline state. The specific heat capacity (C_p) of metallic glasses at room temperature is close to 3R but slightly higher than that the competing crystalline phase(s). The C_p of a metallic glass measured by using differential scanning calorimetry increases gradually with increasing temperature. In the glass-transition region the slope of the plot changes quickly and the specific heat reaches a maximum in the supercooled liquid state (Fig. 1.6). The specific heat remains virtually constant in the supercooled liquid state, and decreases to a low value (similar to that of the glassy state) on crystallization. Visible decrease in C_p starting at about 480 K is connected with the process of structural relaxation. The glass transition and structural relaxation processes will be discussed in the subsequent Chapters.

Fig. 1.6. C_p of the $Cu_{44}Ag_{15}Zr_{36}Ti_5$ metallic glass and liquid phases measured on heating at 0.33 K/s. $C_{p,g}$ and $C_{p,l}$ are C_p of the glassy and liquid state, respectively.

Owing to their non-crystalline structure, bulk metallic glasses exhibit an electrical resistivity (1-3 $\mu\Omega \cdot m$), which is higher than that of the corresponding crystalline alloys. Some intermetallic compounds, like FeAl or Fe_3Al also have high resistivity [77] but, in general, lower than that of metallic glasses.

The thermal conductivity and diffusivity of BMGs and supercooled liquids of bulk metallic glass formers has been measured together with electrical conductivity and the results are shown in Table 1.1 [78,79,80]. The thermal diffusivity, thermal and electrical conductivity of the metallic glasses and amorphous solid were weakly temperature-dependent, with slightly positive temperature coefficients.

Table 1.1. Room temperature electrical conductivity, thermal diffusivity and thermal conductivity of bulk metallic glasses (at 300 K).

Alloy	Thermal diffusivity (mm^2/s)	Thermal conductivity (W/m·K)	Electrical conductivity ($\mu\Omega \cdot m$)$^-$ [1]	Ref.
$Pd_{40}Ni_{10}Cu_{30}P_{20}$	1.62	5.11	0.541	78
$Zr_{55}Al_{10}Ni_5Cu_{30}$	2.2	5.02	0.53	79

Many metallic glasses exhibit moderately negative temperature dependence of the electrical resistivity (Fig. 1.7) though still exhibit metallic type bonding. However, it was shown that the temperature coefficient of resistivity in the $Pd_{40}Ni_{40-x}Cu_xP_{20}$ glassy alloys changes from negative to positive with increase in Cu content [81]. A similar change from the negative to the positive value was reported to occur as the P content is reduced in the Pd-Ni-P alloy system [82]. The Ziman theory was found to explain the negative temperature dependence of the electrical resistivity of some liquid and amorphous alloys [83]. A negative thermal coefficient of resistivity is found to occur if the Fermi sphere diameter ($2k_F$) to the wave number corresponding to the first peak K_p is close to unity, otherwise it is found to be positive. The electronic structure and properties of other metallic glasses were also studied [84].

Fig. 1.7. Inverse temperature dependence of the logarithmic electrical conductivity of the Si-Al-TM (TM-transition metals) alloys.

Atomic diffusion in metallic glasses and supercooled liquids has been studied for many years. The diffusion measurements of Zr-Ti-Cu-Ni-Be BMGs demonstrate that two distinct processes contribute to long-range transport in the supercooled liquid state: single-atom hopping and collective motion which is the dominant process. A change in the slope of diffusion coefficient at T_g is connected with a transition to the liquid state [85]. However, it was shown that even on heating the atomic transport mechanism of the $Pd_{40}Cu_{30}Ni_{10}P_{20}$ metallic glass is thermally activated hopping in a wide temperature range above the glass transition temperature measured by a calorimetric method [86]. Also, unlike diffusion in crystals, however, hopping in the metallic glassy alloys was suggested to be a highly collective process, involving many atoms.

Equilibrium viscosity of the melts increases several orders in magnitude on cooling and reaches 10^{12} Pa·s at T_g on cooling. It suggested that the glass-forming ability parameters/indicators of the metallic glasses can be deduced from their viscosity [87]. Nonequilibrium viscosity of metallic glasses has been measured [88] but it is significantly strain rate dependent.

9

References

[1]   A.L. Greer, New horizons for glass formation and stability, Nat. Mater. 14 (2015) 542. https://doi.org/10.1038/nmat4292

[2]   W. Klement, R.H. Willens and P. Duwez, Non-crystalline structure in solidified gold–silicon Alloys, Nature 187 (1960) 869. https://doi.org/10.1038/187869b0

[3]   H.S. Chen and D. Turnbull, Thermal evidence of a glass transition in gold-silicon alloy, Applied Physics Letters 10 (1967) 284. https://doi.org/10.1063/1.1754813

[4]   H.J. Fecht, J.H. Perepezko, M.C. Lee and W.L. Johnson, Thermodynamic properties and crystallization kinetics of glass - forming undercooled liquid Au - Pb - Sb alloys, Journal of Applied Physics, 68 (1990) 4494. https://doi.org/10.1063/1.346200

[5]   A. Inoue, K. Ohtera, K. Kita and T. Masumoto, New amorphous Mg-Ce-Ni alloys with high strength and good ductility, Japan. J. Appl. Phys. 27 (1988) L2248. https://doi.org/10.1143/JJAP.27.L2248

[6]   V.V. Molokanov and V.N. Chebotnikov, Glass forming ability, structure and properties of Ti and Zr-intermetallic compound based alloys, Key Engineering Materials, 40-41 (1990) 319. https://doi.org/10.4028/www.scientific.net/KEM.40-41.319

[7]   W. Buckel and R. Hilsch, Einfluß der Kondensation bei tiefen temperaturen auf den elektrischen Widerstand und die supraleitung für verschiedene Metalle, Z. Physik, 138 (1954) 109. https://doi.org/10.1007/BF01337903

[8]   N. Chen, D.V. Louzguine-Luzgin and K.F. Yao, A new class of non-crystalline materials: Nanogranular metallic glasses, Journal of Alloys and Compounds 707 (2017) 371. https://doi.org/10.1016/j.jallcom.2016.11.304

[9]   S.V. Ketov, R. Joksimovic, G. Xie, A. Trifonov, K. Kurihara and D.V. Louzguine-Luzgin, Formation of nanostructured metallic glass thin films upon sputtering, Heliyon, 3 (2017) e00228. https://doi.org/10.1016/j.heliyon.2016.e00228

[10]   H.J. Fecht, Nanostructure formation by mechanical attrition, Nanostructured Materials 6 (1995) 33. https://doi.org/10.1016/0965-9773(95)00027-5

[11]   A.W. Weeber and H. Bakker, Amorphization by ball milling. A review, Physica B 153 (1988) 93. https://doi.org/10.1016/0921-4526(88)90038-5

[12]   R.Z. Valiev, Structure and mechanical properties of ultrafine-grained metals, Materials Science and Engineering A, 234-236 (1997) 59. https://doi.org/10.1016/S0921-5093(97)00183-4

[13]   V.M. Segal, Materials processing by simple shear, Mater. Sci. Eng A, 197 (1995) 157. https://doi.org/10.1016/0921-5093(95)09705-8

[14]   A.Inoue and H.M. Kimura, High strength Al- and Mg-based alloys with nanocrystalline or nanoquasicrystalline phase, J. Metast. Nanocr. Mater. 9 (2001) 41. https://doi.org/10.4028/www.scientific.net/JMNM.9.41

[15]   S. Li, G.Q. Xie, D.V. Louzguine-Luzgin, M. Sato and A. Inoue, Microwave-induced sintering of Cu-based metallic glass matrix composites in a single-mode 915-MHz applicator, Metallurgical and Materials Transactions A, 42A (2011) 1463. https://doi.org/10.1007/s11661-010-0346-8

[16]   G. Xie, W. Zhang, D.V. Louzguine-Luzgin, H.M. Kimura and A. Inoue, Fabrication of porous Zr-Cu-Al-Ni bulk metallic glass by spark plasma sintering process, Scripta Mater 55 (2006) 687. https://doi.org/10.1016/j.scriptamat.2006.06.034

[17]   G. Xie, D.V. Louzguine-Luzgin, H.M. Kimura and A. Inoue, Nearly full density $Ni_{52.5}Nb_{10}Zr_{15}Ti_{15}Pt_{7.5}$ bulk metallic glass obtained by spark plasma sintering of gas atomized powders, Applied Physics Letters, 90 (2007) 241902. https://doi.org/10.1063/1.2748102

[18]   J.K. Lee, H.J. Kim, T.S. Kim, S.Y. Shin, Y.C. Kim and J.C. Bae, Deformation behavior of Ni-based bulk metallic glass synthesized by spark plasma sintering, Journal of Materials Processing Technology, 187 (2007) 801. https://doi.org/10.1016/j.jmatprotec.2006.11.206

[19]   A. Inoue and H.M. Kimura, Fabrications and mechanical properties of bulk amorphous, nanocrystalline, nanoquasicrystalline alloys in aluminum-based system, J. Light Met. 1 (2001) 31. https://doi.org/10.1016/S1471-5317(00)00004-3

[20]   T. Yamasaki, P. Schlossmacher, K. Ehrlich and Y. Ogino, Formation of amorphous electrodeposited Ni-W alloys and their nanocrystallization, Nanostructured Materials 10 (1998) 375. https://doi.org/10.1016/S0965-9773(98)00078-6

[21]   J. Ahmad, K. Asami, A. Takeuchi, D.V. Louzguine and A. Inoue, High strength Ni-Fe-W and Ni-Fe-W-P alloys produced by electrodeposition, Materials Transactions, 44 (2003) 1942. https://doi.org/10.2320/matertrans.44.1942

[22] Y. Shen, Y. Li, C. Chen and H.L. Tsai, 3D printing of large, complex metallic glass structures, Materials and Design, 117 (2017) 213. https://doi.org/10.1016/j.matdes.2016.12.087

[23] K. Lu, The future of metals, Science, 328 (2010) 319. https://doi.org/10.1126/science.1185866

[24] D.V. Louzguine-Luzgin, A. Vinogradov, S. Li, A. Kawashima, G. Xie, A.R. Yavari and A. Inoue, Deformation and fracture behavior of metallic glassy alloys and glassy-crystal composites, Metallurgical and Materials Transactions A, 42A (2011) 1504. https://doi.org/10.1007/s11661-010-0391-3

[25] A.M. Glezer, N.A. Shurygina, S.G. Zaichenko and I.E. Permyakova, Interaction of deformation shear bands with nanoparticles in amorphous-nanocrystalline alloys, Russian Metallurgy (Metally), 4 (2013) 235. https://doi.org/10.1134/S0036029513040022

[26] D.V. Louzguine-Luzgin and A. Inoue, Bulk metallic glasses. formation, structure, properties, and applications, Handbook of Magnetic Materials, Edited by K.H.J. Buschow, Elsevier 21 (2013) 131.

[27] D.V. Louzguine-Luzgin, A.I. Bazlov, A.Yu. Churyumov, K. Georgarakis and A.R. Yavari, Comparative analysis of the structure of palladium-based bulk metallic glasses prepared by treatment of melts with flux, Physics of the Solid State, 55 (2013) 1985. https://doi.org/10.1134/S1063783413100211

[28] D.V. Louzguine-Luzgin, L.V. Louzguina-Luzgina and A.Yu. Churyumov, Mechanical properties and deformation behavior of bulk metallic glasses, Metals, 3 (2013) 1. https://doi.org/10.3390/met3010001

[29] D.V. Louzguine-Luzgin, H.K. Nguyen, K. Nakajima, S.V. Ketov and A.S. Trifonov, A study of the nanoscale and atomic-scale wear resistance of metallic glasses, Materials Letters, 185 (2016) 54. https://doi.org/10.1016/j.matlet.2016.08.035

[30] S. Pang, T. Zhang, K. Asami and A. Inoue, Bulk glassy Ni(Co–)Nb–Ti–Zr alloys with high corrosion resistance and high strength, Materials Science and Engineering A, 375 (2004) 368. https://doi.org/10.1016/j.msea.2003.10.152

[31] A. Inoue and A. Takeuchi, Recent progress in bulk glassy alloys, Materials Trans. 43 (2002) 1892. https://doi.org/10.2320/matertrans.43.1892

[32] A. Inoue, T. Zhang and T. Masumoto, Al-La-Ni amorphous alloys with a wide supercooled liquid region, Mater. Trans. Japan. Inst. Metals 30 (1989) 965.

[33] T. Zhang, A. Inoue and T. Masumoto, Amorphous Zr-Al-TM (TM=Co, Ni, Cu) alloys with significant supercooled liquid region of over 100 K, Mater. Trans., JIM 32 (1991) 1005.

[34] H.S. Chen, Thermodynamic considerations on the formation and stability of metallic glasses, Acta Metall. 22 (1974) 1505. https://doi.org/10.1016/0001-6160(74)90112-6

[35] H.W. Kui, A.L. Greer and D. Turnbull, Formation of bulk metallic glass by fluxing, Appl. Phys. Lett. 45 (1984) 615. https://doi.org/10.1063/1.95330

[36] A. Inoue, High strength bulk amorphous alloys with low critical cooling rates, Mater. Trans. JIM 36 (1995) 866. https://doi.org/10.2320/matertrans1989.36.866

[37] W.L. Johnson, Bulk glass-forming metallic alloys: science and technology, MRS Bull. 24 (1999) 42. https://doi.org/10.1557/S0883769400053252

[38] A. Inoue, Stabilization of metallic supercooled liquid and bulk amorphous alloys, Acta Mater. 48 (2000) 279. https://doi.org/10.1016/S1359-6454(99)00300-6

[39] C. Suryanarayana and A. Inoue, Bulk Metallic Glasses, CRC Press, Boca Raton, FL (2010) 12.

[40] W. Zhang, F. Jia and A. Inoue, Formation and properties of new La-based bulk glassy alloys with diameters up to centimeter order, Materials Transactions, 48 (2007) 68. https://doi.org/10.2320/matertrans.48.68

[41] T. Zhang and A. Inoue, Thermal and mechanical properties of Ti-Ni-Cu-Sn amorphous alloys with a wide supercooled liquid region before crystallization, Materials Transactions, JIM, 39 (1998) 1001.

[42] K. Jin and J.F. Löffler, Bulk metallic glass formation in Zr–Cu–Fe–Al alloys, Appl. Phys. Lett., 86 (2005) 1. https://doi.org/10.1063/1.1948513

[43] Q.S. Zhang, W. Zhang, G.Q. Xie, D.V. Louzguine-Luzgin and A. Inoue, Stable flowing of localized shear bands in soft bulk metallic glasses, Acta Materialia, 58 (2010) 904. https://doi.org/10.1016/j.actamat.2009.10.005

[44] A. Inoue, A. Kato, T. Zhang, S. Kim and T. Masumoto, Mg-Cu-Y amorphous alloys with high mechanical strengths produced by metallic mold casting method, Mater. Transactions JIM, 32 (1991) 609.

[45] E.S. Park and D.H. Kim, Formation of Ca–Mg–Zn bulk glassy alloy by casting into cone-shaped copper mold, J. Mater. Res., 19 (2004) 685. https://doi.org/10.1557/jmr.2004.19.3.685

[46] O. Senkov, D. Miracle, V. Keppens et al., Development and characterization of low-density Ca-based bulk metallic glasses. An overview, Metall and Mater. Trans. A 39 (2008) 1888. https://doi.org/10.1007/s11661-007-9334-z

[47] D.H. Xu, G. Duan and W.L. Johnson, Unusual glass-forming ability of bulk amorphous alloys based on ordinary metal copper, Phys. Rev. Lett., 92 (2004) 245504. https://doi.org/10.1103/PhysRevLett.92.245504

[48] H. Men, S.J. Pang and T. Zhang, Effect of Er doping on glass-forming ability of $Co_{50}Cr_{15}Mo_{14}C_{15}B_6$ alloy, J. Mater. Res., 21 (2006) 958. https://doi.org/10.1557/jmr.2006.0109

[49] T. Zhang, Q. Yang, J.I. Yun, L.I. Ran, S.J. Pang, J.F. Wang and T. Xu, Centimeter-scale-diameter Co-based bulk metallic glasses with fracture strength exceeding 5000 MPa, Chinese Science Bulletin, 56 (2011) 3972. https://doi.org/10.1007/s11434-011-4765-8

[50] K. Amiya and A. Inoue, Fe-(Cr, Mo)-(C, B)-Tm bulk metallic glasses with high strength and high glass-forming ability, Rev. Adv. Mater. Sci. 18 (2008) 27.

[51] Q.K. Jiang, G.Q. Zhang, L. Yang, X.D. Wang, K. Saksl, H. Franz, R. Wunderlich, H. Fecht, and J.Z. Jiang, La-based bulk metallic glasses with critical diameter up to 30 mm, Acta Mater., 55 (2007) 4409. https://doi.org/10.1016/j.actamat.2007.04.021

[52] K. Amiya and A. Inoue, Preparation of bulk glassy $Mg_{65}Y_{10}Cu_{15}Ag_5Pd_5$ alloy of 12 mm in diameter by water quenching, Materials Transactions, 42 (2001) 543. https://doi.org/10.2320/matertrans.42.543

[53] A. Inoue and T. Zhang, Thermal stability and glass-forming ability of amorphous Nd-Al-TM (TM = Fe, Co, Ni or Cu) alloys, Mater. Sci. Eng. A, 226–228 (1997) 393. https://doi.org/10.1016/S0921-5093(97)80050-0

[54] Y. Zeng, N. Nishiyama, T. Yamamoto and A. Inoue, Ni-rich bulk metallic glasses with high glass-forming ability and good metallic properties, Mater. Trans. JIM, 50 (2009) 2441. https://doi.org/10.2320/matertrans.MRA2008453

[55] N. Nishiyama, K. Takenaka, H. Miura, N. Saidoh, Y. Zeng and A. Inoue, The world's biggest glassy alloy ever made, Intermetallics, 30 (2012) 19. https://doi.org/10.1016/j.intermet.2012.03.020

[56] J. Schroers and W.L. Johnson, Gold based bulk metallic glass, Appl. Phys. Lett., 84 (2004) 3666. https://doi.org/10.1063/1.1738945

[57]   M.Q. Tang, H.F. Zhang, Z.W. Zhu, H.M. Fu, A.M. Wang, H. Li and Z.Q. Hu, TiZr-base bulk metallic glass with over 50 mm in diameter, J. Mater. Sci. Technol., 26 (2010) 481. https://doi.org/10.1016/S1005-0302(10)60077-1

[58]   A. Peker and W.L. Johnson, A highly processable metallic glass: $Zr_{41.2}Ti_{13.8}Cu_{12.5}Ni_{10.0}Be_{22.5}$, Appl. Phys. Lett. 63 (1993) 2342. https://doi.org/10.1063/1.110520

[59]   W.F. Gale and T.C. Totemeier, editors. Smithells Metals Reference Book 8-th Edition, Elsevier Butterworth-Heinemann Ltd., Oxford UK (2004) 11.

[60]   A. Inoue, N. Matsumoto and T. Masumoto, Al–Ni–Y–Co amorphous alloys with high mechanical strengths, wide supercooled liquid region and large glass-forming capacity, Mater. Trans. JIM, 31 (1990) 493. https://doi.org/10.2320/matertrans1989.31.493

[61]   A. Inoue, K. Ohtera, A.P. Tsai and T. Masumoto, Glass transition behavior of Al-Y-Ni and Al-Ce-Ni amorphous alloys, Jpn. J. Appl. Phys., 27 (1988) 736. https://doi.org/10.1143/JJAP.27.L736

[62]   A. Inoue, K. Ohtera, A.P. Tsai and T. Masumoto, Aluminum-based amorphous alloys with tensile strength above 980 MPa (100 kg/mm^2), Jpn. J. Appl. Phys., 27 (1988) 280. https://doi.org/10.1143/JJAP.27.L280

[63]   D.V. Louzguine and A. Inoue, Proceedings of the ISMANAM 2001 International Symposium on Metastable, Mechanically Alloyed and Nanocrystalline Materials, June 24-29, 2001, Ann Arbor Michigan, USA, Materials Science Forum, 386-388 (2002) 117.

[64]   D.V. Louzguine, Y. Kawamura and A. Inoue, Proceedings of Nagaoka Workshop "Magnesium Platform Science and Engineering 2000" July 27-29, 2000, Nagaoka, Japan. Materials Science Forum, Trans Tech. Publ. Switzerland, 350-351 (2000) 123.

[65]   B.S. Murty and K. Hono, Formation of nanocrystalline particles in glassy matrix in melt-spun Mg-Cu-Y based alloys, Mater. Trans. JIM 41 (2000) 1538. https://doi.org/10.2320/matertrans1989.41.1538

[66]   D.V. Louzguine, A.R. Yavari and A. Inoue, Mischmetal as an alloying addition to amorphous materials and glass formers, Journal of Non-Crystalline Solids, 316 (2003) 255. https://doi.org/10.1016/S0022-3093(02)01602-2

[67]   J.Q. Wang, Y.H. Liu, S. Imhoff, N. Chen, D.V. Louzguine-Luzgin, A. Takeuchi, M.W. Chen, H. Kato, J.H. Perepezko and A. Inoue, Enhance the thermal stability

and glass forming ability of Al-based metallic glass by Ca minor-alloying, Intermetallics 29 (2012) 35. https://doi.org/10.1016/j.intermet.2012.04.009

[68] B.J. Yang, J.H. Yao, J. Zhang, H.W. Yang, J.Q. Wang and E. Ma, Al-rich bulk metallic glasses with plasticity and ultrahigh specific strength, Scripta Materialia 61 (2009) 423. https://doi.org/10.1016/j.scriptamat.2009.04.035

[69] J. Yin, H. Cai, X.W. Cheng and X.Q. Zhang, Al-based bulk metallic glass with large plasticity and ultrahigh strength, Journal of Alloys and Compounds, 648 (2015) 276. https://doi.org/10.1016/j.jallcom.2015.05.285

[70] S.V. Madge, P. Sharma, D.V. Louzguine-Luzgin, A.L. Greer and A. Inoue, New La-based glass–crystal ex situ composites with enhanced toughness, Scripta Materialia, 62 (2010) 210. https://doi.org/10.1016/j.scriptamat.2009.10.029

[71] G. Xie, D.V. Louzguine-Luzgin and A. Inoue, Characterization of interface between the particles in NiNbZrTiPt metallic glassy matrix composite containing SiC fabricated by spark plasma sintering, Journal of Alloys and Compounds, 483 (2009) 239. https://doi.org/10.1016/j.jallcom.2008.07.226

[72] G.Q. Xie, D.V. Louzguine-Luzgin, A. Okubo, H.M. Kimura and A. Inoue, Spark plasma sintering of Al_2O_3 particulate dispersed $Zr_{55}Cu_{30}Al_{10}Ni_5$ metallic glassy matrix composite, Ceramic Transactions, 198 (2007) 39.

[73] S. Li, D.V. Louzguine-Luzgin, G. Xie, M. Sato and A. Inoue, Development of novel metallic glass/polymer composite materials by microwave heating in a separated H-field, Materials Letters, 64 (2010) 235. https://doi.org/10.1016/j.matlet.2009.10.017

[74] M.Yu. Zadorozhnyy, D.I. Chukov, M.N. Churyukanova, M.V. Gorshenkov, V.Yu. Zadorozhnyy, A.A. Stepashkin, A.A. Tsarkov, D.V. Louzguine-Luzgin and S.D. Kaloshkin, Investigation of contact surfaces between polymer matrix and metallic glasses in composite materials based on high-density polyethylene, Materials and Design, 92 (2016) 306. https://doi.org/10.1016/j.matdes.2015.12.031

[75] D.V. Louzguine, H. Kato and A. Inoue, High-strength Cu-based crystal-glassy composite with enhanced ductility, Applied Physics Letters, 84 (2004) 1088. https://doi.org/10.1063/1.1647278

[76] J.S.C. Jang, J.B. Li, S.L. Lee, Y.S. Chang, S.R. Jian, J.C. Huang and T.G. Nieh, Prominent plasticity of Mg-based bulk metallic glass composites by ex-situ spherical Ti particles, Intermetallics, 30 (2012) 25. https://doi.org/10.1016/j.intermet.2012.03.038

[77] Y. Terada, K. Ohkubo, T. Mohri and T. Suzuki, Thermal conductivity of intermetallic compounds with metallic bonding, Materials Transactions, 43 (2002) 3167. https://doi.org/10.2320/matertrans.43.3167

[78] U. Harms, T.D. Shen, R.B. Schwarz, Thermal conductivity of $Pd_{40}Ni_{40-x}Cu_xP_{20}$ metallic glasses, Scripta Materialia 47 (2002) 411. https://doi.org/10.1016/S1359-6462(02)00160-4

[79] M. Yamasaki, S.Kagao and Y. Kawamura, Thermal diffusivity and conductivity of $Zr_{55}Al_{10}Ni_5Cu_{30}$ bulk metallic glass, Scripta Materialia 53 (2005) 63. https://doi.org/10.1016/j.scriptamat.2005.03.021

[80] M. Yamasaki, S. Kagao, Y. Kawamura and K. Yoshimura, Thermal diffusivity and conductivity of supercooled liquid in $Zr_{41}Ti_{14}Cu_{12}Ni_{10}Be_{23}$ metallic glass, Appl. Phys. Lett. 84 (2004) 4653. https://doi.org/10.1063/1.1759768

[81] H.M. Kimura, A. Inoue, N. Nishiyama, K. Sasamori, O. Haruyama and T. Masumoto, Thermal, mechanical, and physical properties of supercooled liquid in Pd–Cu–Ni–P amorphous alloy, Sci. Rep. RITU A43 (1997) 101.

[82] B.Y. Boucher, Influence of phosphorus on the electrical properties of Pd-Ni-P amorphous alloys, J. Non-Cryst. Solids 7 (1972) 277. https://doi.org/10.1016/0022-3093(72)90028-2

[83] J.M. Ziman, A theory of the electrical properties of liquid metals. I: The monovalent metals, Phil. Mag., 6 (1961) 1013. https://doi.org/10.1080/14786436108243361

[84] U. Mizutani, Electronic structure of metallic glasses, Progress in Materials Science, 28 (1983) 97. https://doi.org/10.1016/0079-6425(83)90001-4

[85] X.P. Tang, U. Geyer, R. Busch, W. L. Johnson and Y. Wu, Diffusion mechanisms in metallic supercooled liquids and glasses. Nature 402, (1999) 160. https://doi.org/10.1038/45996

[86] V. Zollmer, K. Ratzke, F. Faupel, A. Rehmet and U. Geyer, Evidence of diffusion via collective hopping in metallic supercooled liquids and glasses, Physical Review B, 65, (2002) 220201. https://doi.org/10.1103/PhysRevB.65.220201

[87] A. Takeuchi and A. Inoue, Calculations of dominant factors of glassforming ability for metallic glasses from viscosity, Mater. Sci. Eng. A 375–377 (2004) 449. https://doi.org/10.1016/j.msea.2003.10.199

[88] O.P. Bobrov, V.A. Khonik, S.A. Lyakhov, K. Csach, K. Kitagawa and H. Neuhäuser, Shear viscosity of bulk and ribbon glassy $Pd_{40}Cu_{30}Ni_{10}P_{20}$ well below and mear the glass transition, J. Appl. Phys. 100 (2006) 033518. https://doi.org/10.1063/1.2226984

Chapter 2

Glass-transition process

Although, the glass-transition phenomenon in metallic glasses has been studied extensively there are still considerable uncertainties in this field of materials science. There are some gaps in obtaining a clear picture of the glass transition. Two general mechanisms of glass formation by limiting crystal nucleation and by limiting their growth will be discussed in the present Chapter.

Contents

2.1   Crystal nucleation controlled mechanism

It is easy to cool a liquid below its liquidus temperature (T_l) owing to kinetic reasons related to difficulties in a crystal nucleation and growth at low temperature and if the crystal nucleation is suppressed then it will vitrify forming a glassy phase [1,2,3]. On the other hand, significant superheating of a solid above its melting temperature and a liquid above its boiling temperature is less likely because the diffusion processes in solids close to the melting temperature, and especially in liquids, are very fast providing no such kinetic limitations to the phase transformation, except for the thermodynamic energy barrier for nucleation of the high-temperature liquid or gaseous phase, respectively.

From the melt metallic glasses (glassy alloys) were first formed by using a rapid solidification technique at a very high solidification rate of 10^6 K/s [4]. For a long time Pd-Cu-Si and Pd-Ni-P were known to be the best metallic glass formers. The bulk glassy alloys possess three common features summarized by Inoue [5], i.e., belong to multicomponent systems (three or more components, in general), have significant atomic size ratios above 12% (but below some critical value) and exhibit negative heats of mixing among the constituent elements. All of them represent indispensable conditions leading to good glass-forming ability (GFA) [1] and relatively high thermal stability against crystallization [6]. The role of size ratio, strong chemical bonding, and valence-electron concentration was discussed in Ref. [7]. The first principle (the number of components) leads to the formation of dense packed structures, "deep" ternary and quaternary eutectics by elemental "confusion" on solidification.

Recently good correlation was found between the alloy system complexity (the number of alloying elements) and critical diameter/size of the glassy sample (D_c) [8]. The glass-forming ability is shown to increase with increasing number of the alloying elements. Statistical analysis shows that the difference in D_c among binary, ternary and quaternary alloys is meaningful. In ternary alloys the GFA is maximized in case of the $A_{70}B_{20}C_{10}$, $A_{65}B_{25}C_{10}$, $A_{65}B_{20}C_{15}$, $A_{56}B_{32}C_{12}$, $A_{55}B_{28}C_{17}$, $A_{44}B_{43}C_{13}$ and $A_{44}B_{38}C_{18}$ alloy compositions (A, B and C symbols represent the constituent elements). Quaternary and quinary bulk glass-forming alloys with a large critical diameter, in general, have a larger supercooled liquid region than those of ternary alloys, so that the addition of a fourth alloying element not only enhances GFA but also stabilizes the supercooled liquid region. For example, Fe-based BMGs [9] were obtained in various metal-metalloid multicomponent alloy systems including Fe–(Al, Ga)–(P, C, B, Si) [10], Fe-Cr-Mo-Nb-C-B, Fe-Cr-Mo-Ta-C-B [11], Fe–Cr–Mo–P–C–B [12], Fe–Ni–P–B [13] and other systems. B, P and C are well-known additions stimulating formation of the glassy phase in Fe-based alloys. Mo and Cr are the effective additions slowing down diffusive phase transformations in steels [14]. Multicomponent equiatomic bulk glassy alloys without a major component, such as $Ti_{20}Zr_{20}Hf_{20}Cu_{20}Ni_{20}$ [15], $Pd_{20}Pt_{20}Cu_{20}Ni_{20}P_{20}$ [16] and other system alloys [17] were also produced by analogy with multiprincipal element alloys also called high-entropy alloys. Bulk glassy alloys were reported to be obtained not only in the ternary but even in some binary systems like Ni-Nb, Ca-Al, Pd-Si, Cu-(Zr or Hf) [18,19,20]. However, addition of the third element significantly enhances the glass-forming ability [21].

Role of the second principle (atomic size ratio) has been rationalized by Egami and Waseda [22,23] who investigated the minimum atomic concentration necessary to destabilize terminal solid solution phases. Later, Miracle and Senkov developed the topological criterion for glass formation [24]. However, the concept of atomic size has some deficiencies. In metallic glasses all bounds are considered to be non-directional, though the existence of some degree of covalent or even partly ionic bonding in the intermetallic compounds (and thus in metallic glasses) have been discussed for years [25,26]. The atomic size of an element derived from the bond length with different elements varies owing to different bond character, charge transfer, etc... Significant bond shortening has been observed in case of Al-Fe-Ce [27] and Au-Al alloys [28].

The third principle (large negative mixing enthalpy) is responsible for the formation of eutectics and dense packing of atoms of different kind. The importance of an efficient atomic packing for the formation of metallic glasses was shown in a number of works [29,30]. One should note that even, pure metals, for example, Ni, Fe can be made amorphous at a high enough cooling rate estimated at 10^8-10^9 K/s [31,32] which are, however, not stable at room temperature unless separated into nanometer scale spheres [33].

As has already been mentioned alloys having a low GFA can be prepared in an amorphous state by condensation from a vapor phase [34] or by electrodeposition in a solution [35] without using the molten state. Some alloys can be produced by a solid state reaction using mechanical attrition [36], in particular ball milling and later compacted making a bulk sample [37] or by severe plastic deformation [38].

The specific volume (V) of a liquid phase decreases faster (while its density (ρ) increases faster) with temperature than that of a competing crystal. According to several studies, the linearity of $d\rho/dT$ slope maintains not only in the equilibrium liquid but in the supercooled liquid region as well [39,40]. Fig. 2.1 shows a typical specific volume versus temperature (T) diagram for a pure substance. The glass-transition region is related to the area of change in the volumetric thermal expansion coefficient from a relatively high value characteristic of a liquid phase to a lower value for a solid glassy phase as shown in Fig. 2.1. Such a change of the thermal expansion coefficient on glass-transition was clearly demonstrated in many works, see for example, Ref. [41].

By analogy with the well-known Kauzmann's entropy crisis [42] illustrated in Fig. 2.2 [43] one can suggest that a liquid metal should not have a lower volume (or a higher density), at a given temperature, than its crystalline counterpart provided that it contracts upon solidification and there are no changes in the chemical bond character. Face-centered cubic (FCC) and hexagonal close packed (HCP) lattices are the densest packing structures for crystalline pure metals, and thus, a liquid or a glassy monoatomic metallic glassy phase should not become denser than them.

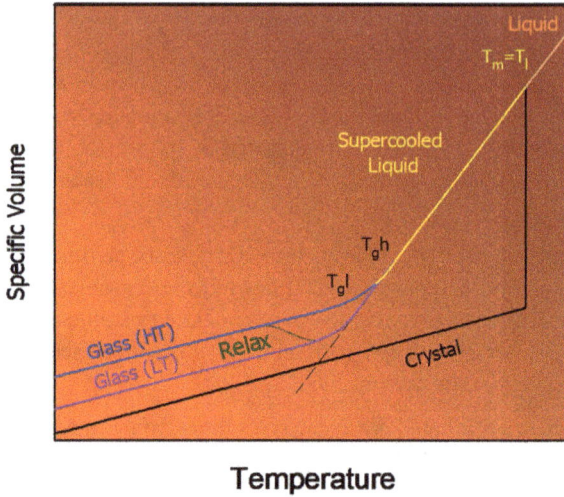

Fig. 2.1. Schematic specific volume versus temperature diagram for a pure substance without the solidification temperature interval. T_g^l and T_g^h illustrate T_g obtained at low and high cooling rates, respectively. The behavior is opposite to the first order phase transitions during crystallization when faster cooling induces higher supercooling.

Fig. 2.2. C_p of the $Au_{49}Cu_{26.9}Ag_{5.5}Pd_{2.3}Si_{16.3}$ alloy in liquid, glassy and crystalline state as a function of temperature measured in the step scan mode. The Kauzmann temperature is 340 K. Reproduced from [43] with permission of Elsevier.

The glass-transition phenomenon is also characterized by the specific heat capacity (C_p) change as one can observe in Fig. 2.2. The glass-transition phenomenon in the $Au_{49}Cu_{26.9}Ag_{5.5}Pd_{2.3}Si_{16.3}$ metallic glass studied using the step-scan calorimetry method on heating showed the existence of two different slopes within the glass-transition region indicating two glass-transition processes one starting at a low temperature around 340 K and the other at a high temperature around 380 K. This phenomenon is likely related to the different diffusion coefficients of the alloying elements in this alloy.

An arbitrary glass-transition (T_g) point is defined as a temperature at which viscosity of a liquid reaches 10^{12} Pa·s on cooling, while the transition takes place in a temperature interval [44]. For some materials this value belongs to the glass-transition region defined by C_p measurement, but it is not the case for other materials [45]. T_g is high for Ferrous-metals based BMGs but for the Ca-Li-based BMGs it is close to room temperature as low as 308 K for the $Ca_{65}Li_{14.54}Mg_{12.46}Zn_8$ one [46]. It is also found that the compressibility of glass-forming liquids and the difference in compressibility between the liquid and glassy phases decreases towards T_g [47].

The glass-transition has similarities with a first-order type diffusionless phase transformation which takes place within a temperature interval provided that not all but a certain volume of the matrix phase is transformed at a definite temperature between the beginning (T_{bg}) and finish (T_{fg}) glass-transition temperatures. The diffusionless character of this transformation implies that the atoms are shifted to the distances smaller than the interatomic distances [48]. One can suppose that the supercooled liquid regions are gradually transformed to the glassy regions upon cooling. The diffusionless transformation can be a natural source of the excess volume and stresses [49].

The reduced glass-transition temperature $T_{rg}=T_g(\beta)/T_l(\beta)$ (T_g glass-transition temperature, T_l liquidus temperature, both can be dependent of cooling or heating rate (β) but usually T_l is extrapolated to the infinitely slow cooling) [50] is one of the most widely known physical-mathematical parameters indicating GFA, derived assuming homogeneous nucleation. Upon crystallization homogeneous nucleation rate depends on the activation energy for transferring atoms across the surface (Q_N) of the nucleus and the nucleation barrier (ΔG^*) the free energy required to form a nucleus of critical size while growth rate depends upon the activation energy for growth (Q_g) and the thermodynamic driving force ΔG. Data in the literature often contain non-equilibrium T_l and non Kauzmann T_g values determined at different heating rates. However, the equilibrium T_l values obtained from the corresponding phase diagrams should be used. One should also mention that T_g increases with increasing cooling rate during solidification from the liquid state [51] and alloys solidified at different cooling rate have different fictive T_g. One should also note that the glass-transition does not occur at a discrete temperature and T_g is defined as an inflection point either on heating (Fig. 2.3) upon devitrification or on cooling during vitrification (see Fig. 2.1).

Fig. 2.3. Typical DSC (a) and differential isothermal calorimetry (b) traces of metallic glasses. T_{br}-temperature indicates the beginning of structural relaxation. An incubation period of about 650 s can be clearly seen in (b).

It has been shown that the width of the supercooled liquid region (ΔT_x) on heating of the glassy sample defined as $T_x(\beta)-T_g(\beta)$ where $T_x(\beta)$ is the onset crystallization temperature as an indicator of the stability of the supercooled liquid against devitrification also correlates quite well with the glass-forming ability [52]. However, the crystallization temperature $T_x(\beta)$ is not purely intrinsic as it depends not only on the heating/cooling rate (β) but also on the nucleation reaction type, the as-solidified structure of the glassy alloy, its purity, existence of heterogeneous nucleation sites [53], and thus, cannot be regarded as purely intrinsic parameter. The same speculation is applied to the γ parameter $\gamma = T_x/(T_g+T_l)$ [54], which takes into account both T_{rg} and ΔT_x criteria as well as the γ_m parameter [55]. Other parameters which take into account both T_{rg} and the fragility index m of the supercooled liquid were introduced [56,57]. However, the T_{rg} concept works well when the alloys of similar fragility are treated as shown in Table 2.1 [58] and plotted in Fig. 2.4. The critical diameter is used for a cylindrical sample as an indicator of the GFA but for a plate-shaped sample it is its thickness.

Table 2.1. The T_{rg} and γ parameters, the critical diameter (d_{cr}) and its natural logarithm of the Cu-Zr-based alloys.

Alloy	T_g/T_l	γ	d_{cr}, mm	$\ln(d_{cr}, mm)$
$Cu_{55}Zr_{45}$	0.59	0.399	1.5	0.41
$Cu_{50}Zr_{50}$	0.59	0.380	2.0	0.69
$Cu_{45}Zr_{45}Ag_{10}$	0.60	0.412	6.0	1.79
$Cu_{35}Zr_{45}Ag_{20}$	0.56	0.404	1.5	0.41
$Cu_{55}Zr_{30}Ti_{10}Ag_5$	0.60	0.397	2.0	1.1
$Cu_{50}Zr_{45}Al_5$	0.61	0.416	9.0	0.69
$Cu_{45}Zr_{45}Al_5Ag_5$	0.62	0.425	10.0	2.20
$Cu_{44}Ag_{15}Zr_{36}Ti_5$	0.63	0.404	25.0	2.30
$Cu_{36}Zr_{48}Al_8Ag_8$	0.63	0.433	3.0	3.22

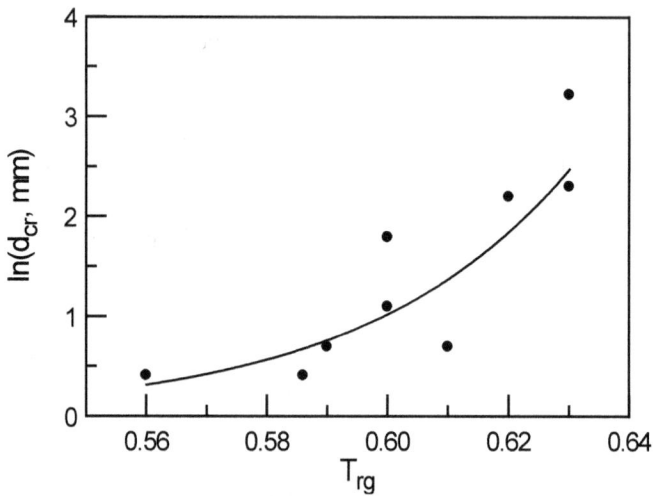

Fig. 2.4. Ln(d_{cr}) of the Cu-Zr-based alloys presented in Table 2.1 as a function of T_{rg}.

The intrinsic α parameter equal to $(T_e(\beta \to 0)-T_g(\beta))/(T_l(\beta \to 0)-T_g(\beta))$ [59] (here T_e is the eutectic temperature T_e) or $(T_s(\beta \to 0)-T_g(\beta))/(T_l(\beta \to 0)-T_g(\beta))$ in a more general form, takes into account the difference T_l-T_s, and thus, $\alpha*T_g/T_l$ criterion has better correspondence with the glass-forming ability of the alloy than T_g/T_l if T_s is significantly lower than T_l. One should note that T_l should be properly measured and extrapolated to $\beta \to 0$ (Fig. 2.5). It was also shown that the liquid structure and the glass-forming ability depend on superheating of the melt [60]. As a result strong influence of the casting temperature on the critical sample diameter of the $Fe_{50}Cr_{15}Mo_{14}C_{15}B_6$ was observed [61].

Fig. 2.5. DTA curves of the $Cu_{36}Zr_{48}Al_8Ag_8$ alloy, heating at 20 K/min and cooling at different rate. Arrows indicate the points of beginning of melting (nonequilibrium solidus temperature) on heating and solidification on cooling (nonequilibrium liquidus temperature).

Another purely intrinsic parameter, ΔT_l [62] indicates lowering of the melting temperature compared to the hypothetical ideal solution. Although, in general, compositions of metallic glasses follow deep eutectics [63], Ni-Si alloys containing from 45 to 58 at.% Si [64] with drastically reduced liquidus temperature and Au-Al alloys [28] could be treated as good glass-formers, which they are not mainly due to their low atomic size ratio. The liquidus temperature of some Au-Al system alloys is as low as about 798 K, which is significantly lower than those for the pure elements. It indicates that this thermodynamically stable liquid is significantly under-cooled compared to that of pure elements. The most common mechanism of deep eutectic formation is due to the enthalpy and size effects. However, the difference in the effective size between Al and Au is small, 4% at most. Therefore the deep depression of the liquidus in Al-Au on the Au-rich side most likely originates from the instability of crystalline phases on the Au-rich side. The local structure of the $AlAu_4$ phase, with icosahedral clusters and strong local distortions, must be very similar to that of the liquid phase. The example of Al-Au also indicates that if the heat of formation is too strongly negative it is harmful to glass formation, because it leads to formation of stable intermetallic compounds. Formation of strong local chemical clusters stabilizing crystals may also be detrimental to glass formation.

Notwithstanding on the formation of a deep eutectic and low stability of oP8 PdSi compound, $Si_{52}Pd_{48}$ alloy also did not produce a glassy phase, though Pd-Si alloys are good glass formers and even bulk glass-formers when fluxed [65,66]. It indicates that glass-formation is hampered in metalloid-rich areas likely owing to unfavorable atomic size ratios. The number of glassy alloys with $R_B/R_A < 1$ (solute to solvent atomic size ratio) is three times larger than with $R_B/R_A > 1$.

Although eutectic alloys have the lowest T_l values the best glass-forming compositions are not found at the equilibrium eutectic point but somewhat shifted usually towards a more refractory phase [67], while T_g is not significantly different in the observed range. This is most likely due to the shift of the eutectic point with supercooling at high enough cooling rate as casting conditions of bulk glassy samples are far from the equilibrium conditions. This may be a result of deep supercooling or existence of the competing crystalline phases in the system. An influence of the kinetic and thermodynamic parameters was also shown [68,69].

Another important intrinsic parameter is thermal conductivity of a molten alloy (λ_l) or its thermal diffusivity. A reasonably good correlation between the initial cooling rate (β_i) for four different alloys ($Cu_{60}Zr_{30}Ti_{10}$, $Zr_{65}Al_{7.5}Ni_{10}Pd_{17.5}$, $Ni_{53}Nb_{20}Ti_{10}Zr_8Co_6Cu_3$ and $Au_{49}Ag_{5.5}Pd_{2.3}Cu_{26.9}Si_{16.3}$) and λ_l values for pure metals was shown [70]. The cooling rates obtained for the $Zr_{55}Cu_{30}Ni_5Al_{10}$, $Ti_{41.5}Zr_{2.5}Hf_5Cu_{42.5}Ni_{7.5}Si_1$, $Pd_{40}Ni_{10}Cu_{30}P_{20}$, and $Ni_{59.5}Nb_{33.6}Sn_{6.9}$ bulk glass-forming alloys using the cooling curves (Fig. 2.6) were also found to scale with the thermal conductivities of these glassy alloys (Table. 2.2., Fig. 2.7) [71]. Thus, contrary to the melt spinning procedure when thermal gradient within the liquid is low the thermal conductivity is among the important factors influencing the GFA upon casting of bulk glassy alloys.

Fig. 2.6. The derivative dT/dt as a function of T for the $Zr_{55}Cu_{30}Ni_5Al_{10}$, $Ti_{41.5}Zr_{2.5}Hf_5Cu_{42.5}Ni_{7.5}Si_1$, $Pd_{40}Ni_{10}Cu_{30}P_{20}$, and $Ni_{59.5}Nb_{33.6}Sn_{6.9}$ bulk glass-forming alloys. Reproduced from [71] with permission of Cambridge University Press.

Table 2.2. The cooling rate (β) of the studied alloys as a function of temperature and their thermal conductivity ($λ_g$) at room temperate. Reproduced from [71] with permission of Cambridge University Press.

System alloy	Cu-Zr-Ti	Zr-Cu-Ni-Al	Ti-Zr-Hf-Cu-Ni-Si	Pd-Ni-Cu-P
β, K/s at 1173 K	4800	3700	3500	-
β, K/s at 1073 K	2850	2900	2400	-
β, K/s at 973 K	2150	2050	1800	1850
β, K/s at 873 K	1500	1350	1200	1200
β, K/s at 773 K	900	750	650	900
β, K/s at 723 K	650	650	500	700
Average $\Sigma(\beta_i/T_i)/N$, s^{-1} (1173-723 K; N=6)	2.95	2.64	2.31	-
Average $\Sigma(\beta_i/T_i)/N$, s^{-1} (973-723 K; N=4)	2.20	2.03	1.75	2.00
λ_g, W/mK	9	5	4.5	7

Fig. 2.7. The average cooling rate the $Zr_{55}Cu_{30}Ni_5Al_{10}$, $Ti_{41.5}Zr_{2.5}Hf_5Cu_{42.5}Ni_{7.5}Si_1$, $Pd_{40}Ni_{10}Cu_{30}P_{20}$, and $Ni_{59.5}Nb_{33.6}Sn_{6.9}$ bulk glass-forming alloys divided by temperature as a function of the thermal conductivities of these alloys. Reproduced from [71] with permission of Cambridge University Press.

One should also point out the importance of the melting/liquidus temperature of the alloy. The high liquidus temperature alloy being cast will cause s higher melt-mould temperature gradient, and thus, a higher initial cooling rate. The nose of the TTT diagram of such an alloy is also likely located at higher temperature. On the other hand, as expected the casting temperature was found to have little effect on the cooling rate of a single alloy with constant liquidus temperature (Fig. 2.8) [72].

Fig. 2.8. The cooling rate of the bulk $Cu_{44}Ag_{15}Zr_{36}Ti_5$ sample (3 mm diameter) at T_g as a function of casting temperature. Reproduced from [72] with permission of Cambridge University Press.

The thermal diffusivity ($\lambda_l/C_p\rho_l$) accounts for the specific volume heat capacity $C_p(T)$ and density $\rho_l(T)$ of the alloy being cast are also useful characteristics. These parameters differ within one order of magnitude (in general 2-3 times) for different molten metals.

Other factors influencing the GFA of alloys are related to the electronic structure [73,74]. It is shown that, the addition of 1 at.% of Cu or Pd to Al-Y-Ni-Co alloy [75,76] or Cu to Al-Y-Fe alloys [77,78] drastically reduces their GFA and causes precipitation of the primary α-Al particles while non of the other parameters such as the average atomic size or overall mixing enthalpy are affected. Replacing Ni in the $Al_{85}Ni_9La_6$ alloy with Cu also decreases its thermal stability [79]. It is known that Al forms readily intermetallic compounds with Cu as well as with Fe, Ni or Y. Thus, the effect cannot be connected with possible repulsive interatomic interactions. Small amount of Si affects crystallization of the $Al_{86}Ni_2Co_{5.8}Gd_{5.7}$ alloy [80]. A comparison study of the GFAs of (Si,Ge)-Ni-Nd alloys [64, 81] and later work [82] also showed that the principles for achieving good GFA known so far are important necessary conditions but not sufficient.

Stability of the supercooled liquid or metallic glass against crystallization is another very important parameter. For example, the alloys which form congruently melting intermetallic compounds like Mg-Yb-Cu [83] have lower GFA compared to alloys with incongruently melting compounds like Mg-Y-Cu even though the liquidus and eutectic temperature for the Mg-Yb alloys are lower than those for the Mg-Y alloys.

Packing density for non-crystalline structures, as a geometrical factor influencing GFA, has been verified using a hard spheres model [84,85]. An importance of topology [86] and the efficient atomic packing for the formation of metallic glasses was shown [87,30]. Actually the formation of dense structures in glassy alloys derives from a low density difference between the bulk glassy alloys and the crystallized counterparts [88,89]. Geometrical aspects of the atomic packing have been considered in binary metallic glasses. It was found that those having the solute-to-solvent radius ratios ranging from 0.60 to 1.45, produce structure-forming, solute-centered clusters with coordination numbers varied from 7 to 20. Also a strong preference was found for special radius ratios, that give efficient local atomic packing in the first coordination shell [90]. There is also a theory of the atomic level stresses in solid solutions and the stress criteria for the instability of the solid solution [91]. However, local distortions may exist even in crystalline pure metals and are quite common in equilibrium intermetallic compounds. The atomic size difference is found even in cI58 α-Mn between atoms of the same sort [92,93].

Fig. 2.9 shows the number of metallic glassy alloys as a function of the atomic size ratio with at least three clear maxima at 0.7, 0.8 and 1.25, a small number of alloys with atomic size ratios from 0.9-1.1 and absence of alloys having R_B/R_A exceeding 1.45 [82]. Similar data were also obtained earlier [94]. Also, surprisingly low atomic size ratio of 1.14 is found for the Ni-Nb system in which a bulk glass-former was produced [95].

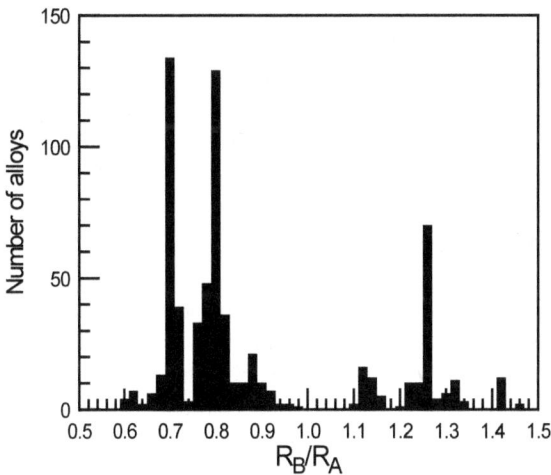

Fig. 2.9. The number of metallic glassy alloys produced by melt spinning as a function of the atomic size ratio of solute to solvent atom (R_B/R_A). Reproduced from [82] with permission of Springer.

It was shown that the electronegativity difference between the constituent elements is an important factor influencing glass formation and the temperature interval of the supercooled liquid region of the glass-forming alloys. The supercooled liquid region in the $Al_{85}RE_8Ni_5Co_2$ alloys was found to increase almost linearly with an increase in the electronegativity of the constituent RE metal [96]. Nearly the same results have been obtained for other Al alloys [97]. High crystallization temperatures (T_x) (compared to the $Al_{85}Y_8Ni_5Co_2$ one) lead to a large supercooled liquid regions of 43 and 48 K for the $Al_{85}Y_6Ni_5Co_2Zr_2$ and $Al_{84}Y_6Ni_4Co_2Sc_4$ glassy alloys, respectively [98].

Thus, upon homogeneous nucleation intrinsic parameters and factors indicating high GFA include T_{rg}, αT_{rg}, δ, λ_1, $\lambda_1/C_p\rho_1$, efficient cluster packing as well as a topological criterion λ [22,99], ΔT_1 [54], σ [100] liquid fragility [56,57,101] which have been summarized and separated for intrinsic and extrinsic factors [102]. ΔT_x and γ can be considered only as nearly intrinsic, because T_x is dependent on other factors such as heterogeneous vs. homogeneous nucleation [103 , 104]. The influence of extrinsic factors can also explain why semi-empirical γ parameter better describes the experimental data within ternary systems than T_{rg} because γ is sensitive to both extrinsic and intrinsic factors.

Nevertheless, there is difficult to create a unique parameter indicating the glass-forming ability because some factors such as T_{rg} deal with the dynamics of liquid (through its viscosity) towards glass transition while others like T_x indicate stability of the liquid versus crystallization: thermodynamic factors in terms of the driving force and kinetic factors in terms of nucleation and growth mechanisms. Moreover, T_x is measured on heating of a glass from room temperature while the high-temperature part where the nose of the TTT diagram can be located remains untreated.

The ideal equivolume glass-transition temperature (temperature at which volumes of liquid and solid phase become equal) was also derived from the thermal expansion data for corresponding liquids and solids [41]. The analysis of the thermal expansion curves allowed to derive a new parameter $\delta = \alpha_l * \rho_l (T_m - 298) / \Delta \rho_{s-l}$ [105,106] where α_l is the volumetric thermal expansion coefficient of the liquid alloy, ρ_m is the density of a liquid measured at a certain temperature, usually close to the melting temperature (T_m) and $\Delta \rho_{s-l}$ is the density difference between the solid and liquid phases. The influence of thermal expansion coefficient of a solid α_s is much smaller than that of the liquid and is excluded for convenience. The δ criterion indicates how fast a metal/alloy can reach the glass transition temperature, while it does not provide information on the stability of the resulted glass against crystallization [107]. One should also note that as predicted [108] the smallest density changes on crystallization of Cu-Zr glasses [109] were found at compositions close to those of the alloys with the best glass-forming ability. Molecular dynamics (MD) simulation resuls of the liquid viscosity [110] and direct viscosity measurements [111] indicate strong liquid character of the best glass-forming liquids. $Cu_{64}Zr_{36}$ alloy, one of the best glass-formers in the binary Cu-Zr system, is characterized by a pronounced icosahedral short range order in the liquid and glassy states.

By using the equivolume (isochoric) criterion the ideal glass-transition temperature (T_g) of Ni was found to be at about 1000-1200 K [112] depending on the thermal expansion coefficient of liquid Ni which is not very precisely defined (Fig. 2.1). The coordination numbers in the first coordination shell at 1800 and 1000 K were found to be 11.23 and 11.86, respectively. The former value is typical for a liquid metal [113] while the later one likely represents a glass. An empirical criterion for the glass transition based on the reduced radial distribution function or pair distribution function (PDF(R)) described by: $\Re = PDF(R)_{min} / PDF(R)_{max}$ where max and min indicate the first maximum and minimum of PDF(R), respectively, [114] also indicated the glass-transition temperature of about 1200 K.

Thus, the ideal T_{rg} of pure Ni may be as high as $\sim$0.6-0.7, which is an extremely high value for a pure metal. In general, bulk glassy alloys are formed at the compositions with high T_g/T_l (T_g glass-transition temperature, T_l liquidus temperature) ratio exceeding approximately 0.6 [50]. However, as has been mentioned earlier poor GFA of pure metals even if they have high T_{rg} is related to their large fragility.

It is also interesting to note that T_g predicted by the point of equal volume for liquid and crystalline Fe of about 1200 K [105] is very close to that found in the MD simulation. The results of MD simulation of liquid Fe preformed by LAMMPS package [115,116] at constant pressure with an embedded atom potential [117] are shown in Fig. 2.10. Boiling was suppressed by using the periodic boundary conditions. The same T_g temperature of 1200 K was found by monitoring the dependence of the size of the largest cluster formed by mutually penetrating and contacting icosahedrons [118].

Fig. 2.10. MD simulation of liquid Fe preformed by LAMMPS package at nearly constant pressure with an embedded atom potential: density as a function of temperature (blue curve). Black curves represent the experimental values for liquid (L) and crystalline (C) iron.

If we assume that the viscosity [119] of a liquid should rise steeply when the liquid Fe volume approaches that of the crystal, take 1200 K as T_g when the viscosity reaches about 10^{12} Pa·s and fit the viscosity of the equilibrium liquid Fe together with the point at T_g by the famous Vogel-Fulcher-Tammann-Hesse (VFTH) equation [120,121,122]:

$$\eta = \eta_0 \cdot \exp[D^* T_0 / (T - T_0)]) \tag{2.1}$$

where η_0, D and T_0 are the fitting parameters, we obtain the fragility parameter D=2.1 [107]. Thus, liquid Fe, as well as Ni [105] should be more fragile than any metallic glass-former known so far. The most fragile metallic glasses have D~6 [123]. It indicates that pure metals, as very fragile liquids, should be very unstable against crystallization.

Nevertheless, as can be seen from Table 2.3, metals, which were found to vitrify easily in the experiment [33], i.e. Ni, Fe, Co, V have a relatively large δ parameter and a large activation energy for the temperature dependence of viscosity (E) in the equilibrium liquid state. Most of the data is from Ref. [40].

Table 2.3. Physical properties of several elements in the liquid state. T_m and T_l are the melting and liquidus temperatures, respectively; Q is the activation energy for temperature dependence of viscosity, dV/V_l is the volume change upon melting. The state: crystalline or amorphous obtained upon rapid solidification in Ref. [33]. Dash character indicates that the parameter is not known.

El.	Lattice at RT and at melting	$T_m = T_l$, K	Q, kJ/mol	dV/V_l, %	δ	State
Ti	HCP, BCC	1958	-	1.9	0.99	Cr
Zr	HCP, BCC	2123	-	3.7	0.99	Cr
V	BCC	2185	-	2.1	1.33	Am
Fe	BCC, FCC*, BCC	1809	41.4	3.5	1.59	Am
Co	HCP, FCC	1766	44.4	3.5	1.46	Am
Ni	FCC	1727	50.2	5.5	1.71	Am
Pd	FCC	1825	-	6	1.25	Cr
Pt	FCC	2042	-	6.7	1.94	Cr
Cu	FCC	1358	30.5	4	0.88	Cr
Ag	FCC	1234	22.2	4	0.73	-
Au	FCC	1336	15.9	5	0.8	Cr

*intermediate state

The plot scaling liquid viscosity by the reduced temperature at ambient pressure introduced by Laughlin-Uhlmann [124] and extended by Angell [125] shows the temperature dependence of an equilibrium liquids' viscosity (η) normalized by T_g (Fig. 2.11) [129]. Not all of the substances reach the equilibrium viscosity of 10^{12} Pa·s at the glass-transition temperature monitored by the heat capacity variation, and, in general this value ranges from about 10^{10} to 10^{12} Pa·s [126,127]. The viscosity is proportional to the relaxation time [128]. Moreover, using the relation of $\eta=\sigma/3\dot{\varepsilon}$, where η is the viscosity, σ is the flow stress and $\dot{\varepsilon}$ is the strain rate; is used to calculate viscosity from the strain rate.

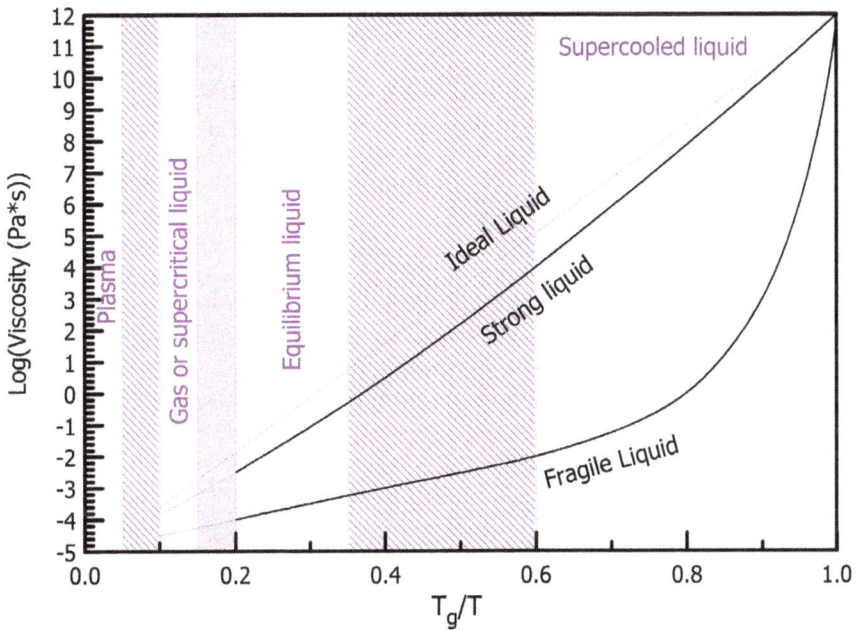

Fig. 2.11. A schematic plot of the viscosity as a function of temperature for strong and fragile liquids together with the corresponding phase regions. The "ideal liquid" shows the Arrhenius behavior while none of the known substances follows this law in the entire range from T_g to boiling temperature. Filled areas represent the temperature regions of liquidus and boiling temperature depending on the alloy composition.

This plot differentiates "strong" and "fragile" liquids [129,130] which exhibit different degrees of deviation from the Arrhenius law:

$$\eta = \eta_0 \exp(-Q/RT) \qquad (2.2)$$

(where Q is the activation energy for viscous flow and R is the gas constant) in the supercooled liquid state down to T_g. Such a deviation is generally called "fragility" or the dynamic fragility. The correlations between the thermodynamic (the difference in specific heat capacity C_p between a liquid and a glass) and dynamic fragilities were discussed [131]. For a numeric representation of $\eta F(T)$ dependence below T_l the VFTH equation (eq. 2.1) is usually used [132,133]. Above T_l the equilibrium viscosity of liquid pure metals generally obeys the Arrhenius law in a wide temperature range where the measurements were made [134,135]. Fragility (here and below dynamic) can also be defined using the parameter

$$m = d(\log(\eta))/d(T_g/T) \qquad (2.3)$$

at T_g [136].

Another representation of fragility is by assuming that the activation energy for viscous flow is not constant but changes as the structure of a liquid varies with temperature [137,138].

$$R_D = Q_H/Q_L \qquad (2.4)$$

The correlation observed between the critical cooling rate for glass formation R_c, liquid fragility index m or D, and T_{rg} [56] explains the low glass-forming ability (GFA) of a pure metal even if it has a high T_g. This is intuitively clear because according to the plot (Fig. 2.7) [129] strong liquids, as a rule, have a higher viscosity above T_g, and thus, should be less prone to crystallization for kinetic reasons. It, however, does not include the thermodynamic aspects related to the driving force in terms of the Gibbs free energy difference between the liquid and crystalline phases [82]. As lower viscosity of the melt promotes both nucleation and growth processes of crystals [139] even at the same values of the reduced glass-transition temperature ($T_{rg}=T_g/T_l$) a more fragile liquid is going to have a lower viscosity in the entire range between T_l and T_g (see Fig. 2.11).

The origin of fragile behavior should be connected with structural changes in the supercooled liquid leading to the variation in the activation energy for viscous flow. For example, intensification of the covalent bonding between the metallic atoms and P was found in Pd-Cu-Ni-P melts in-situ cooled down to the glass-transition region [140] which was responsible for the changes in the structure of a liquid (clustering [141]), and thus, "fragile" behavior of this melt. Similar changes were observed in Zr-Cu [142] and Zr-Cu-Al [143] liquids on cooling. Although fragile liquids are generally predisposed to have a lower GFA compared to strong liquids Pd-based alloys still have a good GFA owing to their exceptionally high T_{rg} values approaching 0.7 [59].

A structural parameter, γ, based on the shift of the first peak in the structure factor, $S(q_1)$ is used to characterize the liquid fragility through the rate of structural ordering with decreasing temperature [144]. The link between fragility and glass-forming ability arises from the coupling of fragility with structure and dynamics. Low fragility of the Zr-Pt metallic melts [145] is attributed to a combination of localized, highly polar, sd-sd bonds between Zr and Pt atoms [146].

It is also found that the reduced Arrhenius crossover temperature $\theta_A = T_A/T_g$, a crossover in the mean diffusion coefficient from the high-temperature Arrhenius to low-temperature non-Arrhenius behavior in metallic glass-forming liquids was found to take place approximately at $\theta_A = 2$ in the equilibrium liquid state [147]. In case of fragile molecular liquids, this crossover occurs at $\theta_A = 1.4$. The θ_A values for strong network liquids are higher than 2. It is suggested that the reduced crossover temperature depends strongly on the liquid fragility, and can be observed either in the supercooled state (molecular glass former) or in the equilibrium liquid state (metallic and network glass formers). The effective activation barrier of the high-temperature Arrhenius behavior was found to be a nearly universal value of $11k_B T_g$ for nonpolar molecular and metallic liquids. Such correlations between the low and high-temperature parameters imply that T_g can be estimated from the high-temperature activation barrier E_∞ and the fragility m can be estimated from the reduced crossover temperature [148].

Also, a liquid, in order to be able to attain a high temperature and retain its liquid state boiling must be avoided. This is possible provided that the liquid is under an overcritical external pressure [149]. Below the critical point (the point above which one cannot distinguish between the liquid and gaseous phase) the high-temperature part of the viscosity plot cannot exist. Above this point the liquid/gaseous phase can reach high temperatures without undergoing boiling phase transformation before it transforms to plasma (Fig. 2.11).

The temperature dependence of the equilibrium and non-equilibrium viscosity below T_g has been studied extensively [150]. Due to long relaxation times a non-equilibrium viscosity is often measured leading to a Non-Newtonian flow and at room temperature can be as low as 10^{17} Pa·s [151]. The extrapolated equilibrium viscosity at room temperature should be very high and it is not clear is there a change in the slope of the equilibrium viscosity at T_g.

Following the variable activation energy of viscous flow $Q(T)$ in the equation of the equilibrium viscosity (eq. 2.2) [138] has two constant asymptotes – high Q_h for low temperature (supercooled) liquid and low Q_l for high temperature (equilibrium) liquid (Fig. 2.12). Q_g corresponds to that of a glass but in this range the viscosity is strongly strain rate sensitive and it is not clear what the behavior of equilibrium viscosity in this range is. More generally the universal temperature relationship for the activation energy of viscous flow of liquids is:

$$Q(T)=Q_l+RT\cdot\ln[1+\exp(-S_d/R)\exp((Q_h-Q_l)/RT)] \tag{2.5}$$

which depends on the asymptotic energies and on the entropy S_d.

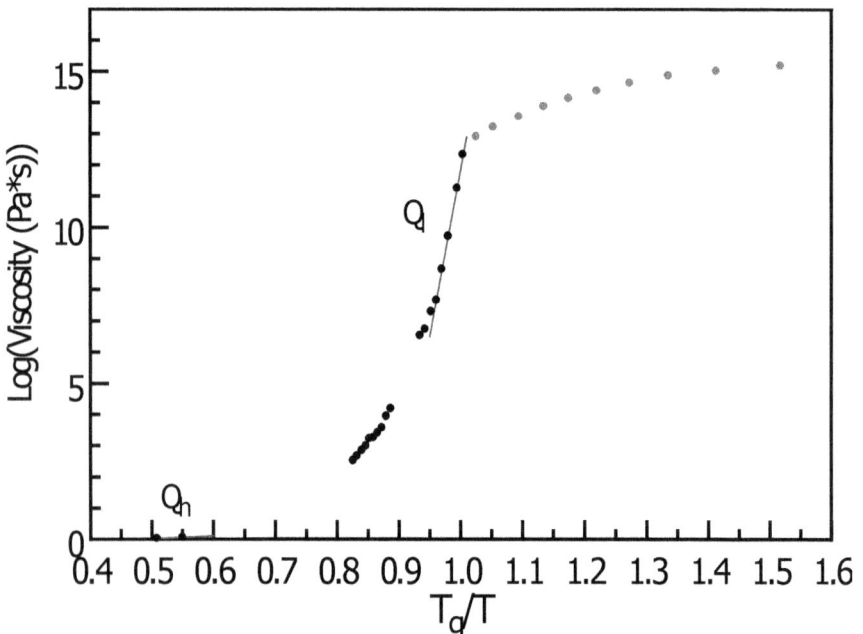

Fig. 2.12. Summarized temperature dependence of viscosity of the $Pd_{40}Ni_{40}P_{20}$ and $Pd_{40}Cu_{30}Ni_{10}P_{20}$ alloys with similar properties in the liquid, supercooled liquid and metallic glassy state. The data is taken from Refs. [152,153,154]. The values below T_g are for non-equilibrium viscosity.

There was also shown that there is a correlation between the Poisson's ratio of BMGs and the fragility index of the corresponding liquids: the higher v is, the higher is the fragility index m [155].

In situ measurements of the vibrational spectra of strong and fragile metallic glasses in the glass, liquid and crystalline phases were performed using inelastic neutron scattering. The similarity of the phonon density of states curves above and below T_g indicates that the excess vibrational entropy is unchanged on the glass transition. The excess entropy of the liquid and glass is dominated by configurational entropy in these metallic glasses while the excess vibrational entropy in the $Cu_{50}Zr_{50}$ and $Cu_{46}Zr_{46}Al_8$ alloy was found to be very small [156]. To the contrary molecular and network glasses indicate vibrational contribution to the excess liquid/glass entropy [157]. The configurational entropy increases with increasing temperature but at higher temperatures anharmonicity may cause thermodynamically significant changes in the vibrational dynamics of the liquid as well.

2.2 Crystal growth controlled mechanism

The $Fe_{48}Cr_{15}Mo_{14}C_{15}B_6Tm_2$ and $Fe_{48}Cr_{15}Mo_{14}C_{15}B_6Y_2$ alloys are the example of crystal growth mechanism limited BMGs because they contain pre-existed nuclei [158]. GFA of the Fe-Cr-Mo-C-B alloys is drastically improved by the addition of rare-earth (RE) metals (Fig. 2.13 (a)) [159]. The additions of Y [160,161] and Gd [162,163] were reported to shift the chemical composition closer to eutectic and to combine oxygen into the yttrium oxide. $(Fe,Co)_{50-x}Cr_{15}Mo_{14}C_{15}B_6Tm_x$ bulk metallic glasses showed exceptionally high glass-forming ability: the critical thickness of cast samples reached 1 cm [164,165]. The addition of Tm is more effective than Y in improving the glass-forming ability of Fe-based alloy as it leads to a lower T_l^* temperature, and thus to higher T_{rg} and γ parameters. Also, a strong influence of the casting temperature on the critical sample diameter of the $Fe_{50}Cr_{15}Mo_{14}C_{15}B_6$ alloy was observed [166].

Fig. 2.13. (a) High-resolution TEM image of the $Fe_{48}Cr_{15}Mo_{14}C_{15}B_6Tm_2$ BMG alloy in the as-solidified state. (b) Bright -field TEM image of the $Fe_{48}Cr_{15}Mo_{14}C_{15}B_6Tm_2$ glassy alloy annealed at 893 K for 420 s.

The nanoscale size particles of the χ-$Fe_{36}Cr_{12}Mo_{10}$ phase were also found to precipitate on heating from the supercooled liquid state in the $Fe_{48}Cr_{15}Mo_{14}C_{15}B_6Y_2$ alloy. TEM of the $Fe_{48}Cr_{15}Mo_{14}C_{15}B_6Tm_2$ alloy annealed at 893 K for 420 s shown in Fig. 2.13 (b) indicates the particles of about 10 nm size, while a wide (330) X-ray diffraction peak corresponds to a significantly smaller particles size of 3.1 nm which indicates a large amount of the internal strain in this phase.

The energy-dispersive X-ray (EDX) spectra obtained indicated that the distribution of Tm content is close to Gaussian. This indicates that Tm as well as Y [158] is nearly equally present in both the χ-$Fe_{36}Cr_{12}Mo_{10}$ phase and the residual glassy matrix. At the same time χ-$Fe_{36}Cr_{12}Mo_{10}$ phase has a higher Fe content than the glassy matrix. The diffusion coefficient (D) of Mo (D_{Mo}) in Fe at 873 of $2.0*10^{-20}$ m^2/s is very close to that for the self diffusion of Fe (D_{Fe}) of $1.9*10^{-20}$ m^2/s [167] while RE metals have a much lower D value of about $4.2*10^{-23}$ m^2/s at 900 K [168] consistent with their large atomic size. Such a slow diffusion of RE metals drastically slows down the growth rate of χ-$Fe_{36}Cr_{12}Mo_{10}$ phase and explains good glass-forming ability of the $Fe_{48}Cr_{15}Mo_{14}C_{15}B_6Tm_2$ BMG.

The differential isothermal calorimetry investigation (Fig. 2.14) was carried out below T_x measured on continuous heating. Crystallization in the $Fe_{48}Cr_{15}Mo_{14}C_{15}B_6RE_2$ (RE=Tm and Y) alloys started without an incubation period. No incubation period was observed in Fig. 2.11 while the χ-$Fe_{36}Cr_{12}Mo_{10}$ phase precipitated. Also, the existence of crystal-like clusters which can act as pre-existing nuclei in the $Fe_{48}Cr_{15}Mo_{14}C_{15}B_6Tm_2$ alloy was observed earlier [169]. On the contrary $Fe_{50}Cr_{15}Mo_{14}C_{15}B_6$, $Co_{48}Cr_{15}Mo_{14}C_{15}B_6Tm_2$, $Fe_{50}Cr_{15}Mo_{14}C_6B_{15}$ and $Fe_{48}Cr_{15}Mo_{14}C_6B_{15}Tm_2$ glassy alloys showed nucleation and growth behavior and crystallization started after the incubation period as usual nucleation rate controlled BMGs [170]. Destabilization of the competing crystalline phases, as it was found for marginal glass-formers [171,172] is considered to be the dominant reason for the improved GFA of these BMGs by the RE metals.

Another example of crystal growth-controlled BMG is the $Zr_{65}Al_{7.5}Ni_{10}Pd_{17.5}$ bulk glassy alloy [173] in which an icosahedral quasicrystalline phase is also formed without the incubation period. Some of Al-RE-TM (TM-transition metals) [174] system alloys (their critical thickness is about 0.5 mm) produced by rapid solidification techniques also showed growth controlled mechanism, especially those containing Cu [175]. Partial replacement of Ni by Cu in the $Al_{85}Y_8Ni_5Co_2$ metallic glass caused formation of the nanoscale Al particles, resulted in sharp drops in tensile strength, hardness, crystallization temperature and disappearance of the supercooled liquid prior to crystallization. Nevertheless, the absolute number of such growth-controlled BMGs is low and formation of the most of BMGs is nucleation-controlled.

Fig. 2.14. Differential isothermal calorimetry traces of $Fe_{48}Cr_{15}Mo_{14}C_{15}B_6Tm_2$ at 868 K and $Fe_{50}Cr_{15}Mo_{14}C_{15}B_6$ at 848 K. The data is taken from Ref. [170] with permission of Elsevier.

2.3   Extrinsic factors influencing glass-forming ability

First group of purely extrinsic factors influencing the GFA are [102]: factors that modify the melt-mould heat transfer coefficient (h). They include mould surface quality and cleanliness, mould surface temperature, thermal conductivity of the mould, casting temperature, casting pressure, liquid metal turbulence during casting as well as the chamber pressure and atmosphere [176]. The presence or absence of turbulence and the degree of turbulence can have a strong influence on heat transfer and GFA. These factors are typically not reported in the literature. These factors may play a central role in the poor reproducibility in critical thickness for a given metallic glass composition that is often apparent in the literature. The influence of these factors is often not known quantitatively, and only a qualitative discussion is given below.

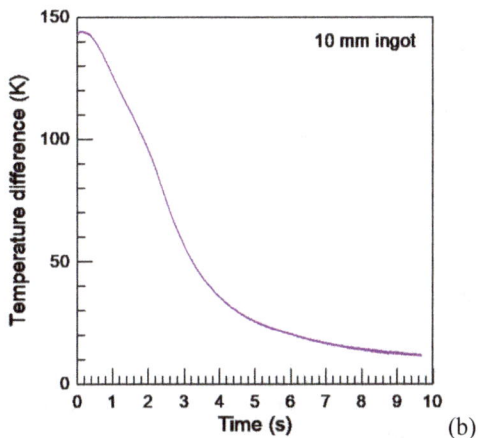

Fig. 2.15. (a) Schematic representation of temperature gradients close to the mould-melt interface depending upon the Biot number. Reproduced from [102] with permission of WILEY-VCH Verlag GmbH & Co. (b) The temperature difference between the center of the $Cu_{44}Ag_{15}Zr_{36}Ti_5$ ingot of 10 mm diameter and half distance to the mold upon casting of the melt.

The mold surface cleanliness and roughness affect the melt-mould heat transfer coefficient and the Biot number $Bi=hL/\lambda_l$, (where L is a characteristic length scale equal to radius of the cast cylindrical rod, and λ_l is the thermal conductivity of the molten alloy being cast). It indicates whether heat transfer through the melt or heat transfer through the melt/mould interface dominates in cooling (see Fig. 2.15 (a)). The value of the Biot number calculated for typical Cu- and Zr-based glass-forming alloys is about unity which implies that the temperature gradients inside the melt are non-negligible in accordance with Fig. 2.15 (b). Thus, both heat transfer through the melt and heat transfer through the melt/mould interface influence the cooling rate [177] of bulk metallic glasses.

The heat flow Q_f is proportional to the temperature gradient:

$$\frac{dQ_f}{dt} = -\lambda A \frac{dT}{dx} \tag{2.6}$$

where Q_f is the heat flow, t is time, λ is a temperature dependent thermal conductivity, A is the transversal surface area, dT is the temperature gradient through which the heat is being transferred, dx is the thickness of the body of matter through which the heat is passing.

An ideal time-temperature relation for a cooling object is Newton's exponential decay law of cooling:

$$T(t) = T_s + (T_0 - T_s)e^{-t/\tau} \tag{2.7}$$

(where T is the body temperature, $T_0=T$ at t=0, T_s is the temperature of the surrounding matter (mould), and $-1/\tau$ (or K) is a constant). Plotting $ln(T-T_s)$ versus t should therefore give a straight line with a slope $-1/\tau$ if cooling is influenced only by one kind of process. However, this law fulfils only if the Biot number is low enough. The cooling curve of the $Cu_{58.8}Zr_{26.2}Ti_{15}$ bulk glassy alloy (Fig. 2.16) can't be fit well with a single exponential decay function fitting (violet line). However, it can be fit very well (Fig. 2.16) if an exponential decay equation with two exponents (second order) is used:

$$T(t) = T_0 + A_1 e^{-t/\tau_1} + A_2 e^{-t/\tau_2} \tag{2.8}$$

where A_1, A_2, τ_1 and τ_2 are the constants at $T_0=423$ K, $A_1=356$ K, $\tau_1=0.074$ s, $A_2=571$ K, $\tau_2=0.478$ s. Thus, the cooling of the melt is influenced by both processes: the heat transfer through the melt as well as through the melt/mould interface.

Fig. 2.16. The cooling curve of the $Cu_{58.8}Zr_{26.2}Ti_{15}$ bulk glassy alloy and $Ni_{59.5}Nb_{33.6}Sn_{6.9}$ alloy which crystallized. Violet line – the best fit with the first order exponential decay function. Green line – fit with the second order exponential decay function. Reproduced from [71] with permission of Cambridge University Press.

The second group of factors takes into account heterogeneous nucleation as a potent source for decreasing GFA. Possible sources of heterogeneous nucleation include irregularities of the mould and melt cleanliness. An external alternating electromagnetic field [178] can also influence the nucleation process. Thus, for successful bulk glass formation one should take into account both intrinsic and extrinsic factors. For example, an alloy with high intrinsic GFA may not form a fully amorphous product due to limiting extrinsic GFA parameters related to purity of the alloying elements and the processing methodology.

2.4   Effect of flux treatment

Contamination by impurities or inclusions can be removed by flux treatment [179,180]. Fluxing by removing metallic oxide inclusions suppresses heterogeneous nucleation and thus inhibits crystallization. It favors homogeneous nucleation rather than heterogeneous nucleation [181] or reduces the nucleation rate in case of heterogeneous nucleation and drastically improves the critical diameter of Pd-based bulk glassy samples [182,183]. Pd-Ni-Cu-P bulk metallic glasses [184,185] are the best glass-formers known to date. Their GFA is higher than that of ternary Pd-Ni-P [186]. Similar to the Pd-Ni-Cu-P alloys some of Fe-based BMG were also successfully subjected to flux treatment used to improve their glass-forming ability [187,188]. The $Pd_{40}Ni_{20}Cu_{20}P_{20}$ glassy alloy was prepared in a bulk form with a diameter up to 72 mm [189]. A cyclic heating-cooling treatment with fluxing can significantly promote the purification effects, implying that impurities within the melting alloy can be driven to the surface by moving the solidification front [190]. Consequently, the critical cooling rate for glass formation can be drastically reduced.

Fig. 2.17 shows the DTA traces of the un-fluxed and fluxed $Pd_{40}Ni_{40}Si_4P_{16}$ bulk metallic glasses. The differently processed samples exhibit almost the same glass transition temperature T_g and liquidus (nonequilibrium liquidus) temperature T_{lh} determined by the heating curves [191]. Both T_g and T_{lh} are important material characteristics that are sensitive to the alloy composition. The similar behaviors in glass transition and melting (in addition to the EDX compositional analysis) confirmed the compositional identity of the fluxed and not fluxed $Pd_{40}Ni_{40}Si_4P_{16}$ alloys. The supercooled liquid regions for both not fluxed and fluxed samples exceeds 100 K. From Fig. 2.18 and Table 2.4 one can see that crystallization of the fluxed $Pd_{40}Ni_{40}Si_4P_{16}$ alloy is much more sluggish than that of the not fluxed sample upon cooling. On continuous cooling crystallization of the fluxed sample started at 769 K while the crystallization of the not fluxed sample started at 872 K. The difference is 103 K. In addition, the supercooling ΔT (T_{lh}-T_{lc}, where T_{lc} is the temperature at which solidification by cooling starts) of the fluxed sample reaches 202 K. The supercooling ΔT (T_{lh}-T_{lc}) determines stability of a liquid, which is of importance to evaluate the glass-forming ability of alloys.

Table 2.4. Characteristic temperatures derived from the DTA traces shown in Fig. 2.17.

Alloys	T_g (±2 K)	T_x (±2 K)	T_{lh} (± 2K)	T_{lc} (± 2K)	ΔT (K)
fluxed	579	688	971	769	202
un-fluxed	579	683	971	872	99

Fig. 2.17. DTA traces of (a) the fluxed and (b) the un-fluxed $Pd_{40}Ni_{40}Si_4P_{16}$ BMGs. Reproduced from [191] with permission of Elsevier.

Fig. 2.18 shows the crystalline eutectic colonies presenting a spherical morphology formed in not fluxed and fluxed samples. The density of precipitates related to the number of nucleation events is clearly different in two cases.

Fig. 2.18. SEM images of the polished cross section of not fluxed (a) and fluxed (b) samples annealed at 673 K for 9 minutes. Reproduced from [191] with permission of Elsevier.

The effect of small additions of Dy, Gd and Er to $Zr_{62.5}Cu_{22.5}Fe_5Al_{10}$ was also studied recently [192]. Ingots of the alloy without RE elements had a mixed glassy-crystalline structure and fully crystalline structure for the samples of 3 and 5 mm in diameter, respectively. The addition of 0.5 at.% Dy and Er did not promote formation of the amorphous structure, however the samples containing 1 at.% Dy and Er exhibited improved GFA leading to fully amorphous structure in both samples: 3 and 5 mm in diameter (Table 2.5). Gd containing samples with high γ parameter demonstrated amorphous structure at both diameters. The oxide particles are present in the structure of the alloys containing RE elements. RE elements formed oxides and inhibited oxygen-induced crystallization during cooling. Dy_2O_3, Gd_2O_3, Er_2O_3 with the average size of about 0.5 μm were uniformly distributed in the glassy phase. Their average volume fraction is 0.4 % in case of 0.5 at.% RE content and 0.7 % in case of 1.0 at. % RE content.

Table 2.5. The values of T_g, T_x, T_s, T_l^* (approximate T_l) and GFA criteria for the studied alloys. Reproduced from [192] with permission of Elsevier.

Alloy	T_g (K)	T_x (K)	T_s (K)	T_l^* (K)	$T_{rg}=T_g/T_l$	$\gamma=T_x/(T_g+T_l)$
Base alloy	681	763	1143	1168	0.583	0.413
Dy 0.5 %	663	755	1066	1168	0.567	0.413
Dy 1 %	656	753	1070	1153	0.569	0.416
Gd 0.5 %	663	759	1067	1162	0.571	0.416
Gd 1 %	661	760	1085	1171	0.565	0.415
Er 0.5 %	662	758	1086	1161	0.570	0.416
Er 1 %	657	747	1084	1175	0.559	0.408

$Fe_{66}Nb_4B_{30}$ ferromagnetic bulk glassy alloy with high B content was also produced by fluxing and casting [193].

References

[1]   C.A. Angell, Structural instability and relaxation in liquid and glassy phases near the fragile liquid limit, J. Non-Cryst. Solids 102 (1988) 205. https://doi.org/10.1016/0022-3093(88)90133-0

[2]   T. Egami, Atomistic mechanism of bulk metallic glass formation, Journal of Non-Crystalline Solids 317 (2003) 30. https://doi.org/10.1016/S0022-3093(02)02003-3

[3]   M.I. Ojovan, Ordering and structural changes at the glass–liquid transition, Journal of Non-Crystalline Solids, 382 (2013) 79. https://doi.org/10.1016/j.jnoncrysol.2013.10.016

[4]   H.S. Chen and D. Turnbull, Evidence of a glass–liquid transition in a gold–germanium–silicon alloy, J. Chem. Phys. 48 (1968) 2560. https://doi.org/10.1063/1.1669483

[5]   A. Inoue, High strength bulk amorphous alloys with low critical cooling rates, Mater. Trans. JIM 36 (1995) 866. https://doi.org/10.2320/matertrans1989.36.866

[6]   D.V. Louzguine and A. Inoue, Comparison of the long-term thermal stability of various metallic glasses under continuous heating, Scripta Mater 47 (2002) 887. https://doi.org/10.1016/S1359-6462(02)00268-3

[7]   J. Hafner, Theory of the formation of metallic glasses, Phys. Rev. B 21 (1980) 406. https://doi.org/10.1103/PhysRevB.21.406

[8]   D.V. Louzguine-Luzgin, D.B. Miracle, L. Louzguina-Luzgina and A. Inoue, Comparative analysis of glass-formation in binary, ternary and multicomponent alloys, Journal of Applied Physics, 108 (2010) 103511. https://doi.org/10.1063/1.3506687

[9]   A. Inoue, A. Takeuchi and T. Zhang, Ferromagnetic bulk amorphous alloys, Metallurgical and Materials Transactions A, 29 (1998) 1779. https://doi.org/10.1007/s11661-998-0001-9

[10]   A. Inoue, Y. Shinohara and J.S. Gook, Thermal and magnetic properties of bulk Fe-Based Glassy alloys prepared by copper mold casting, Mater Trans JIM, 36 (1995) 1427. https://doi.org/10.2320/matertrans1989.36.1427

[11]   S.J. Pang, T. Zhang, K. Asami and A. Inoue, New Fe-Cr-Mo-(Nb, Ta)-C-B glassy alloys with high glass-forming ability and good corrosion resistance, Mater. Trans. JIM, 42 (2001) 376. https://doi.org/10.2320/matertrans.42.376

[12]   X.J. Gu, S.J. Poon, G.J. Shiflet and M. Widom, Ductility improvement of amorphous steels Roles of shear modulus and electronic structure, Acta Materialia 56 (2008) 88. https://doi.org/10.1016/j.actamat.2007.09.011

[13]   T.D. Shen and R.B. Schwarz, Bulk ferromagnetic glasses in the Fe-Ni-P-B system, Acta Mater., 49 (2001) 837. https://doi.org/10.1016/S1359-6454(00)00365-7

[14]   R.D. Doherty, Diffusive phase transformations, in Physical Metallurgy, 4th Edition by R.W. Cahn, P. Haasen, North Holland (1996) 2740.

[15] L. Ma, L. Wang, T. Zhang and A. Inoue, Bulk glass formation of Ti-Zr-Hf-Cu-M (M= Fe, Co, Ni) alloys Mater. Trans., 43 (2002) 277. https://doi.org/10.2320/matertrans.43.277

[16] A. Takeuchi, N. Chen, T. Wada, Y. Yokoyama, H. Kato A.Inoue and J.W.Yeh, $Pd_{20}Pt_{20}Cu_{20}Ni_{20}P_{20}$ high-entropy alloy as a bulk metallic glass in the centimeter, Intermetallics, 19 (2011) 1546. https://doi.org/10.1016/j.intermet.2011.05.030

[17] A. Inoue, Z. Wang, D.V. Louzguine-Luzgin, Y. Han, F.L. Kong, E. Shalaan and F. Al-Marzouki, Effect of high-order multicomponent on formation and properties of Zr-based bulk glassy alloys, J. Alloys Compd., 638 (2015) 197. https://doi.org/10.1016/j.jallcom.2015.03.078

[18] D. Xu, B. Lohwongwatana, G. Duan, W. L. Johnson and C. Garland, Bulk metallic glass formation in binary Cu-rich alloy series - $Cu_{100-x}Zr_x$ (x=34, 36, 38.2, 40 at.%) and mechanical properties of bulk $Cu_{64}Zr_{36}$ glass, Acta Mater. 52 (2004) 2621. https://doi.org/10.1016/j.actamat.2004.02.009

[19] D. Wang, Y. Li, B.B. Sun, M.L. Sui, K. Lu and E. Ma, Bulk metallic glass formation in the binary Cu–Zr system, Appl. Phys. Lett. 84 (2004) 4029. https://doi.org/10.1063/1.1751219

[20] A. Inoue, W. Zhang and J. Saida, Synthesis and fundamental properties of Cu-based bulk glassy alloys in binary and multi-component systems, Mater. Trans. 45 (2004) 1153. https://doi.org/10.2320/matertrans.45.1153

[21] P.J. Desré, On the effect of the number of components on glass-forming ability of alloys from the liquid state: application to the new generation of multicomponent bulk glasses, Materials Transactions, JIM 38 (1997) 583.

[22] T. Egami and Y. Waseda, Atomic size effect on the formability of metallic glasses, J. Non-Cryst Solids 64 (1984) 113. https://doi.org/10.1016/0022-3093(84)90210-2

[23] T. Egami, The atomic structure of aluminum based metallic glasses and universal criterion for glass-formation, J. Non-Cryst. Solids 205–207 (1996) 575. https://doi.org/10.1016/S0022-3093(96)00277-3

[24] D.B. Miracle and O.N. Senkov, Topological criterion for metallic glass formation, Mater. Sci. Eng. A 347 (2003) 50. https://doi.org/10.1016/S0921-5093(02)00579-8

[25] L. Pauling, Atomic radii and interatomic distances in metals, J. Am. Chem. Soc. 69 (1947) 542. https://doi.org/10.1021/ja01195a024

[26] R. Eiblert, J. Redinger and A. Neckel, Electronic structure, chemical bonding and spectral properties of the intermetallic compounds FeTi, CoTi and NiTi, J. Phys. F. Met. Phys. 17 (1987) 1533. https://doi.org/10.1088/0305-4608/17/7/011

[27] T. Egami, Universal criterion for metallic glass formation, J. Non-Cryst. Solids 205-207 (1996) 575. https://doi.org/10.1016/S0022-3093(96)00277-3

[28] T. Egami, M. Ojha, D.M. Nicholson, D.V. Louzguine-Luzgin, N. Chen and A. Inoue, Glass formability and the Al-Au system, Philos. Mag. 92 (2012) 655. https://doi.org/10.1080/14786435.2011.630692

[29] D.B. Miracle, The efficient cluster packing model. An atomic structural model for metallic glasses, Acta Mater. 54 (2006) 4317. https://doi.org/10.1016/j.actamat.2006.06.002

[30] H.W. Sheng, W.K. Luo, F.M. Alamgir, J.M. Bai and E. Ma, Atomic packing and short-to- medium-range order in metallic glasses, Nature 439 (2006) 419. https://doi.org/10.1038/nature04421

[31] K. Tamura and H. Endo, Ferromagnetic properties of amorphous nickel, Phys. Lett. A 29 (1969) 52. https://doi.org/10.1016/0375-9601(69)91025-1

[32] H.A. Davies, Metallic glass formation, in Amorphous Metallic Alloys (ed) F.E. Luborsky, London, Butterworths (1983) 8. https://doi.org/10.1016/B978-0-408-11030-3.50007-8

[33] Y.W. Kim, H.M. Lin and T.F. Kelly, Amorphous solidification of pure metals in submicron spheres, Acta Metall. 37 (1989) 247. https://doi.org/10.1016/0001-6160(89)90283-6

[34] P.G. Le Comber, A. Madan and W.E. Spear, Electronic and structural properties of amorphous semiconductors, Ed. by P.G. Le Comber and J. Mort, Academic press, London and New York (1973) 373.

[35] T. Yamasaki, P. Schlossmacher, K. Ehrlich and Y. Ogino, Formation of amorphous electrodeposited Ni-W alloys and their nanocrystallization, Nanostructured Materials 10 (1998) 375. https://doi.org/10.1016/S0965-9773(98)00078-6

[36] H.J. Fecht, Nanostructure formation by mechanical attrition, Nanostructured Materials 6 (1995) 33. https://doi.org/10.1016/0965-9773(95)00027-5

[37] D.V. Louzguine, Y. Kawamura and A. Inoue, Production of $Si_{55}Al_{20}Fe_{10}Ni_5Cr_5Zr_5$ bulk amorphous alloy by hot pressing, Materials Science and Technology, 15 (1999) 583. https://doi.org/10.1179/026708399101506139

[38] R.Z. Valiev, Structure and mechanical properties of ultrafine-grained metals, Materials Science and Engineering A 234-236 (1997) 59. https://doi.org/10.1016/S0921-5093(97)00183-4

[39] J. Brillo and I. Egry, Density determination of liquid copper, nickel and their alloys, International Journal of Thermophysics 24 (2003) 1155. https://doi.org/10.1023/A:1025021521945

[40] W.F. Gale and T.C. Totemeier, editors. Smithells Metals Reference Book 8-th Edition (Elsevier Butterworth-Heinemann Ltd., Oxford UK, 2004) 14.

[41] I.-R. Lu, G.P. Gorler, H.J. Fecht and R. Willnecker, Investigation of specific volume of glass-forming Pd-Cu-Ni-P alloy in the liquid, vitreous and crystalline state, Journal of Non-Crystalline Solids, 312–314 (2002) 547. https://doi.org/10.1016/S0022-3093(02)01767-2

[42] W. Kauzmann, The nature of the glassy state and the behavior of liquids at low temperatures, Chem. Rev. 43 (1948) 219. https://doi.org/10.1021/cr60135a002

[43] D.V. Louzguine-Luzgin, I. Seki, S.V. Ketov, L.V. Louzguina-Luzgina, V.I. Polkin, N. Chen, H. Fecht, A.N. Vasiliev and H. Kawaji, Glass-transition process in an Au-based metallic glass, Journal of Non-Crystalline Solids, 419 (2015) 12. https://doi.org/10.1016/j.jnoncrysol.2015.03.018

[44] C. Suryanarayana and A. Inoue, Bulk metallic glasses, Taylor and Francis Group, LLC (2011) 17.

[45] L.-M. Martinez and C.A. Angell, A thermodynamic connection to the fragility of glass-forming liquids, Nature 410 (2001) 663. https://doi.org/10.1038/35070517

[46] W.H. Wang, Bulk metallic glasses with functional physical properties, Adv. Mater., 21 (2009) 4524. https://doi.org/10.1002/adma.200901053

[47] I.S. Klein and C.A. Angell, Excess thermodynamic properties of glassforming liquids. The rational scaling of heat capacities and the thermodynamic fragility dilemma resolved, Journal of Non-Crystalline Solids 451 (2016) 116. https://doi.org/10.1016/j.jnoncrysol.2016.06.006

[48] C.M. Wayman, R.W. Cahn and P. Haasen, Eds, Phase transformations nondiffusive in Physical Metallurgy, third-ed, Elsevier BV (1983) 1032.

[49] H.S. Chen, H. Kato and A. Inoue, A fictive stress model calculation of nonlinear viscoelastic behaviors in a Zr-based glassy alloy: stress growth and relaxation, Jpn. J. Appl. Phys. 39 (2000) 5184. https://doi.org/10.1143/JJAP.39.5184

[50] D. Turnbull and M.H. Cohen, Free-volume model of the amorphous phase: glass transition, J. Chem. Phys. 34 (1961) 120. https://doi.org/10.1063/1.1731549

[51] P.G. Debenedetti and F.H. Stillinger, Supercooled liquids and the glass transition, Nature 410 (2001) 259. https://doi.org/10.1038/35065704

[52] A. Inoue, Stabilization of metallic supercooled liquid and bulk amorphous alloys, Acta Mater. 48 (2000) 279. https://doi.org/10.1016/S1359-6454(99)00300-6

[53] K.F. Kelton, Transient nucleation in glasses, Mater. Sci. Eng. B 32 (1995) 145. https://doi.org/10.1016/0921-5107(95)80023-9

[54] Z.P. Lu and C.T. Liu, A new glass-forming ability criterion for bulk metallic glasses, Acta Materialia 50 (2002) 3501. https://doi.org/10.1016/S1359-6454(02)00166-0

[55] X.H. Du, J.C. Huang, C.T. Liu and Z.P. Lu, New criterion of glass forming ability for bulk metallic glasses, J. Appl. Phys. 101 (2007) 086108. https://doi.org/10.1063/1.2718286

[56] O.N. Senkov, Correlation between fragility and glass-forming ability of metallic alloys, Phys. Rev. B 76 (2007) 104202. https://doi.org/10.1103/PhysRevB.76.104202

[57] W.L. Johnson, J.H. Na and M.D. Demetriou, Quantifying the origin of metallic glass formation, Nat. Commun. 7 (2016) 10313. https://doi.org/10.1038/ncomms10313

[58] D.V. Louzguine-Luzgin, G. Xie, Q. Zhang and A. Inoue, Devitrification behavior and crystal-glassy mixed-phase structures observed in partially crystallized Cu-based glassy alloys, Ceramic Transactions, 219 (2010) 3.

[59] N. Nishiyama and A. Inoue, Direct comparison between critical cooling rate and some quantitative parameters for evaluation of glass-forming ability in Pd-Cu-Ni-P alloys, Mater. Trans., 43 (2002) 1913. https://doi.org/10.2320/matertrans.43.1913

[60] L. Zhang, Y. Wu, X. Bian, H. Li, W. Wang and S. Wu, Short-range and medium-range order in liquid and amorphous $Al_{90}Fe_5Ce_5$ alloys, J. Non-Cryst. Sol., 262 (2000) 169.

[61] K.S. Filippov, Correlation between hereditary structures and properties of an $Fe_{50}Cr_{15}Mo_{14}C_{15}B_6$ bulk amorphous alloy in the solid and liquid states, Russian Metallurgy (Metally) 5 (2010) 379. https://doi.org/10.1134/S0036029510050034

[62] I.W. Donald and H.A. Davies, Prediction of glass-forming ability for metallic systems, J. Non-Cryst. Solids, 30 (1978) 77. https://doi.org/10.1016/0022-3093(78)90058-3

[63] A.R. Yavari, Solving the hume-rothery eutectic puzzle using miracle glasses, Nature Mater. 4 (2005) 1. https://doi.org/10.1038/nmat1289

[64] D.V. Louzguine, L.V. Louzguina and A. Inoue, Factors influencing glass formation in rapidly solidified Si,Ge–Ni and Si,Ge–Ni–Nd alloys, Appl. Phys. Lett., 80 (2002) 1556. https://doi.org/10.1063/1.1457522

[65] K.F. Yao, N. Chen, Pd-Si binary bulk metallic glass, Sci. China. Ser. G 51 (2008) 414. https://doi.org/10.1007/s11433-008-0051-4

[66] N. Chen, H.A. Yang, A. Caron, P.C. Chen, Y.C. Lin, D.V. Louzguine-Luzgin, K.F. Yao, M. Esashi and A. Inoue, Glass-forming ability and thermoplastic formability of a $Pd_{40}Ni_{40}Si_4P_{16}$ glassy alloy, Journal of Materials Science, 46 (2011) 2091. https://doi.org/10.1007/s10853-010-5043-x

[67] H. Tan, Y. Zhang, D. Ma, Y.P. Feng and Y. Li, Optimum glass formation at off-eutectic composition and its relation to skewed eutectic coupled zone in the La based La–Al–(Cu,Ni) pseudo ternary system, Acta Materialia 51 (2003) 4551. https://doi.org/10.1016/S1359-6454(03)00291-X

[68] A. Zhu, G. Shiflet and D.B. Miracle, Glass forming ranges of Al–rare earth metal alloys: thermodynamic and kinetic analysis, Scripta Mater., 50 (2004) 987. https://doi.org/10.1016/j.scriptamat.2004.01.019

[69] S. Gorsse, G. Orveillon, O.N. Senkov and D.B. Miracle, Thermodynamic analysis of glass-forming ability in a Ca-Mg-Zn ternary alloy system, Phys. Rev. B, 73 (2006) 224202. https://doi.org/10.1103/PhysRevB.73.224202

[70] D.V. Louzguine-Luzgin, A.D. Setyawan, H. Kato and A. Inoue, Thermal conductivity of an alloy in relation to the observed cooling rate and glass-forming ability, Phil. Mag., 87 (2007) 1845. https://doi.org/10.1080/14786430601096928

[71] D.V. Louzguine-Luzgin, T. Saito, J. Saida and A. Inoue, Thermal conductivity of metallic glassy alloys and its relationship to the glass forming ability and the observed cooling rates, J. Mater. Res., 23 (2008) 2283. https://doi.org/10.1557/JMR.2008.0286

[72] D.V. Louzguine-Luzgin, T. Saito, J. Saida and A. Inoue, Influence of cooling rate on the structure and properties of a Cu–Zr–Ti–Ag glassy alloy, J. Mater. Res., 23 (2008) 515. https://doi.org/10.1557/JMR.2008.0066

[73] S.R. Nagel and J. Taue, Nearly-free-electron approach to the theory of metallic glass alloys, Phys. Rev. Lett. 35 (1975) 380. https://doi.org/10.1103/PhysRevLett.35.380

[74] P. Hiussler, Interrelations between electronic and ionic structure in metallic glasses, Z. Phys. B Cond.-Matt. 53 (1983) 15. https://doi.org/10.1007/BF01578243

[75] D.V. Louzguine and A. Inoue, Investigation of structure and properties of the Al-Y-Ni-Co-Cu metallic glasses, J Mater Res.17 (2002) 1014. https://doi.org/10.1557/JMR.2002.0149

[76] D.V. Louzguine-Luzgin and A. Inoue, Structure and transformation behaviour of a rapidly solidified Al-Y-Ni-Co-Pd alloy, J. Alloys Compd. 399 (2005) 78. https://doi.org/10.1016/j.jallcom.2005.02.018

[77] J.H. Perepezko and K. Hildal, Metallic glass formation reactions and interfaces, Mater. Sci and Eng. B, 148 (2008) 171. https://doi.org/10.1016/j.mseb.2007.09.058

[78] J.H. Perepezko, S.D. Imhoff and R.J. Hebert, Nanostructure development during devitrification and deformation, J. Alloys Comp. 495 (2010) 360. https://doi.org/10.1016/j.jallcom.2009.10.051

[79] Z. Huang, J. Li, Q. Rao and Y. Zhou, Thermal stability and primary phase of Al–Ni(Cu)–La amorphous alloys, J. Alloys Compd. 463 (2008) 328. https://doi.org/10.1016/j.jallcom.2007.09.005

[80] V.I. Tkatch, S.G. Rassolov, V.V. Popov, V.V. Maksimov, V.V. Maslov, V.K. Nosenko, A.S. Aronin, G.E. Abrosimova and O.G. Rybchenko, Complex crystallization mode of amorphous/nanocrystalline composite $Al_{86}Ni_2Co_{5.8}Gd_{5.7}Si_{0.5}$, Journal of Non-Crystalline Solids 357 (2011) 1628. https://doi.org/10.1016/j.jnoncrysol.2011.02.029

[81] D.V. Louzguine-Luzgin and A. Inoue, A glance on the glass-transition phenomenon from the viewpoint of devitrification, J. Alloys Comp., 434-435 (2007) 121. https://doi.org/10.1016/j.jallcom.2006.08.146

[82] D.V. Louzguine-Luzgin, N. Chen, A.Yu. Churymov, L.V. Louzguina-Luzgina, V. I. Polkin, L. Battezzati and A.R. Yavari, Role of different factors in the glass-forming ability of binary alloys, Journal of Materials Science, 50 (2015) 1783. https://doi.org/10.1007/s10853-014-8741-y

[83] A.A. Tsarkov, E.N. Zanaeva, A.Yu. Churyumov, S.V. Ketov and D.V. Louzguine-Luzgin, Crystallization kinetics of Mg–Cu–Yb–Ca–Ag metallic glasses,

Materials Characterization, 111 (2016) 75.
https://doi.org/10.1016/j.matchar.2015.10.034

[84]   J.D. Bernal, The structure of liquids, Proc. R. Soc. A., 280 (1964) 299.
https://doi.org/10.1098/rspa.1964.0147

[85]   W.M. Visscher and M. Bolsterli, Random packing of equal and unequal spheres in
two and three dimensions, Nature 239 (1972) 504.
https://doi.org/10.1038/239504a0

[86]   O.N. Senkov, D.B. Miracle and H.M. Mullens, Topological criteria for
amorphization based on a thermodynamic approach, J. Appl. Phys. 97 (2005)
103502. https://doi.org/10.1063/1.1896434

[87]   D.B. Miracle, W.S. Sanders and O.N. Senkov, The influence of efficient atomic
packing on the constitution of metallic glasses, Phil. Mag. 83 (2003) 2409.
https://doi.org/10.1080/1478643031000098828

[88]   A. Inoue, T. Negishi, H.M. Kimura, T. Zhang and A.R. Yavari, High packing
density of Zr- and Pd-based bulk amorphous alloys, Mater. Trans. JIM 39 (1998)
318. https://doi.org/10.2320/matertrans1989.39.318

[89]   D.V. Louzguine-Luzgin, A.R. Yavari, M. Fukuhara, K. Ota, G. Xie, G. Vaughan
and A. Inoue, Free volume and elastic properties changes in Cu–Zr–Ti–Pd bulk
glassy alloy on heating, J. Alloys and Comp. 431 (2007) 136.
https://doi.org/10.1016/j.jallcom.2006.05.069

[90]   D.B. Miracle, D.V. Louzguine-Luzgin, L.V. Louzguina-Luzgina and A. Inoue, An
assessment of binary metallic glasses correlations between structure, glass forming
ability and stability, International Materials Reviews 55 (2010) 218.
https://doi.org/10.1179/095066010X12646898728200

[91]   T. Egami, Universal criterion for metallic glass formation, Mater. Sci. Eng. A
226–228 (1997) 261. https://doi.org/10.1016/S0921-5093(97)80041-X

[92]   D. Hobbs, J. Hafner and D. Spisak, Understanding the complex metallic element
Mn. I. Crystalline and noncollinear magnetic structure of α-Mn, Phys. Rev. B 68
(2003) 014407. https://doi.org/10.1103/PhysRevB.68.014407

[93]   D.V. Louzguine-Luzgin, A.R. Yavari, G. Vaughan and A. Inoue, Clustered
crystalline structures as glassy phase approximants, Intermetallics 17 (2009) 477.
https://doi.org/10.1016/j.intermet.2008.12.008

[94] D.B. Miracle, W.S. Sanders and O.N. Senkov, The influence of efficient atomic packing on the Constitution of metallic glasses, Philos Mag 83 (2003) 2409. https://doi.org/10.1080/1478643031000098828

[95] L. Xia, W.H. Li, S.S. Fang, B.C. Wei and Y.D. Dong, Binary Ni–Nb bulk metallic glasses, J Appl. Phys. 99 (2006) 026103. https://doi.org/10.1063/1.2158130

[96] D.V. Louzguine and A. Inoue, Electronegativity of the constituent rare-earth metals as a factor stabilizing the supercooled liquid region in Al-based metallic glasses, Appl. Phys. Lett. 79 (2001) 3410. https://doi.org/10.1063/1.1420781

[97] C.S. Ma, J. Zhang, X.C. Chang, W.L. Hou and J.Q. Wang, Electronegativity difference as a factor for evaluating the thermal stability of Al-rich metallic glasses, Philosophical Magazine Letters 88 (2008) 917. https://doi.org/10.1080/09500830802526596

[98] D.V. Louzguine-Luzgin, A. Inoue and W.J. Botta, Reduced electronegativity difference as a factor leading to the formation of Al-based glassy alloys with a large supercooled liquid region of 50 K, Appl. Phys. Lett. 88 (2006) 011911. https://doi.org/10.1063/1.2159420

[99] T. Egami, Universal criterion for metallic glass formation, Mater. Sci. Eng. A., 226-228 (1997) 261. https://doi.org/10.1016/S0921-5093(97)80041-X

[100] E.S. Park, D.H. Kim and W.T. Kim, Parameter for glass forming ability of ternary alloy systems, Appl. Phys. Lett., 86 (2005) 061907. https://doi.org/10.1063/1.1862790

[101] K. Song, X.F. Bian, X.Q. Lu, M.T. Xie and R. Jia, The correlations among the fragility of supercooled liquids, the fragility of superheated melts and the glass-forming ability for marginal metallic glasses, J. Appl. Phys. 105 (2009) 024304. https://doi.org/10.1063/1.3043632

[102] D.V. Louzguine-Luzgin, D.B. Miracle and A. Inoue, Intrinsic and extrinsic factors influencing the glass-forming ability of alloys, Adv. Eng. Mater. 10 (2008) 1008. https://doi.org/10.1002/adem.200800134

[103] R.W. Cahn, Materials science: Melting and the surface, Nature, 323 (1986) 668. https://doi.org/10.1038/323668a0

[104] J.H. Perepezko, Nucleation in undercooled liquids, Mater. Sci. Eng., 65 (1984) 125. https://doi.org/10.1016/0025-5416(84)90206-4

[105] D.V. Louzguine-Luzgin and A. Inoue, An extended criterion for estimation of glass-forming ability of metals, J. Mater. Res., 22 (2007) 1378. https://doi.org/10.1557/jmr.2007.0167

[106] D.V. Louzguine-Luzgin and A. Inoue, Erratum: An extended criterion for estimation of glass-forming ability of metals, J. Mater. Res., 22 (2007) 2352. https://doi.org/10.1557/JMR.2007.0167e

[107] D.V. Louzguine-Luzgin, Vitrification and devitrification processes in metallic glasses, Proceedings of International Symposium on Metastable, Amorphous and Nanostructured Materials ISMANAM 2012, June 18-22, Moscow, Journal of Alloys and Compounds, 586 (2014) S2. https://doi.org/10.1016/j.jallcom.2012.09.057

[108] A.R. Yavari, Small volume change on melting as a new criterion for easy formation of metallic glasses, Phys. Lett. 95A (1983) 165. https://doi.org/10.1016/0375-9601(83)90825-3

[109] Y. Li, Q. Guo, J.A. Kalb and C.V. Thompson, Matching glass-forming ability with the density of the amorphous phase, Science 322 (2008) 1816. https://doi.org/10.1126/science.1163062

[110] N. Jakse and A. Pasturel, Local order and dynamic properties of liquid and undercooled Cu_xZr_{1-x} alloys by ab-initio molecular dynamics, Phys. Rev. B 78 (2008) 214204. https://doi.org/10.1103/PhysRevB.78.214204

[111] K. Russew, L. Stojanova, S. Yankova, E. Fazakas and L.K. Varga, Thermal behavior and melt fragility number of $Cu_{100-x}Zr_x$ glassy alloys in terms of crystallization and viscous flow, Journal of Physics Conference Series, 144 (2009) 012094. https://doi.org/10.1088/1742-6596/144/1/012094

[112] D.V. Louzguine-Luzgin, R. Belosludov, M. Saito, Y. Kawazoe and A. Inoue, Glass-transition behavior of Ni. Calculation, prediction and experiment, J. Appl. Phys. 104 (2008) 123529. https://doi.org/10.1063/1.3042240

[113] Y. Waseda, in Liquid Metals, edited by R. Evans and D.A. Greenwood, The Institute of Physics, Bristol, United Kingdom (1977) 230.

[114] H.H. Wendt and F.F. Abraham, Empirical criterion for the glass transition region based on monte carlo simulations, Phys. Rev. Lett., 41 (1978) 1245. https://doi.org/10.1103/PhysRevLett.41.1244

[115] http //lammps.sandia.gov/

[116] S. Plimpton, Fast parallel algorithms for short-range molecular dynamics, J. Comp. Phys, 117 (1995) 1. https://doi.org/10.1006/jcph.1995.1039

[117] M.I. Mendelev, S. Han, D.J. Srolovitz, G.J. Ackland, D.Y. Sun and M. Asta, Development of new interatomic potentials appropriate for crystalline and liquid iron, Phil. Mag., 83 (2003) 3977. https://doi.org/10.1080/14786430310001613264

[118] A.V. Evteev, A.T. Kosilov and E.V. Levchenko, Atomic mechanisms of pure iron vitrification, Journal of Experimental and Theoretical Physics, 99 (2004) 522. https://doi.org/10.1134/1.1809680

[119] R. Busch, The thermophysical properties of bulk metallic glass-forming liquids, JOM, 52 (2000) 39. https://doi.org/10.1007/s11837-000-0160-7

[120] H. Vogel, Das temperaturabhängigkeitsgesetz der viskosität von flüssigkeiten, Physik. Z. 22 (1921) 645

[121] G.S. Fulcher, Analysis of recent measurements of viscosity of glass, J. Am. Ceram. Soc. 8 339 (1925) 789. https://doi.org/10.1111/j.1151-2916.1925.tb18582.x

[122] G. Tammann and H. Hesse, Die Abhängigkeit der Viskosität von der Temperatur bei unterkühlten Flüssigkeiten, Z. Anorg. Allg. Chem. 156 (1926) 245. https://doi.org/10.1002/zaac.19261560121

[123] S. Mukherjee, J. Schroers, Z. Zhou, W.L. Johnson and W.-K. Rhim, Viscosity and specific volume of bulk metallic glass-forming alloys and their correlation with glass forming ability, Acta Materialia 52 (2004) 3689. https://doi.org/10.1016/j.actamat.2004.04.023

[124] W.T. Laughlin and D.R. Uhlmann, Viscous flow in simple organic liquids, J. Phys. Chem. 76 (1972) 2317. https://doi.org/10.1021/j100660a023

[125] C.A. Angell, Spectroscopy simulation and scattering and the medium range order problem in glass, J. Non-Cryst. Solids 73 (1985) 1. https://doi.org/10.1016/0022-3093(85)90334-5

[126] C.A. Angell, Glass formers and viscous liquid slowdown since David Turnbull enduring puzzles and new twists, MRS Bulletin 33 (2008) 544. https://doi.org/10.1557/mrs2008.108

[127] J.C. Mauro, Y.Z. Yue, A.J. Ellison, P.K. Gupta and D.C. Allan, Viscosity of glass-forming liquids, Proc. Natl. Acad. Sci. USA 106 (2009) 19780. https://doi.org/10.1073/pnas.0911705106

[128] W.L. Johnson, J. Lu and M.D. Demetriou, Deformation and flow in bulk metallic glasses and deeply undercooled glass forming liquids—a self consistent dynamic free volume model, Intermetallics, 10 (2002) 1039. https://doi.org/10.1016/S0966-9795(02)00160-7

[129] C.A. Angell, Formation of glasses from liquids and biopolymers, Science, 267 (1995) 1924. https://doi.org/10.1126/science.267.5206.1924

[130] P.G. Debenedetti and F.H. Stillinger, Supercooled liquids and the glass transition, Nature 410 (2001) 259. https://doi.org/10.1038/35065704

[131] D.-H. Huang and G.B. McKenna, New insights into the fragility dilemma in liquids, J. Chem. Phys., 114 (2001) 5621. https://doi.org/10.1063/1.1348029

[132] M.H. Cohen and D. Turnbull, Molecular transport in liquids and glasses, J. Chem. Phys. 31 (1959) 1164. https://doi.org/10.1063/1.1730566

[133] M.H. Cohen and G.S. Grest, Liquid-glass transition, a free-volume approach, Phys. Rev. 20 (1979) 1077. https://doi.org/10.1103/PhysRevB.20.1077

[134] L. Battezzati and A.L. Greer, The viscosity of liquid metals and alloys, Acta Metallurgica 37 (1989) 1791. https://doi.org/10.1016/0001-6160(89)90064-3

[135] W.F. Gale and T.C. Totemeier, editors. Smithells Metals Reference Book 8-th Edition, Elsevier Butterworth-Heinemann Ltd., Oxford UK (2004) 14.

[136] V.N. Novikov, Y. Ding and A.P. Sokolov, Correlation of fragility of supercooled liquids with elastic properties of glasses, Phys. Rev. E 71 (2005) 061501. https://doi.org/10.1103/PhysRevE.71.061501

[137] M.I. Ojovan, Viscosity and glass transition in amorphous oxides, Advances in Condensed Matter Physics, 2008 (2008) 817829. https://doi.org/10.1155/2008/817829

[138] M.I. Ojovan, About activation energy of viscous flow of glasses and melts, Materials Research Society Symposium Proceedings 1757 (2015) 7. https://doi.org/10.1557/opl.2015.44

[139] D.R. Uhlmann, A kinetic treatment of glass formation, Journal of Non-crystalline Solids 7 (1972) 337. https://doi.org/10.1016/0022-3093(72)90269-4

[140] D.V. Louzguine-Luzgin, R. Belosludov, A.R. Yavari, K. Georgarakis, G. Vaughan, Y. Kawazoe, T. Egami and A. Inoue, Structural basis for supercooled liquid fragility established by synchrotron-radiation method and computer simulation, Journal of Applied Physics 110 (2011) 043519. https://doi.org/10.1063/1.3624745

[141] M.I. Ojovan, Ordering and structural changes at the glass–liquid transition, Journal of Non-Crystalline Solids 382 (2013) 79. https://doi.org/10.1016/j.jnoncrysol.2013.10.016

[142] V. Wessels, A.K. Gangopadhyay, K.K. Sahu, R.W. Hyers, S.M. Canepari, J.R. Rogers, M.J. Kramer, A.I. Goldman, D. Robinson, J.W. Lee, J.R. Morris and K.F. Kelton, Rapid chemical and topological ordering in supercooled liquid $Cu_{46}Zr_{54}$, Phys. Rev. B 83 (2011) 094116. https://doi.org/10.1103/PhysRevB.83.094116

[143] K. Georgarakis, L. Hennet, G.A. Evangelakis, J. Antonowicz, G.B. Bokas, V. Honkimaki, A. Bytchkov, M.W. Chen and A.R. Yavari, Probing the structure of a liquid metal during vitrification, Acta Materialia 87 (2015) 174. https://doi.org/10.1016/j.actamat.2015.01.005

[144] K.F. Kelton, Kinetic and structural fragility - A correlation between structures and dynamics in metallic liquids and glasses, Journal of Physics Condensed Matter 29 (2017) 023002. https://doi.org/10.1088/0953-8984/29/2/023002

[145] N.A. Mauro, M. Blodgett, M.L. Johnson, A.J. Vogt and K.F. Kelton, A structural signature of liquid fragility, Nat. Commun. 5 (2014) 4616. https://doi.org/10.1038/ncomms5616

[146] H. Wang and E.A. Carter, Metal-metal bonding in Engel-Brewer intermetallics , anomalous, charge transfer in zirconium-platinum ($ZrPt_3$) J. Am. Chem. SOC 115 (1993) 2357. https://doi.org/10.1021/ja00059a034

[147] T. Iwashita, D. M. Nicholson and T. Egami, Elementary excitations and crossover phenomenon in liquids, Phys. Rev. Lett. 110 (2013) 205504. https://doi.org/10.1103/PhysRevLett.110.205504

[148] N. A. Mauro, M. Blodgett, M. L. Johnson, A. J. Vogt and K. F. Kelton, A structural signature of liquid fragility, Nature Commun. 5 (2014) 1. https://doi.org/10.1038/ncomms5616

[149] D.V. Louzguine-Luzgin, L.V. Louzguina-Luzgina and H. Fecht, On limitations of the viscosity versus temperature plot for glass-forming substances, Materials Letters, 182 (2016) 355. https://doi.org/10.1016/j.matlet.2016.07.006

[150] J.C. Mauro, D.C. Allan and M. Potuzak, Nonequilibrium viscosity of glass, Physical Review B 80 (2009) 094204. https://doi.org/10.1103/PhysRevB.80.094204

[151] A.E. Berlev, O.P. Bobrov, V.A. Khonik, K. Csach, A. Jurıkova, J. Miskuf, H. Neuhauser and M.Y. Yazvitsky, Viscosity of bulk and ribbon Zr-based glasses well

below and in the vicinity of T_g. A comparative study, Physical Review B 68 (2003) 132203. https://doi.org/10.1103/PhysRevB.68.132203

[152] Y. Kawamura and A. Inoue, Newtonian viscosity of supercooled liquid in a $Pd_{40}Ni_{40}P_{20}$ metallic glass, Applied Physics Letters, 77 (2000) 1114. https://doi.org/10.1063/1.1289502

[153] K. Csach O.P. Bobrov, V.A. Khonik and S.A. Lyakhov and K. Kitagawa, Relationship between the shear viscosity and heating rate of metallic glasses below T_g, Physical Review B 73 (2006) 092107. https://doi.org/10.1103/PhysRevB.73.092107

[154] G. Wilde, G.P. Görler, R. Willnecker and H.J. Fecht, Calorimetric, thermomechanical and rheological characterizations of bulk glass-forming $Pd_{40}Ni_{40}P_{20}$, J. Appl. Phys. 87 (2000) 1141. https://doi.org/10.1063/1.371991

[155] V.N. Novikov and A.P. Sokolov, Poisson's ratio and the fragility of glass-forming liquids, Nature 431 (2004) 961. https://doi.org/10.1038/nature02947

[156] H.L. Smith, C.W. Li, A. Hoff, G.R. Garrett, D.S. Kim, F.C. Yang, M.S. Lucas, T. Swan-Wood, J.Y.Y. Lin, M.B. Stone, D.L. Abernathy, M.D. Demetriou and B. Fultz, Separating the configurational and vibrational entropy contributions in metallic glasses, Nature Physics (2017) 4142.

[157] P.D. Gujrati and M. Goldstein, Viscous liquids and the glass transition. 9. Nonconfigurational contributions to the excess entropy of disordered phases, J. Phys. Chem. 84 (1980) 869. https://doi.org/10.1021/j100445a013

[158] D.V. Louzguine-Luzgin, A.I. Bazlov, S.V. Ketov, A.L. Greer and A. Inoue, Crystal growth limitation as a critical factor for formation of Fe-based bulk metallic glasses, Acta Materialia, 82 (2015) 396. https://doi.org/10.1016/j.actamat.2014.09.025

[159] V. Ponnambalam, S.J. Poon and J. Shiflet, Synthesis of iron-based bulk metallic glasses as nonferromagnetic amorphous steel alloys, J. Mater. Res., 19 (2004) 3046. https://doi.org/10.1557/JMR.2004.0374

[160] Z.P. Lu, C.T. Liu and W.D. Porter, Role of yttrium in glass formation of Fe-based bulk metallic glasses, Appl. Phys. Lett. 83 (2003) 2581. https://doi.org/10.1063/1.1614833

[161] T.A. Baser and M. Baricco, Fe-based bulk metallic glasses with Y addition, Journal of Alloys and Compounds, 434–435 (2007) 176. https://doi.org/10.1016/j.jallcom.2006.08.190

[162] M. Iqbal, J.I. Akhter, H.F. Zhang and Z.Q. Hu, Synthesis and characterization of bulk amorphous steels, J. of Non-Ciyst. Solids 354 (2008) 3284. https://doi.org/10.1016/j.jnoncrysol.2008.02.009

[163] A. Bouchare, B. Bendjemil, R. Piccin and M. Baricco, Formation and thermal properties of Fe-based BMG's with Y or Gd addition, Int. J. Nanoelectronics and Materials 3 (2010) 63.

[164] K. Amiya and A. Inoue, Fe-(Cr, Mo)-(C, B)-Tm bulk metallic glasses with high strength and high glass-forming ability, Rev. Adv. Mater. Sci., 18 (2008) 27.

[165] T. Zhang, Q. Yang, J.I. Fei, L. I. Ran, S.J. Pang, J.F. Wang and T. Xu, Centimeter-scale-diameter Co-based bulk metallic glasses with fracture strength exceeding 5000 MPa, Chinese Science Bulletin, 56 (2011) 3972. https://doi.org/10.1007/s11434-011-4765-8

[166] K.S. Filippov, Correlation between hereditary structures and properties of an $Fe_{50}Cr_{15}Mo_{14}C_{15}B_6$ bulk amorphous alloy in the solid and liquid states, Russian Metallurgy (Metally) 5 (2010) 379. https://doi.org/10.1134/S0036029510050034

[167] H. Nitta, K. Miura and Y. Iijima, Self-diffusion in iron-based Fe–Mo alloys, Acta Mater. 54 (2006) 2833. https://doi.org/10.1016/j.actamat.2006.02.020

[168] C. Hin, B.D. Wirth and J.B. Neaton, Formation of Y_2O_3 nanoclusters in nanostructured ferritic alloys during isothermal and anisothermal heat treatment. A kinetic Monte Carlo study, Phys. Rev. B 80 (2009) 134118. https://doi.org/10.1103/PhysRevB.80.134118

[169] A. Hirata, Y. Hirotsu, K. Amiya and A. Inoue, Crystallization process and glass stability of an $Fe_{48}Cr_{15}Mo_{14}C_{15}B_6Tm_2$ bulk metallic glass, Physical Review B 78 (2008) 144205. https://doi.org/10.1103/PhysRevB.78.144205

[170] D.V. Louzguine-Luzgin, A.I. Bazlov, S.V. Ketov and A. Inoue, Crystallization behavior of Fe- and Co-based bulk metallic glasses and their glass-forming ability, Materials Chemistry and Physics, 162 (2015) 197. https://doi.org/10.1016/j.matchemphys.2015.05.058

[171] D. Turnbull, Under what conditions can a glass be formed?, Contemporary Physics 10 (1969) 473. https://doi.org/10.1080/00107516908204405

[172] F. Spaepen and D. Turnbull, Formation of metallic glasses, In Proceedings of 2nd International Conference on Rapidly Quenched Metals, edited by N.J. Grant and B.C. Giessen (MIT Press, Cambridge, MA) (1976) 205.

[173] D.V. Louzguine-Luzgin, Y. Zeng, A.D.H. Setyawan, N. Nishiyama, H. Kato, J. Saida and A. Inoue, Deformation behavior of Zr- and Ni-based bulk glassy alloys, Journal of Materials Research, 22 (2007) 1087. https://doi.org/10.1557/jmr.2007.0126

[174] A. Inoue, K. Ohtera, A.P. Tsai and T. Masumoto, Glass transition behavior of Al-Y-Ni and Al-Ce-Ni amorphous alloys, Jpn. J. Appl. Phys., 27 (1988) 280. https://doi.org/10.1143/JJAP.27.L280

[175] D.V. Louzguine and A. Inoue, Investigation of structure and properties of the Al-Y-Ni-Co-Cu metallic glasses, Journal of Materials Research, 17 (2002) 1014. https://doi.org/10.1557/JMR.2002.0149

[176] H. Kato, J. Saida and A. Inoue, Influence of hydrostatic pressure during casting on as cast structure and mechanical properties in $Zr_{65}Al_{7.5}Ni_{10}Cu_{17.5-x}Pd_x$ (x = 0, 17.5) alloys, Scripta Mater., 51 (2004) 1063. https://doi.org/10.1016/j.scriptamat.2004.08.004

[177] K.J. Laws, B. Gun and M. Ferry, Influence of casting parameters on the critical casting size of bulk metallic glass, Metallurgical and Materials Transactions A 40 (2009) 2377. https://doi.org/10.1007/s11661-009-9929-7

[178] T. Tamura, K. Amiya, R.S. Rachmat, Y. Mizutani and K. Miwa, Electromagnetic vibration process for producing bulk metallic glasses, Nature Mater., 4 (2005) 289. https://doi.org/10.1038/nmat1341

[179] H.W. Kui, A.L. Greer and D. Turnbull, Formation of bulk metallic glass by fluxing, Appl. Phys. Lett., 45 (1984) 615. https://doi.org/10.1063/1.95330

[180] Y. He, R.B. Schwarz and J.I. Archuleta, Bulk glass formation in the Pd–Ni–P system, Appl. Phys. Lett., 69 (1996) 1861. https://doi.org/10.1063/1.117458

[181] G. Wilde, J.L. Sebright and J.H. Perepezko, Bulk liquid undercooling and nucleation in gold, Acta Mater., 54 (2006) 4759. https://doi.org/10.1016/j.actamat.2006.06.007

[182] A. Inoue, N. Nishiyama and T. Matsuda, Preparation of bulk glassy $Pd_{40}Ni_{10}Cu_{30}P_{20}$ alloy of 40 mm in diameter by water quenching, Mater. Trans. JIM, 37 (1996) 181. https://doi.org/10.2320/matertrans1989.37.181

[183] N. Nishiyama and A. Inoue, Glass-forming ability of $Pd_{42.5}Cu_{30}Ni_{7.5}P_{20}Pd_{42.5}Cu_{30}Ni_{7.5}P_{20}$ alloy with a low critical cooling rate of 0.067 K/s, Appl. Phys. Lett., 80 (2002) 568. https://doi.org/10.1063/1.1445475

[184] N. Nishiyama and A. Inoue, Glass-forming ability of bulk $Pd_{40}Ni_{10}Cu_{30}P_{20}$ alloy, Mater Trans JIM 37 (1996) 1531. https://doi.org/10.2320/matertrans1989.37.1531

[185] I.-R. Lu, G. Wilde, G.P. Görler and R. Willnecker, Thermodynamic properties of Pd-based glass-forming alloys, J. Non-Cryst. Solids, 250–252 (1999) 577. https://doi.org/10.1016/S0022-3093(99)00135-0

[186] H.S. Chen, Thermodynamic considerations on the formation and stability of metallic glasses, Acta Metall. 22 (1974) 1505. https://doi.org/10.1016/0001-6160(74)90112-6

[187] K.F. Yao and C.Q. Zhang, Fe-based bulk metallic glass with high plasticity, Appl. Phys. Lett. 90 (2007) 061901. https://doi.org/10.1063/1.2437722

[188] C. Duhamel, K.G. Georgarakis, A. LeMoulec, A.R. Yavari, G. Vaughan, A. Baron and N. Lupu, Influence of fluxing in the preparation of bulk Fe-based glassy alloys, Journal of Alloys and Compounds, 483 (2009) 243. https://doi.org/10.1016/j.jallcom.2008.08.132

[189] N. Nishiyama and A. Inoue, Flux treated Pd-Cu-Ni-P amorphous alloy having low critical cooling rate, Mater Trans JIM. 38 (1997) 464. https://doi.org/10.2320/matertrans1989.38.464

[190] T.D. Shen and R.B. Schwarz, Lowering critical cooling rate for forming bulk metallic glass, Appl. Phys. Lett. 88 (2006) 091903. https://doi.org/10.1063/1.2172160

[191] N. Chen, L. Gu, G.Q. Xie, D.V. Louzguine-Luzgin, A.R. Yavari, G. Vaughan, S.D. Imhoff, J. H. Perepezko, T. Abe and A. Inoue, Flux-induced structural modification and phase transformations in a $Pd_{40}Ni_{40}Si_4P_{16}$ bulk-glassy alloy, Acta Materialia, 58 (2010) 5886. https://doi.org/10.1016/j.actamat.2010.07.003

[192] A.Yu. Churyumov, A.I. Bazlov, A.A. Tsarkov, A.N. Solonin and D.V. Louzguine-Luzgin, Microstructure, mechanical properties and crystallization behavior of Zr-based bulk metallic glasses prepared under a low vacuum, Journal of Alloys and Compounds, 654 (2016) 87. https://doi.org/10.1016/j.jallcom.2015.09.003

[193] M. Stoica, S. Kumar, S. Roth, S. Ram, J. Eckert, G. Vaughan and A.R. Yavari, Crystallization kinetics and magnetic properties of $Fe_{66}Nb_4B_{30}$ bulk metallic glass, Journal of Alloys and Compounds, 483 (2009) 632. https://doi.org/10.1016/j.jallcom.2007.11.150

Chapter 3

Structure

Crystals are characterized by a unit cell and existence of a translational symmetry. Quasicrystals do not have such a cell in three-dimensional space but exhibit a rotational symmetry. The structure of liquids and glasses is less well understood. As the atomic structure of bulk metallic glasses defines their properties, in particular mechanical, research on the structure of glassy alloys is an important topic of modern physical metallurgy.

Contents

3.1   Structure of liquids and glasses

Structure of liquids [1] or a glasses was studied by X-ray diffraction (small and wide angle methods) [2,3], including synchrotron radiation X-ray experiments [4,5], neutron diffraction [6] (especially useful when alloys contain light elements like B, Si, C, P etc...), transmission electron microscopy [7,8] and fluctuation electron microscopy (FEM) [9] experiments. It was also successfully studied by modeling: the reverse Monte Carlo modeling (RMC) [10,11], classical [12] and the quantum mechanics molecular dynamics simulation (MD) [13]. The Lennard-Jones (LJ) potential utilized for simulating the systems with van der Waals interactions [14] is not suitable for modeling metallic glasses. The embedded-atom method (EAM) many-body potential [15] is more efficient and usually applicable for metals and alloys.

The structure of metallic glasses looks completely disordered at first glance as can be seen from the high-resolution TEM image in Fig. 3.1. However, for a glassy structure it is possible to define the degree of topological (TSRO) and chemical short range order (CSRO). CSRO depends on chemical environment of the nearest-neighbor atoms as deviation from the expectation of a random distribution [16]. In general, a certain atom in a metallic glass attracts atoms of different kind as the nearest neighbors. A typical example is related to the metalloid centered clusters observed in metal-metalloid glasses, for example in Ni-P metallic glasses [17].

Fig. 3.1. High-resolution TEM image of a Zr-based BMG. The insert exhibits a nanobeam diffraction pattern.

Conventional X-ray diffraction pattern produces a series of broad maxima (Fig. 3.2 (a)) while selected-area and nanobeam electron diffraction patterns produce a series of diffraction rings (Fig. 3.1 insert).

Fig. 3.2. Scattering intensity (a) and radial distribution function (b) of the $Cu_{60}Zr_{30}Ti_{10}$ alloy.

In order to produce the radial distribution function (RDF) shown in Fig. 3.2 (b) correction for the scattering from the sample container, air scattering, polarization, absorption [18], and Compton scattering [19] must be done. The measured intensity can be converted to electron units per atom with the generalized Krogh-Moe-Norman method [20], using the X-ray atomic scattering factors and anomalous dispersion corrections [21]. The total structure factor S(Q) and the interference function $Qi(Q) = Q[S(Q) – 1]$ $(Q = 4\pi \cdot \sin\theta/\lambda$, θ is the diffraction angle) can be obtained from the coherent scattering intensity by using atomic scattering factors. The radial distribution RDF and pair distribution functions PDF can be subsequently obtained by the Fourier transform of $Qi(Q)$:

$$\text{RDF(R)} = 4\pi r^2 \rho(R) = 4\pi r^2 \rho_0 + 2R/\pi \int_0^{Q_{max}} Q(S(Q) - 1)\sin(QR)dQ \quad (3.1)$$

where $\rho(R)$ is the total radial number density function and ρ_0 is the average number density of the sample. RDF also shows a series of the maxima as a function of the interatomic distance (Fig. 2b) indicative of the short and medium range order.

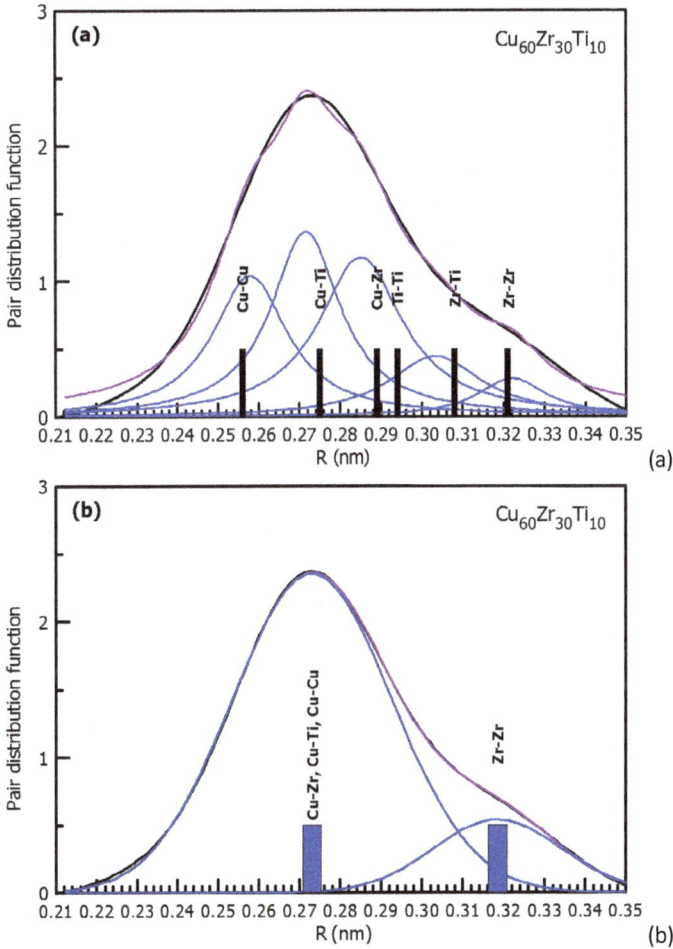

Fig. 3.3. Fitting of the pair distribution function of the $Cu_{60}Zr_{30}Ti_{10}$ alloy with 5 and 2 Gaussian peaks. Reproduced from [22] with permission of Elsevier.

Lorentzian function fitting of the first maximum of the reduced RDF from 0.21 to 0.35 nm of $Cu_{60}Zr_{30}Ti_{10}$ showed good correspondence to the Cu-Cu, Cu-Ti, Cu-Zr, Zr-Ti and Zr-Zr interatomic distances produced a reasonable correspondence with the original PDF(r) data of the $Cu_{60}Zr_{30}Ti_{10}$ alloy (Fig. 3.3) [22]. As the maximum of the left shoulder (0.273 nm) corresponds to lower value than the predicted Cu-Zr interatomic distance (0.289 nm) one can assume that this distance in the studied glassy alloy is lower than that predicted by the Goldschmidt atomic radii due to negative mixing enthalpy (attractive interaction) in this atomic pair which is quite common in the intermetallic compounds, for example, in oC68 $Cu_{10}Zr_7$ phase [23]. A similar decrease in the interatomic distances owing to charge transfer was observed in Zr-based alloys [24,25].

An icosahedral-type medium range order which is expected to be a dominant local configuration in liquids and metallic glasses in terms of the distorted icosahedra [26] may be a good explanation why no short-range order crystal-like atomic arrays are observed in the HRTEM images (Fig. 3.1) of Cu-Zr- and Zr-Cu-based metallic glasses. The structure of Cu-Zr glassy alloys is one of the most widely studied by various methods including MD simulation [27,28,29]. The structure of Zr–Cu binary metallic glasses except for the $Zr_{67}Cu_{33}$ one can be modeled by statistics of Cu–Zr, Cu–Cu, and Zr–Zr nearest neighbor populations in an "ideal solid solution" [30]. The existence of icosahedral structure was found in Cu-Zr metallic glasses [31,32]. On the other hand, the atomic structure of Cu–Zr glasses analyzed by means of the reverse Monte Carlo (RMC), XRD, neutron diffraction and extended X-ray absorption fine structure (EXAFS) methods did not show a dominant special atomic structure [33].

Some of crystalline compounds have the structure resembling that of metallic glasses. However, as in the glassy state there is a wider variation of the interatomic distances than in the crystal one can perform integration of the number of interatomic distances in the crystal. The structure of the $Cu_{60}Zr_{30}Ti_{10}$ glassy alloy in the first coordination shell can be described by that of the oC68 $Cu_{10}Zr_7$ compound. The structure of Zr-based alloys having composition close to $Zr_2(TM,Al)$ might be described as an aggregation of the atomic clusters of a metastable cF96 Zr_2Ni compound (a=1.227 nm; Ni has nearly the same atomic radius as Cu) which is quite common metastable phase formed in Zr-based glassy alloys. The RDF(r)^g of the $Pd_{40}Cu_{30}Ni_{10}P_{20}$ glassy alloy can be described with a complex cI58 α-Mn structure (a=0.89125 nm) [34,35] provided that the interatomic distances are integrated at 0.01 nm step and shifted towards larger r values by a factor of 1.09 in accordance with a larger atomic size of Pd [36].

More detailed structural information can be obtained by using anomalous X-ray scattering (AXS) [37], the X-ray absorption fine structure (XAFS) [38], including the extended X-ray absorption fine structure (EXAFS) [39,40] and X-ray absorption near edge structure (XANES) [41] when environmental RDFs for certain atomic pairs can be obtained. Upon anomalous X-ray scattering experiments energy dependencies of the anomalous dispersion terms should be used for each constituent element. The difference between the scattering intensities for the element A, $\Delta i_A(Q, E_1, E_2)$, measured at two energies of E_1 and E_2 that are usually 300 and 25 eV lower than the energies of the corresponding absorption edges, respectively is a function of the scattering vector Q are:

$$\Delta i_A(Q, E_1, E_2) \cong \frac{\{I_{obs}(Q, E_1) - \langle f^2(Q, E_1) \rangle\} - \{I_{obs}(Q, E_2) - \langle f^2(Q, E_2) \rangle\}}{c_A\{f'_A(E_1) - f'_A(E_2)\}W(Q, E_1, E_2)} =$$

$$= \sum_{m=1}^{n} \frac{c_m \Re\{j_m(Q, E_1) + j_m(Q, E_2)\}}{W(Q, E_1, E_2)}\{a_m(Q) - 1\},$$

(3.2)

where

$$W(Q, E_1, E_2) = \sum_{m=1}^{n} c_m \Re\{f_m(Q, E_1) + f_m(Q, E_2)\}$$

where n is the number of element, the values c_m and f_m are the atomic fraction and the X-ray atomic scattering factor of the m-th element, respectively. $\Re$ denotes the real part of the values in brackets. Energy dependencies of the anomalous dispersion terms can be calculated the relativistic Cromer and Liberman method [42].

The method is illustrated using the research data for the $Ge_{50}Al_{40}Cr_{10}$ alloy [37]. The partial pair distribution functions are shown in Fig. 3.4.

Fig. 3.4. Partial pair distribution functions of the $Ge_{50}Al_{40}Cr_{10}$ alloy. Reproduced from [37] with permission of the Physical Society of Japan.

The results obtained in the present study indicate that the structure of the $Ge_{50}Al_{40}Cr_{10}$ alloy is inhomogeneous at nanoscale. Some medium range order (MRO) zones of about 1 nm were observed in the high-resolution TEM image. A similar structure is formed in the $Ge_{55}Al_{30}Cr_{10}La_5$ metallic glass [43]. Nanoscale zones produce distinct spots in the dark-field TEM images (Figure 3.5), even though their size is estimated at about 1-2 nm. This may indicate separation to Ge-rich and Ge-poor zones producing a double type halo pattern with clearly visible two haloes one corresponding to the scattering vector (Q) equal to 23.5 nm^{-1} another to 30.2 nm^{-1}. One of these regions may be a precursor for further nanocrystallization observed on heating.

Fig. 3.5. (a) bright-field image of the $Ge_{55}Al_{30}Cr_{10}La_5$ metallic glass in as-solidified state with the selected-area electron diffraction pattern as an inset. (b and c) dark-field images, respectively, taken from the first and the second ring of the separated halo pattern (the inset). The first and the second rings are marked as (1) and (2).

The partial pair distribution functions (see Fig. 3.4) indicate that corresponding pairs of the alloying elements have their own preferential interatomic distances. Only the Ge-Ge and Ge-Cr pairs have one sharp peak in the first coordination shell, whereas corresponding functions of the Ge-Al, Cr-Cr, Cr-Al and Al-Al atomic pairs show two subpeaks. The interatomic distance of about 0.255 nm is usual around Ge atoms, whereas partial pair distribution functions around Al atoms show two distances of about 0.246 and 0.280 nm. The partial coordination number in the Ge-Ge pair of 3.4 is not far from that value of 4 typical for amorphous Ge. On the other hand the values for other atomic pairs are closer to that of usual metallic glasses. Thus, Ge forms a framework like a-Ge which is different from the distribution of the other elements. The results obtained are in agreement with the idea of separation over a short range into Ge-rich and Al-rich amorphous phases. The interatomic distances are shortened owing to strong Ge-Cr interactions. Also, in the Al-Fe-Ce system alloys the Al-Fe distances are significantly shortened [44] and a strong interaction between Al and Fe leads to chemical and topological short-range ordering. It indicates that the Al–Fe pairs (and in general Al-TM pairs) involve covalent-like interaction originating from the hybridization of the $3d$-states of Fe with the sp-sates of Al [45]. Local chemical order was studied in Al–Tb alloys and found to be inhomogeneous on nanoscale [46]. MD simulations also showed pronounced chemical SRO in the Al-Mn and Al-Ni alloys [47].

Pair distribution functions can be also obtained by using the selected-area electron diffraction patters (SAED) representing two-dimensional intensity distribution in the reciprocal space. For example, PDF(R) of the $Fe_{48}Cr_{15}Mo_{14}C_{15}B_6Y_2$ glassy alloy in as-cast state is shown in Fig. 3.6. One-dimensional intensity distribution function was obtained by circular integration of the scattered intensity over 180 degrees around the center of the spot produced by the transmitted beam (000).

Fig. 3.6. PDF(R) of the $Fe_{48}Cr_{15}Mo_{14}C_{15}B_6Y_2$ glassy alloy (black curve fitted with 3 Gaussian functions shown in green). The resulted fitting curve is marked in red. The corresponding SAED pattern is shown in the inset.

By using the interatomic forces calculations of the atomic structure and of the thermodynamic properties of the crystalline, liquid, and amorphous phases have been performed since long ago [48]. The interrelation between the formation of topologically close-packed intermetallic compounds, eutectics, and metallic glasses was found to depend on optimal embedding of the neighboring atoms. Recently MD simulation of Fe was preformed by LAMMPS package [49,50] at constant pressure with an embedded atom method potential [51] and used to model the structure of liquid (Fig. 3.7 (a)) and glassy (Fig. 3.7 (b)) iron. Visualization and analysis of the atomistic simulation data was performed with OVITO - the Open Visualization Tool [52]. The atomic structure exhibits no crystals though the positions of the maxima in the reduced radial distribution function also called pair distribution function (PDF) correspond quite well to those of crystalline α-Fe (Fig. 3.7 (c)). Visually the structures Fig. 3.7 (a) and (b) look very similar but the PDFs show that the structure of a glass is more ordered than that of a liquid indicating a stronger PDF peak. Another clear difference is splitting of the second PDF maximum observed in the case of glass. This is a common feature observed in many glasses. The first PDF maximum is also asymmetrical as consisting of two peaks.

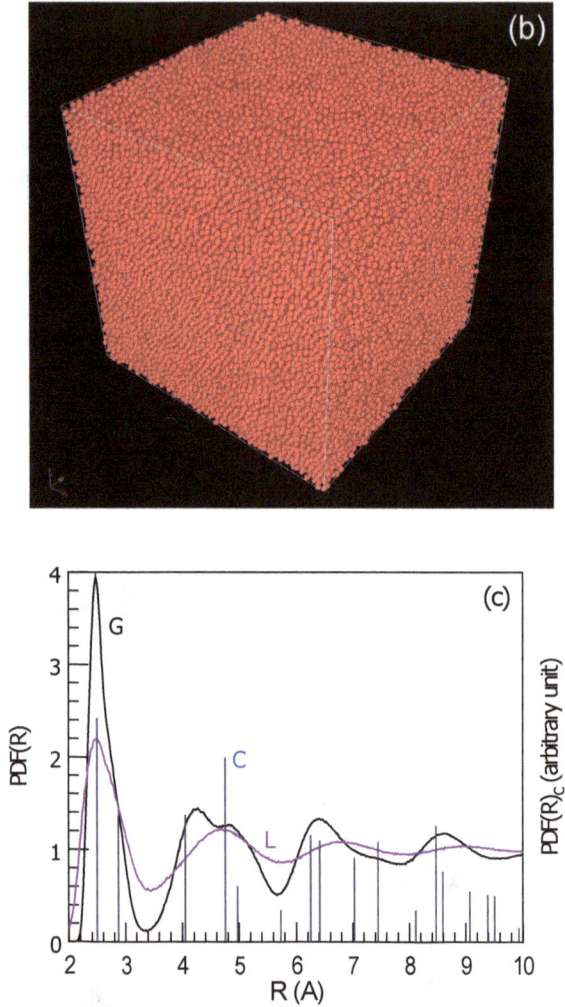

Fig. 3.7. MD simulation of liquid Fe preformed by LAMMPS package at nearly constant pressure with an embedded atom potential. (a) an atomic cell at 2500 K and (b) the same atomic cell at 300 K after cooling. (c) PDFs of crystalline (C), liquid (L) and glassy (G) phase. Note that although (a) and (b) look very similar their PDFs (c) are significantly different.

The coordination number determined from the experimental PDF of the first peak is used as a measure of local order and compared with the results of simulation. The bond-orientational order [53] (recently discussed in detail [54]), the common neighbor [55], the Voronoi polyhedral analysis displayed in Fig. 3.8 (a) [56] and other methods can be used to study the atomic structures of glasses produced by simulation.

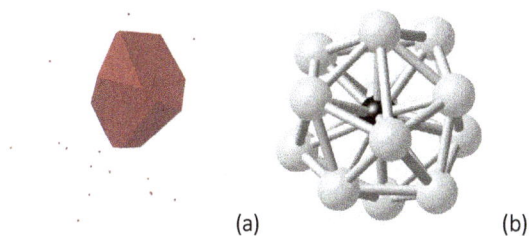

(a)   (b)

Fig. 3.8. A Voronoi cell produced by sections of planes located at the centers between the points in space (a) (courtesy of A. Tsarkov), distorted icosahedron in a metallic glass (b).

The structure of metallic liquids as a package of hard spheres (dense random packing) was proposed by Bernal [57] using a tetrahedron, octahedron, tetragonal dodecahedron, capped trigonal prism and Archimedean capped antiprism. Similar following works have been also published afterwards [58,59]. However, dense random packing method cannot fill out the space efficiently. The maximum packing fraction is significantly lower than the packing fraction of the closed packed crystalline lattices. Moreover, metallic atoms are not "hard" but rather "soft" spheres. Owing to a small density difference between bulk metallic glassy and the corresponding crystalline structure [60] it become clear that the metallic glasses are quite well packed in expense of their short and medium-range order. According to modern viewpoint the structure of metallic glasses consists of the efficiently packed atomic clusters rather than dense random packing of atoms [61]. Cluster-based models of glassy structure have been issued recently [62,63] and the importance of an efficient atomic packing for the formation of metallic glasses is emphasized [64,65]. It explains high relative densities of metallic glasses close to those of the crystalline counterparts [66].

It was shown that Pd- and Ni-based bulk glassy alloys of Pd-Si [67], Pd-Ni-P [68,69,70], Pd-Cu-Si [71] and Ni-Pd-P [72] systems contain clear medium-range order zones. The structure models created for the Fe-B glassy alloys showed existence of the prismatical clusters typical for Fe-boride compounds formed around B atoms [73,74]. Frank-Kasper polyhedra and defective icosahedral clusters (Fig. 3.8 (b)) were found in metallic glasses

[75]. Quasicrystalline-type short-range order found in Zr-based metallic glasses consisting of distorted icosahedra was demonstrated by sub-nanometer beam electron diffraction [76]. The local atomic structures around Pt as well as Zr in the amorphous and quasicrystalline Zr-Al-Ni-Pt alloys determined by the anomalous X-ray scattering method [77] showed possible existence of the chemical short range order clusters in the glassy state similar to those of a quasicrystal. This idea is in a good consistency with the formation of the nanoscale icosahedral phase observed in a Cu-Zr-Ti-Pd alloy on devitrification [78,79]. However, as has been already mentioned there are various metallic glasses, for example, Fe-based and Al-based where icosahedral clusters are not the dominant structure units.

The reduced supercooling (undercooling) before crystallization from the melt was found to be the largest for crystal line phases, smaller for crystal approximants and the smallest for quasicrystals indicating possible structural similarity to metallic glassy phase [80]. It is also reasonable to assume that complex structural units of quasicrystals beyond the first shell [81] also exist in metallic glasses. Such a research topic deserves further investigation.

Recently, a new parameter called flexibility volume ($v_{flex} = <r^2>/a^2 \cdot \Omega_a$, where $<r^2>$ is the mean root squared displacement, a is the interatomic distance and Ω_a is the average atomic volume) was introduced as an universal indicator, to bridge the structural state of metallic glasses with their properties, on both atomic and macroscopic levels [82]. The flexibility volume combines static atomic volume with dynamics information via atomic vibrations that probe local configurational space and interaction between the neighboring atoms. The flexibility volume can quantitatively predict the shear modulus and it correlates strongly with atomic packing topology, and with the activation energy for thermally activated relaxation.

The surface structure of the $Ni_{62}Nb_{38}$ metallic glass was also found to consist of atomic clusters (Fig. 3.9) [83,84]. A simple empirical model for the estimation of the cluster size was suggested, and the value obtained (about 0.8 nm) was in good agreement with the corresponding scanning tunneling microscopy (STM) observation. The results were compared with those obtained by transmission electron microscopy and nano-beam diffraction measurements.

Fig. 3.9. STM image of a Ni-Nb metallic glass indicating a clustered structure. Courtesy of A. Oreshkin.

The as-cast $Au_{49}Cu_{26.9}Ag_{5.5}Pd_{2.3}Si_{16.3}$ [85] BMG rods contained a micrometer-scale crystalline surface layer containing the Au-based solid solution grains embedded in the amorphous phase (with a small part of an intermetallic phase) [86], which was responsible for nice yellow color of the samples surface. Internal part of the ingot is grey metallic owing to relatively high Si content. The origin of such an effect could be surface nucleation and growth of Au crystal layer. It was found that in the liquid state somewhat above the eutectic temperature this Au-based alloy also contains a very thin layer of a crystalline phase [87]. The thickness of this crystalline layer depends on the casting temperature. In the samples cast from 720 K the thickness was 80 μm while when casted from 1570 K the thickness was only about 10 μm. The ribbon samples had no such a layer owing to significantly higher cooling rate (about 10^6 K/s). One should also mention that a liquid-like layer was found to exist on top of lead crystal down to about 40 K below the liquidus temperature as a consequence of the surface effects [88].

Recently-discovered materials called nanoglasses or nano-structured metallic glasses consisting of nanoscale spheroid glassy particles separated by glassy interfaces represent a new class of metallic glasses which demonstrate good mechanical and chemical properties. The former material is produced by the compaction of nanometer-sized amorphous spheres formed by the inert gas condensation [89] while the latter one is formed by magnetron sputtering (Fig. 3.10) [90,91].

Fig. 3.10. SEM image of the $Zr_{62.5}Pd_{37.5}$ nanostructured metallic glass obtained in secondary electrons.

3.2 Thermal expansion detected by X-ray diffraction

The thermal expansion of the glassy alloys and melts upon heating or cooling was studied by Synchrotron radiation X-ray diffraction for the $Fe_{40}Ni_{40}P_{14}B_6$ [92], $Pd_{40}Cu_{30}Ni_{10}P_{20}$ [93], $Zr_{55}Cu_{30}Ni_5Al_{10}$ [94 , 95], various La-based [96 , 97], $Cu_{55}Hf_{25}Ti_{15}Pd_5$ and $Cu_{55}Zr_{30}Ti_{10}Ni_5$ [98] compositions. As an example, shift in the first PDF maximum of the $Cu_{55}Hf_{25}Ti_{15}Pd_5$ glassy alloy with temperature on heating is shown in Fig. 3.11 [99]. The least squares fitting gives the thermal expansion coefficient (β) of $1.6 \cdot 10^{-5}$ K^{-1}. The linear coefficient of thermal expansion measured by thermo-mechanical analysis in the range of 300-530 K was 1.5 ± 0.1 10^{-5} K^{-1} [98].

3.3 Structural changes on glass formation from the liquid state

The evolution of atomic structure of liquids [100, 101] including liquid metals [102,103] on undercooling/supercooling was studied. [104]. For example, a structural change in liquid Ga is discovered at the critical temperature of about 1000 K [105]. Below this temperature, the liquid Ga has a large increasing slope of peak position shift of the first peak of the structure factor. Above this temperature, a small reducing slope of peak-position shift of the first peak of the structural factor and pair-distribution functions. Liquid polymorphism was also reported to be observed in the case of Te-Ge alloys [106] and In [107] as a function of temperature.

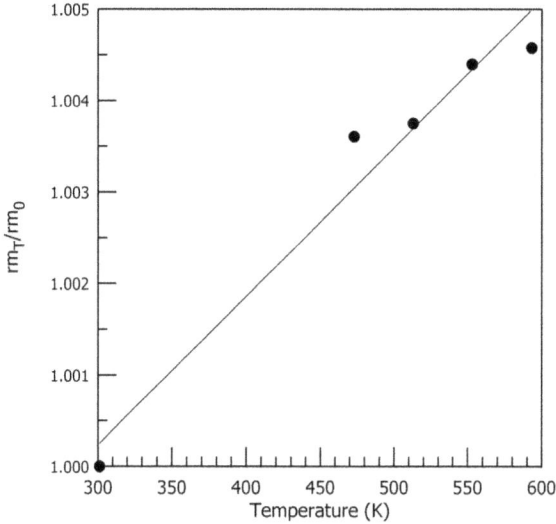

Fig. 3.11. Location of the main peak of PDF(r) rm_T divided by that at room temperature rm_0 as a function of temperature. Reproduced from [99] with permission of AIP Publishing.

The structural changes and vitrification of the $Pd_{42.5}Cu_{30}Ni_{7.5}P_{20}$ liquid alloy on cooling from above the equilibrium liquidus temperature were also studied in-situ by synchrotron radiation X-ray diffraction and compared with the results of ab-initio computer simulation [108,109]. $Pd_{42.5}Cu_{30}Ni_{7.5}P_{20}$ forms a relatively fragile melt (in terms of its temperature dependence of the viscosity). Its fragility index m is close to 60 [110]. The experimental setup is schematically shown in Fig. 3.12. Several sets of heating and cooling experiments were performed. As a result a series of PDF(R) spectra were recorded in-situ at different temperatures (Fig. 3.13).

Fig. 3.12. Schematic representation of the experimental setup. Reproduced from [108] with permission of Elsevier.

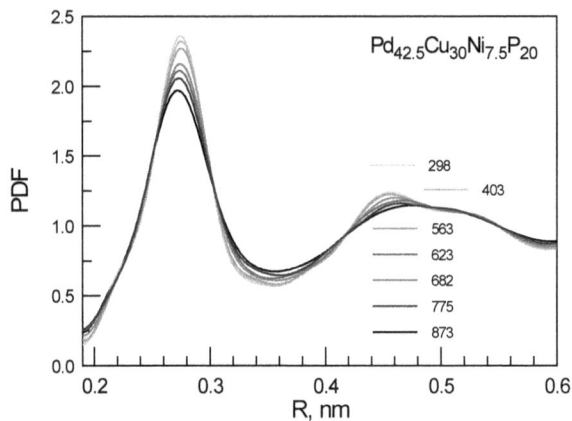

Fig. 3.13. PDF(R) functions of the studied alloy obtained at different temperatures given in Kelvin. Reproduced from [108] with permission of Elsevier.

Intensification of the shoulder in the second coordination peak found in the $Pd_{42.5}Cu_{30}Ni_{7.5}P_{20}$ liquid alloy at about 0.45 nm (Fig. 3.13) on cooling was also observed in the $Fe_{80}B_{20}$ liquid alloy at 0.415 nm (Fig.3.14 (a)) [111].

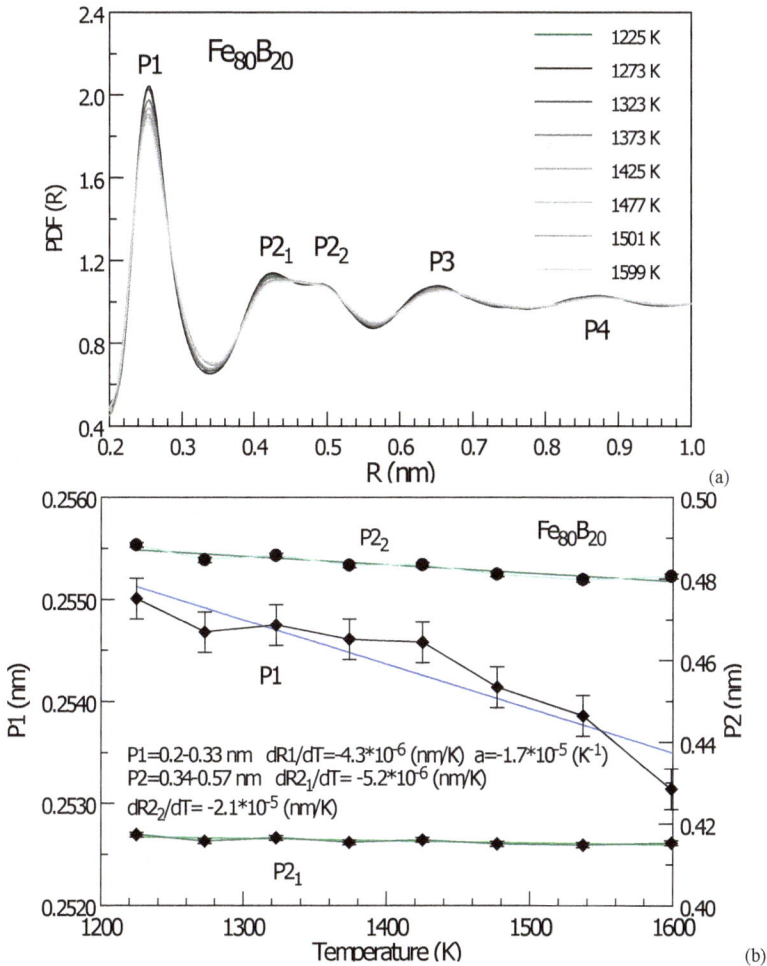

Fig. 3.14. (a) PDF and the corresponding maxima of the $Fe_{80}B_{20}$ liquid alloy in the 1st, 2nd, 3rd and 4th coordination shells. (b) Displacement of P1 and P2 (two subpeaks/shoulders of P2 are marked as $P2_1$ and $P2_2$) with temperature. The curves are drawn through the experimental points, while linear fits were used to calculate the average thermal expansion coefficient only. Reproduced from [111] with permission of Elsevier.

The thermal expansion coefficient is negative in the case of the initial two PDF(R) peaks of the $Fe_{80}B_{20}$ alloy (Fig. 3.14 (b) while for pure Fe it is negative only for the first peak. The positions of peak 3 and 4 of PDF(R) exhibit a positive shift with temperature. This behavior argues for continuous structural changes in the supercooled liquids. Intensive changes in the short and medium range order in the supercooled liquid are also reflected in the different thermal expansion coefficients obtained for the subsequent atomic coordination shell [112,113].

The integration of the first RDF(R) (radial distribution function) maximum of the $Pd_{42.5}Cu_{30}Ni_{7.5}P_{20}$ liquid alloy (both subpeaks from 0.20 nm to 0.35 nm) (not shown here) produced the area under the peak (A) corresponding to the coordination number (CN) in the first coordination shell at room temperature of 12.6 (Fig. 3.15). It indicates formation of rather a densely packed structure.

Fig. 3.15. Coordination number (CN) from the RDF(R) in the first coordination shell (0.2-0.35 nm) as a function of temperature on cooling as well as two fits – linear below T_g and exponential above it. Reproduced from [108] with permission of Elsevier.

Fitting of the first PDF(R) maximum from 0.2 to 0.35 nm (baseline was corrected to make both outermost points be equal to zero) using two Gaussian peak functions produced a reasonable correspondence with the original PDF(R) plot (Fig. 3.16). According to the interatomic distances, the first subpeak (P1) corresponds to nearest Cu-P or Ni-P atomic pairs while the second subpeak (P2) mostly corresponds to the mixture of Pd-Pd, Pd-Cu and Pd-Ni pairs. Radial distribution functions for three different Pd-based alloys were compared in Ref. [114] and their comparative analysis was performed.

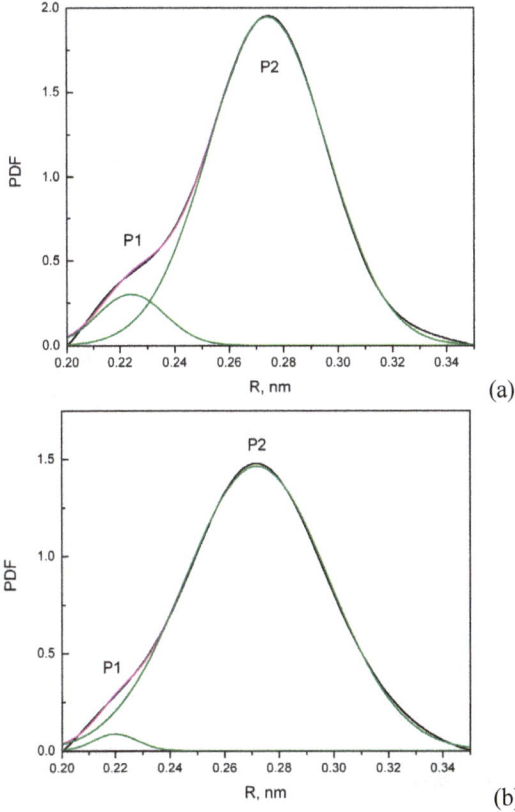

Fig. 3.16. Fitting of the first PDF(R) maximum (black curve) from 0.2 to 0.35 nm using two Gaussian functions related to P1 (left) and P2 (right). The curves in (a) correspond to 298 K while the curves in (b) to 873 K. The baseline was corrected prior to fitting. Reproduced from Ref. [109] with permission from the American Institute of Physics.

Fig. 3.17 shows the area under the subpeak P1 divided by the area under the subpeak P2 obtained on cooling. A drastic change in this ratio above T_g is observed.

Fig. 3.17. (a) The area under the subpeak P1 (A(P1)) divided by the area under P2 (A(P2)) in the first coordination shell obtained on cooling (solid circles) as well as the viscosity of the studied alloy (solid line) as a function of temperature. (b) The same plot using the reciprocal temperature scale normalized by T_g. Reproduced from Ref. [109] with permission of the American Institute of Physics.

A high degree of the short-range order was observed consistent with models which predict that metallic glasses are not made of random packing of atoms but consists of dense packing of clusters [61,62] which is also consistent with the small volume difference between glassy and the corresponding crystalline phases [115,116].

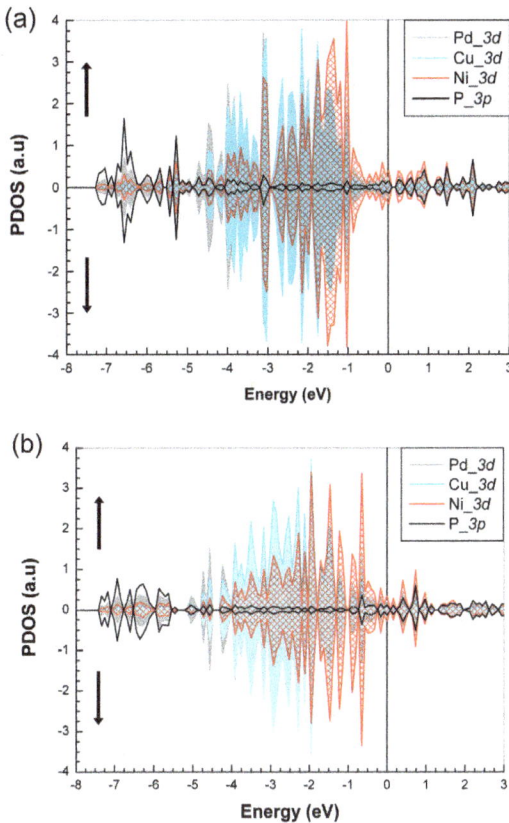

Fig. 3.18. The calculated partial densities of states for the spin-up (↑) and spin-down (↓) 3d electrons of the Pd (gray), Cu (blue), Ni (red) and 3p electron P (black) atoms, at T= 950K (a) and at T= 550K (b) respectively. The Fermi level (vertical line) has been chosen as zero energy. Reproduced from Ref. [109] with permission of the American Institute of Physics.

Partial PDF(R) for the Ni-P and Cu-P atomic pairs obtained by MD simulation also indicated an increase in the number of such atomic bonds on cooling in agreement with the experimental results. Analysis of the atomic and electronic structure of the alloy in the liquid and glassy states reveals rapid formation of chemical bonds with *p-d* hybridization between P and metal atoms due to charge transfer from a metal to phosphorous in the supercooled liquid close to the glass-transition region. The calculated partial densities of states projected on the $3p$ state of P atoms and the $3d$ and $4d$ states of Ni, Cu and Pd atoms show noticeable changes in the electronic structures of the $Pd_{42.5}Cu_{30}Ni_{7.5}P_{20}$ alloy towards a chemical short-range ordering in the glassy state (Fig. 3.18). Thus covalently bonded Ni,Cu-P clusters are formed on cooling towards T_g.

As shown in Fig. 3.17 with a decrease in temperature the integrated area under P2 (AP2) changes slightly while that under P1 (AP1) increases significantly and the ratio AP1/AP2 increases about 5 times on cooling from the melt to T_g. It indicates that the number of Ni-P and Cu-P atomic pairs, and thus chemical short range order (CSRO) around P, increases on cooling the melt and then remains nearly constant in the glassy state.

The temperature dependence of viscosity of a liquid can be written as [117]:

$$\eta = (Nh/Vm)\exp(Q/RT) \tag{3.3}$$

where Q is the activation energy for viscous flow, V_m is molar volume, N is the Avogadro's number, h is the Planck's constant and R is the universal gas constant. The activation energy Q depends on the structure of the liquid and supports the idea that the structural changes affect the viscosity. This fact is responsible for a relatively high fragility of the Pd-Ni-Cu-P liquid in the supercooled state compared to other metallic glasses as the temperature dependence of the viscosity of this bulk metallic glass-former strongly deviates from the Arrhenius behavior (Fig. 3.17). This is also reflected in its high fragility index m [118] (additional to D which is also used) [119]. The link between the activation energy for viscous flow and structure of the melt is discussed in [120]. Such an argument was also used for interpretation of the glass-formation in Ni [121].

As has been also mentioned in Chapter 2 Ni-Nb alloys behave in a similar way [122,123]. Larger values of γ parameter related to the shift in the peak in the structure factor $S(q_1)$, introduced in Ref. [124], indicate a more fragile liquid. For the fragile $Ni_{62}Nb_{38}$ liquid γ is equal to 8.63, while for the stronger Zr-based Vit 106a liquid it is 3.85. Significant structural changes in metallic glasses in the supercooled liquid state have been also observed in the $Ti_{39.5}Zr_{39.5}Ni_{21}$ liquid [125] and predicted in Cu-Zr-Al glasses by MD computer simulation [126]. A link between the structure and viscosity of the $Cu_{64}Zr_{36}$ alloy was also suggested in Ref [127].

The structural changes on heating leading to crystallization will be discussed in detail in Chapter 4.

3.4 Structural changes on mechanical loading

Anelastic and viscoplastic strains can have distinct, even opposite, effects on properties. For example, relaxation enthalpy was found to decrease after indentation [128] and in an as-cast Pd-based BMG subjected to shot peening [129]. On the other hand, shot peening induced rejuvenation of the annealed glass. In a Zr-based BMG cold-rolled to thickness reductions up to 50%, relaxation enthalpy first increased and then decreased [130]. A similar effect was observed in the case of a Zr-based BMG deformed in compression [131,132].

The first demonstration that elastic-cycling effects extend to macroscopic loading, and are not limited to nanoindentation, was by Caron et al. [133]. They loaded a Zr-based bulk metallic glass (BMG) in uniaxial compression at $(37\pm15)\%$ of the yield stress σ_y. As a result of this cycling, the MG had a lower relaxation enthalpy as measured in DSC, indicating a more relaxed state, consistent with the observed decreases in internal friction. Load-free samples showed a residual contraction along the loading axis, taken to indicate a densification, though there was no measurable change in the Young's modulus (measured by the ultrasonic pulse-echo technique). The structural changes underlying these effects were attributed to an accumulation of anelastic strains associated with the cycle periods being much shorter than the anelastic recovery time. The cyclic loading was found to induce partial nanocrystallization of the metallic glass (Fig. 3.19) forming a crystalline phase. Further work on the same BMG at significantly higher stress values confirmed that cyclic loading can induce crystallization, but of a metastable phase [134]. There was a detectable shift of the first peak in X-ray diffraction patterns of the MG, and it was noted that this could be associated with anisotropic structural change. Cyclic loading was found to lead to a 0.5% decrease in density of the MG and increase in the crystallization enthalpy, opposite to the change expected for relaxation. This could be the result of initial nanoscale heterogeneity in the BMG sample [135].

Fig. 3.19. High-resolution TEM images of cube sample in (a) as-cast and (b) after mechanical cycling at 586 ±242 MPa stress. The inserts represent corresponding NBD patterns. Reproduced from Ref. [133] with permission of the American Institute of Physics.

As noted above, elastic static loading in nanoindentation has no detectable effect on MGs though the loading times are short. Static loading can be easily maintained for long time. BMG samples elastically loaded for a long time period in uniaxial compression show significant changes in properties [136,137], even though there is typically only a very small residual contraction along the loading axis. The relaxation enthalpy, measured in DSC, is increased, indicating that the loading has induced rejuvenation (i.e. the opposite of relaxation and ageing). This should be associated with a decrease in density, and such a decrease was later confirmed by direct measurement [138]. The rejuvenated samples show increases in plasticity and decreases in the elastic bulk and shear moduli [136,137,138].

It is clear from the works reviewed above that cyclic elastic loading has many effects on BMGs [139]. Although the stress states in nanoindentation and in macroscopic uniaxial compression are very different, there are some analogous effects of cycling in the two cases. MD simulations which, as noted above, can reproduce the observed effects of cycling in nanoindentation, have failed to detect any clear changes in the glassy structure [139] though such changes are clearly observed within the shear bands on plastic deformation [140]. The property changes indicate that the cycling induces both relaxation (consistent with observed hardening) or rejuvenation (consistent with lowered density and the increased mobility). It is possible, that mechanical treatments such as elastic cycling induce structural changes that are qualitatively different from those normally associated with relaxation and rejuvenation (such as would be achieved by annealing or by forming the glass from the liquid by quenching at different rates). For example, the relative changes in topological and in chemical short-range order [141] may be different in the mechanical and thermal treatments. These effects will also be discussed in Chapter 5.

3.5 Pressure effects, polyamorphism

The structure of a metallic liquid was studied using molecular dynamics simulation. It was found that isothermal compression slows down dynamics of the metallic liquid leading to glass transition [142]. The increase of the kinetic fragility with increasing pressure is accompanied by structural changes. With increasing pressure the liquid dynamics was found to become more heterogeneous. Local ordering at higher pressure and lower temperature leads to the increased five-fold symmetry.

The phenomenon of liquid-liquid transition between two thermodynamically distinct phases – a low-density liquid and high-density liquid below the critical temperature [143,144] and polyamorphism which is well known in the case of the amorphous ice [145,146] is also observed in inorganic solids [147], the $Ce_{55}Al_{45}$ metallic glass [148] and the $La_{32}Ce_{32}Al_{16}Ni_5Cu_{15}$ bulk metallic glass [149] as a function of hydrostatic pressure. Significant increase of density was found on compression of the $Ce_{55}Al_{45}$ metallic glass while the $La_{32}Ce_{32}Al_{16}Ni_5Cu_{15}$ one exhibited a sudden change in compressibility at about 14 GPa. According to MD simulations and X-ray absorption experiments in a $Ce_{75}Al_{25}$ [150] the origin of this transition is the delocalization of $4f$ electrons in Ce. After delocalization the bonds become shorter than those between the Ce atoms with localized $4f$ electrons.

References

[1]   J.L. Finney, Random packing and the structure of simple liquids. Proc. R. Soc. A 319 (1970) 479. https://doi.org/10.1098/rspa.1970.0189

[2]   K. Suzuki, Proceedings of the 4th International Conference on Rapidly Quenched Metals, edited by T. Masumoto and K. Suzuki Japan Institute of Metals, Sendai, Vol. 1 (1982) 309.

[3]   Y. Waseda and H.S. Chen, A structural study of metallic glasses containing boron (Fe-B, Co-B, and Ni-B), Phys. Stat. Solidi. 49 (1978) 387. https://doi.org/10.1002/pssa.2210490149

[4]   N. Mattern, H. Hermann, S. Roth, J. Sakowski, M.P. Macht, P. Jovari and J. Jiang, Structural behavior of $Pd_{40}Cu_{30}Ni_{10}P_{20}$ bulk metallic glass below and above the glass transition, Appl. Phys. Lett., 82 (2003) 2589. https://doi.org/10.1063/1.1567457

[5]   A.R. Yavari, N. Nikolov, N. Nishiyama, T. Zhang, A. Inoue, J.L. Uriarte and G. Heunen, The glass transition of bulk metallic glasses studied by real-time diffraction in transmission using high-energy synchrotron radiation, Mater. Sci. Eng. A 375–377 (2004) 709. https://doi.org/10.1016/j.msea.2003.10.224

[6]   P. Lamparter, Structure of metallic glasses, Physica Scripta, T57 (1995) 45. https://doi.org/10.1088/0031-8949/1995/T57/008

[7]   A. Hirata, Y. Hirotsu, T. Ohkubo, N. Tanaka and T.G. Nieh, Local atomic structure of Pd–Ni–P bulk metallic glass examined by high-resolution electron microscopy and electron diffraction, Intermetallics 14 (2006) 903. https://doi.org/10.1016/j.intermet.2006.01.007

[8]   D.V. Louzguine-Luzgin, J. Antonowicz, K. Georgarakis, G. Vaughan, A.R. Yavari and A. Inoue, Real-space structural studies of Cu-Zr-Ti glassy alloy, Journal of Alloys and Compounds, 466 (2008) 106. https://doi.org/10.1016/j.jallcom.2007.11.039

[9]   P.M. Voyles, J.M. Gibson and M.M.J. Treacy, Fluctuation microscopy a probe of atomic correlations in disordered materials, J. Electron Microsc., 49 (2000) 259. https://doi.org/10.1093/oxfordjournals.jmicro.a023805

[10]   R.L. McGreevy, Reverse Monte Carlo modelling, J. Phys. – Condens Matter, 13 (2001) R877. https://doi.org/10.1088/0953-8984/13/46/201

[11]   J. Hwang, A.M. Clausen, H. Cao and P.M. Voyles, Reverse Monte Carlo structural model for a zirconium-based metallic glass incorporating fluctuation microscopy medium-range order, data J. Mater. Res., 24 (2009) 3121. https://doi.org/10.1557/jmr.2009.0386

[12]   D.C. Rapaport, The art of molecular dynamics simulation, Cambridge University Press, The Edinburgh Building, Cambridge, UK 2004.

[13]   G. Kresse and J. Hafner, Ab initio molecular dynamics for liquid metals, Phys. Rev. B 47 (1993) 558. https://doi.org/10.1103/PhysRevB.47.558

[14]   J.E. Lennard-Jones, On the determination of molecular fields, Proc. R. Soc. Lond. A, 106 (1924) 463.

[15]   M.S. Daw and M.I. Baskes, Embedded-atom method – derivation and application to impurities, surfaces, and other defects in metals, Phys. Rev. B, 29 (12) (1984) 6443. https://doi.org/10.1103/PhysRevB.29.6443

[16] G.S. Cargill and F. Spaepen, Description of chemical ordering in amorphous-alloys, J. Non-Cryst. Solids, 43 (1) (1981) 91. https://doi.org/10.1016/0022-3093(81)90174-5

[17] H.W. Sheng, W.K. Luo, F.M. Alamgir, J.M. Bai and E. Ma, Atomic packing and short-to-medium-range order in metallic glasses, Nature, 439 (2006) 419. https://doi.org/10.1038/nature04421

[18] C.N.J. Wagner, Direct methods for the determination of atomic-scale structure of amorphous solids (X-ray, electron, and neutron scattering) J. Non-Cryst. Solids, 31 (1978) 1. https://doi.org/10.1016/0022-3093(78)90097-2

[19] D.T. Cromer, Compton scattering factors for aspherical free atoms, J. Chem. Phys. 47 (1969) 4857. https://doi.org/10.1063/1.1670980

[20] Y. Waseda, The structure of non-crystalline materials, McGraw-Hill, New York, 1980, p 670.

[21] International tables for x-ray crystallography, ed. J.A. Ibers and W.C. Hamilton, Kynoch, Birmingem, 1974, Vol. 4, p.1.

[22] D.V. Louzguine-Luzgin, J. Antonowicz, K. Georgarakis, G. Vaughan, A.R. Yavari and A. Inoue, Real-space structural studies of Cu-Zr-Ti glassy alloy, Journal of Alloys and Compounds, 466 (2008) 106. https://doi.org/10.1016/j.jallcom.2007.11.039

[23] J.L.C. Daams, P. Villars and J.H.N. van Vucht, Atlas of Crystal Structure Types (ASM International, Materials Park, 1991) Vol. 2, p. 3051.

[24] E. Matsubara, T. Tamura, Y. Waseda, A. Inoue, T. Zhang and T. Masumoto, Structural study of $Zr_{60}Al_{15}Ni_{25}$ amorphous alloys with a wide supercooled liquid region by the anomalous x-ray scattering (AXS) Method, Materials Transactions, JIM, 33 (1992) 873.

[25] S. Sato, T. Sanada, J. Saida, M. Imafuku, E. Matsubara and A. Inoue, Effect of Al on local structures of Zr-Ni and Zr-Cu metallic glasses, Materials Transactions, 46 (2005) 2893. https://doi.org/10.2320/matertrans.46.2893

[26] K.F. Kelton, G.W. Lee, A.K. Gangopadhyay, R.W. Hyers, T.J. Rathz, J.R. Rogers, M.B. Robinson and D.S. Robinson, First X-ray scattering studies on electrostatically levitated metallic liquids: demonstrated influence of local icosahedral order on the nucleation barrier, Phys. Rev. Lett. 90 (2003) 195504. https://doi.org/10.1103/PhysRevLett.90.195504

[27]   Y.Q. Cheng, H.W. Sheng and E. Ma, Relationship between structure, dynamics, and mechanical properties in metallic glass-forming alloys, Phys. Rev. B 78 (2008) 014207. https://doi.org/10.1103/PhysRevB.78.014207

[28]   V. Wessels, A.K. Gangopadhyay, K. Sahu, R.W. Hyers, S.M. Canepari, J.R.S. Rogers, M.J. Kramer, A.I. Goldman, D. Robinson, J.W. Lee, J. Morris and K.F. Kelton, Rapid chemical and topological ordering in supercooled liquid $Cu_{46}Zr_{54}$, Phys. Rev. B 83 (2011) 094116. https://doi.org/10.1103/PhysRevB.83.094116

[29]   N. Mattern, J. Bednarcik, M. Stoica, J. Eckert, Temperature dependence of the short-range order of $Cu_{65}Zr_{35}$ metallic glass, Intermetallics 32 (2013) 51. https://doi.org/10.1016/j.intermet.2012.08.024

[30]   K. Georgarakis, A.R. Yavari, D.V. Louzguine-Luzgin, J. Antonowicz, M. Stoica, Y. Li, M. Satta, A. LeMoulec, G. Vaughan and A. Inoue, Atomic structure of Zr-Cu glassy alloys and detection of deviations from ideal solution behavior with Al addition by x-ray diffraction using synchrotron light in transmission, Applied Physics Letters 94 (2009) 191912. https://doi.org/10.1063/1.3136428

[31]   Ch.E. Lekka, A. Ibenskas, A.R. Yavari and G.A. Evangelakis, Tensile deformation accommodation in microscopic metallic glasses via subnanocluster reconstructions, Appl. Phys. Lett., 91 (2007) 214103. https://doi.org/10.1063/1.2816912

[32]   K.W. Park, J.J. Jang, M. Wakeda, Y. Shibutani and J.C. Lee, Atomic packing density and its influence on the properties of Cu–Zr amorphous alloys, Scripta Mater., 57 (2007) 805. https://doi.org/10.1016/j.scriptamat.2007.07.019

[33]   N. Mattern, P. Jóvári, I. Kaban, S. Gruner, A. Elsner, V. Kokotin, H. Franz, B. Beuneu and J. Eckert, Short-range order of Cu–Zr metallic glasses, J. Alloys Compd., 485 (2009) 163. https://doi.org/10.1016/j.jallcom.2009.05.111

[34]   A.J. Bradley and J. Thewlis, The crystal structure of alpha-Manganese, Proceedings of the Royal Society of London, A 115 (1927) 456. https://doi.org/10.1098/rspa.1927.0103

[35]   C.P. Gazzara, R.M. Middleton, R.J. Weiss and E.O. Hall, A refinement of the parameters of α-manganese, Acta Crystallogr. 22 (1967) 859. https://doi.org/10.1107/S0365110X67001689

[36]   D.V. Louzguine-Luzgin, A.R. Yavari, G. Vaughan and A. Inoue, Clustered crystalline structures as glassy phase approximants, Intermetallics, 17 (2009) 477. https://doi.org/10.1016/j.intermet.2008.12.008

[37]   D.V. Louzguine, M. Saito, Y. Waseda and A. Inoue, Structural study of amorphous Ge$_{50}$Al$_{40}$Cr$_{10}$ alloy, Journal of the Physical Society of Japan, 68 (1999) 2298. https://doi.org/10.1143/JPSJ.68.2298

[38]   J. Antonowicz, A. Pietnoczka, K. Pękała, J. Latuch and G.A. Evangelakis, Local atomic order, electronic structure and electron transport properties of Cu-Zr metallic glasses, J. Appl. Phys. 115 (2014) 203714. https://doi.org/10.1063/1.4879903

[39]   W.K. Luo and E. Ma, EXAFS measurements and reverse Monte Carlo modeling of atomic structure in amorphous Ni$_{80}$P$_{20}$ alloys, J. Non-Cryst Solids, 354 (2008) 945. https://doi.org/10.1016/j.jnoncrysol.2007.08.028

[40]   J. Antonowicz, A. Pietnoczka, W. Zalewski, R. Bacewicz, M. Stoica, K. Georgarakis and A.R. Yavari, Local atomic structure of Zr–Cu and Zr–Cu–Al amorphous alloys investigated by EXAFS method, J. Alloys Compd. 509S (2011) S34. https://doi.org/10.1016/j.jallcom.2010.10.105

[41]   A.L. Ankudinov, B. Ravel, J.J. Rehr and S.D.Conradson, Real-space multiple-scattering calculation and interpretation of X-ray-absorption near-edge structure, Phys. Rev. B 58 (1998) 7565. https://doi.org/10.1103/PhysRevB.58.7565

[42]   D.T. Cormer and D.A. Liberman, Relativistic calculation of anomalous scattering factors for X-rays, J. Chem. Phys. 53 (1970) 1891. https://doi.org/10.1063/1.1674266

[43]   D.V. Louzguine and A. Inoue, Ge-Al-Cr-La amorphous alloys containing crystalline-like zones, NanoStructured Materials, 11 (1999) 115. https://doi.org/10.1016/S0965-9773(99)00024-0

[44]   H.Y. Hsieh, B.H. Toby, T. Egami, Y. He, S.J. Poon and G.J. Shiflet, Atomic-structure of amorphous Al$_{90}$Fe$_x$Ce$_{10-x}$, J. Mater. Res., 5 (1990) 2807. https://doi.org/10.1557/JMR.1990.2807

[45]   T. Egami, The atomic structure of aluminum based metallic glasses and universal criterion for glass formation, J Non-Cryst Solids, 207 (1996) 575. https://doi.org/10.1016/S0022-3093(96)00277-3

[46]   Y.E. Kalay, I. Kalay, J.W. Hwang, P.M. Voyles and M.J. Kramer, Local chemical and topological order in Al–Tb and its role in controlling nanocrystal formation, Acta Materialia 60 (2012) 994. https://doi.org/10.1016/j.actamat.2011.11.008

[47]   N. Jakse, O. Lebacq and A. Pasturel, Ab initio molecular-dynamics simulations of short-range order in liquid Al$_{80}$Mn$_{20}$ and Al$_{80}$Ni$_{20}$ alloys, Phys. Rev. Lett., 93 (2004) 207801. https://doi.org/10.1103/PhysRevLett.93.207801

[48] J. Hafner, Theory of the formation of metallic glasses, Phys. Rev. B 21 (1980) 406. https://doi.org/10.1103/PhysRevB.21.406

[49] http //lammps.sandia.gov/

[50] S. Plimpton, Fast parallel algorithms for short-range molecular dynamics, J Comp. Phys., 117 (1995) 1. https://doi.org/10.1006/jcph.1995.1039

[51] M.I. Mendelev, S. Han, D.J. Srolovitz, G.J. Ackland, D.Y. Sun and M. Asta, Development of new interatomic potentials appropriate for crystalline and liquid iron, Phil. Mag., 83 (2003) 3977. https://doi.org/10.1080/14786430310001613264

[52] A. Stukowski, Visualization and analysis of atomistic simulation data with OVITO - the Open Visualization Tool, Modelling Simul. Mater. Sci. Eng. 18 (2010) 015012. https://doi.org/10.1088/0965-0393/18/1/015012

[53] P.J. Steinhardt, D.R. Nelson and M. Ronchetti, Icosahedral bond orientational order in supercooled liquids, Phys. Rev. Lett. 47 (1981) 1297. https://doi.org/10.1103/PhysRevLett.47.1297

[54] W. Mickel, S.C. Kapfer, G.E. Schröder-Turk and K. Mecke, Shortcomings of the bond orientational order parameters for the analysis of disordered particulate matter, Journal of Chemical Physics 138 (2013) 044501. https://doi.org/10.1063/1.4774084

[55] J.D. Honeycutt and H.C. Andersen, Molecular dynamics study of mel, ting and freezing of small Lennard-Jones clusters, J. Phys. Chem. 91 (1987) 4950. https://doi.org/10.1021/j100303a014

[56] F. Aurenhammer, Voronoi diagrams – A survey of a fundamental geometric data structure, ACM Computing Surveys, 23 (1991) 345. https://doi.org/10.1145/116873.116880

[57] J.D. Bernal, Geometry of the structure of monatomic liquids, Nature, 185 (1960) 68. https://doi.org/10.1038/185068a0

[58] G.D. Scott, Packing of equal spheres, Nature, 188 (4754) (1960) 908. https://doi.org/10.1038/188908a0

[59] J.L. Finney, Random packings and structure of simple liquids. 1. Geometry of random close packing, Proc. Roy. Soc. Lond. Ser. A – Math. Phys. Sci., 319 (1970) 479.

[60] A. Inoue, T. Negishi, H.M. Kimura, T. Zhang and A.R. Yavari, High packing density of Zr- and Pd-based bulk amorphous alloys, Mater. Trans. JIM, 39 (1998) 318. https://doi.org/10.2320/matertrans1989.39.318

[61]   D.B. Miracle, A structural model for metallic glasses, Nature Mater., 3 (2004) 697. https://doi.org/10.1038/nmat1219

[62]   D.B. Miracle, The efficient cluster packing model – An atomic structural model for metallic glasses, Acta Mater., 54 (2006) 4317. https://doi.org/10.1016/j.actamat.2006.06.002

[63]   H.W. Sheng, W.K. Luo, F.M. Alamgir, J.M. Bai and E. Ma, Atomic packing and short-to-medium-range order in metallic glasses, Nature 439 (2006) 419. https://doi.org/10.1038/nature04421

[64]   D.B. Miracle, W.S. Sanders and O.N. Senkov, The influence of efficient atomic packing on the constitution of metallic glasses, Phil. Mag 83 (2003) 2409. https://doi.org/10.1080/1478643031000098828

[65]   Y.Q. Cheng and E. Ma, Atomic-level structure and structure–property relationship in metallic glasses, Prog. Mater. Sci., 56 (2011) 379. https://doi.org/10.1016/j.pmatsci.2010.12.002

[66]   D.V. Louzguine-Luzgin, A.R. Yavari, M. Fukuhara, K. Ota, G. Xie, G. Vaughan and A. Inoue, Free volume and elastic properties changes in Cu-Zr-Ti-Pd bulk glassy alloy on heating, Journal of Alloys and Compounds 431 (2007) 136. https://doi.org/10.1016/j.jallcom.2006.05.069

[67]   T. Ohkubo and Y. Hirotsu, Electron diffraction and high-resolution electron microscopy study of an amorphous $Pd_{82}Si_{18}$ alloy with nanoscale phase separation, Phys. Rev. B, 67 (2003) 094201. https://doi.org/10.1103/PhysRevB.67.094201

[68]   Y. Hirotsu, T. Ohkubo and M. Matsushita, Study of amorphous alloy structures with medium range atomic ordering, Microsc. Res. Tech. 40 (1998) 284. https://doi.org/10.1002/(SICI)1097-0029(19980215)40:4<284::AID-JEMT5>3.0.CO;2-S

[69]   O. Haruyama, K. Sugiyama, M. Sakurai and Y. Waseda, A local structure change of bulk $Pd_{40}Ni_{40}P_{20}$ glass during full relaxation, J. Non-Crystalline Solids, 353 (2007) 3053. https://doi.org/10.1016/j.jnoncrysol.2007.05.071

[70]   A. Hirata, Y. Hirotsu, T. Ohkubo, N. Tanaka and T.G. Nieh, Local atomic structure of Pd–Ni–P bulk metallic glass examined by high-resolution electron microscopy and electron diffraction, Intermetallics 14 (2006) 903. https://doi.org/10.1016/j.intermet.2006.01.007

[71]   Y. Hirotsu, M. Uehara and, M. Ueno, Microcrystalline domains in amorphous palladium-copper-silicon ($Pd_{77.5}Cu_6Si_{16.5}$) alloys studied by high-resolution electron microscopy, J. Appl. Phys., 59 (1986) 3081. https://doi.org/10.1063/1.336932

[72] D.V. Louzguine-Luzgin, Y.H. Zeng, A.D. Setyawan, N. Nishiyama and H. Kato, J. Saida and, A. Inoue, Deformation behavior of Zr- and Ni-based bulk glassy alloys, Journal of Materials Research, 22 (2007) 1087. https://doi.org/10.1557/jmr.2007.0126

[73] Y. Waseda and H.S. Chen, On the structure of amorphous $Fe_{84}B_{16}$, Solid State Comm., 27 (1978) 809. https://doi.org/10.1016/0038-1098(78)90703-2

[74] A. Hirata, Y. Hirotsu, T. Ohkubo, E. Matsubara, T. Ohkubo and K. Hono, Mechanism of nanocrystalline microstructure formation in amorphous Fe–Nb–B alloys, Phys. Rev. B, 74 (2006) 214206. https://doi.org/10.1103/PhysRevB.74.214206

[75] D.W. Qi and S. Wang, Icosahedral order and defects in metallic liquids and glasses, Phys. Rev. B 44 (1991) 884. https://doi.org/10.1103/PhysRevB.44.884

[76] A. Hirata, L.J. Kang, T. Fujita, B. Klumov, K. Matsue, M. Kotani, A.R. Yavari and M.W. Chen, Geometric frustration of icosahedron in metallic glasses, Science, 341 (2013) 376. https://doi.org/10.1126/science.1232450

[77] E. Matsubara, M. Sakurai, T. Nakamura, M. Imafuku, S. Sato, J. Saida and A. Inoue, environmental structural studies in amorphous and quasicrystalline $Zr_{70}Al_6Ni_{10}Pt_{14}$ alloys, Scripta Mater. 44 (2001) 2297. https://doi.org/10.1016/S1359-6462(01)00895-8

[78] D.V. Louzguine and A. Inoue, Nanoparticles with icosahedral symmetry in Cu-based bulk glass former induced by Pd addition, Scripta Mater. 48 (2003) 1325. https://doi.org/10.1016/S1359-6462(03)00018-6

[79] D.V. Louzguine, A.R. Yavari and A. Inoue, Devitrification behaviour of Cu-Zr-Ti-Pd bulk glassy alloys, Phil. Mag. 83 (2003) 2989. https://doi.org/10.1080/1478643031000151394

[80] K.F. Kelton, Crystallization of liquids and glasses to quasicrystals, J. Non-Cryst. Sol. 334-335 (2004) 253.

[81] S. Ranganathan and A. Inoue, An application of Pettifor structure maps for the identification of pseudo-binary quasicrystalline intermetallics, Acta Materialia 54 (2006) 3647. https://doi.org/10.1016/j.actamat.2006.01.041

[82] J. Ding, Y.Q. Cheng, H. Sheng, M. Asta, R.O. Ritchie and E. Ma, Universal structural parameter to quantitatively predict metallic glass properties, Nature Communications, 7 (2016) 13733. https://doi.org/10.1038/ncomms13733

[83] A.I. Oreshkin, N.S. Maslova, V.N. Mantsevich, S.I. Oreshkin, S.V. Savinov, V.I. Panov and D.V. Louzguine-Luzgin, Metallic glass electronic structure peculiarities revealed by UHV STM/STS, JETP Letters, 94 (2011) 58. https://doi.org/10.1134/S0021364011130157

[84] A.I. Oreshkin, V.N. Mantsevich, S.V. Savinov, S.I. Oreshkin, V.I. Panov, A.R. Yavari, D.B. Miracle and D.V. Louzguine-Luzgin, In situ visualization of Ni–Nb bulk metallic glasses phase transition, Acta Materialia, 61 (2013) 5216. https://doi.org/10.1016/j.actamat.2013.05.014

[85] J. Schroers, B. Lohwongwatana, W.L. Johnson and A. Peker, Gold based bulk metallic glass, Appl. Phys. Lett. 87 (2005) 061912. https://doi.org/10.1063/1.2008374

[86] S.V. Ketov, N. Chen, A. Caron, A. Inoue and D.V. Louzguine-Luzgin, Structural features and high quasi-static strain rate sensitivity of $Au_{49}Cu_{26.9}Ag_{5.5}Pd_{2.3}Si_{16.3}$ bulk metallic glass, Applied Physics Letters, 101 (2012) 241905. https://doi.org/10.1063/1.4770072

[87] S. Mechler, E. Yahel, P.S. Pershan, M. Meron and B. Lin, Crystalline monolayer surface of liquid Au–Cu–Si–Ag–Pd: Metallic glass former, App. Phys. Let. 98 (2011) 251915. https://doi.org/10.1063/1.3599515

[88] J.W.M. Frenken and J.F. van der Veen, Observation of surface melting, Phys. Rev. Lett. 54 (1985) 134. https://doi.org/10.1103/PhysRevLett.54.134

[89] J.X. Fang, U. Vainio, W. Puff, R. Würschum, X.L. Wang, D. Wang, M. Ghafari, F. Jiang, J. Sun, H. Hahn and H. Gleiter, Atomic structure and structural stability of $Sc_{75}Fe_{25}$ nanoglasses, Nano Lett. 12 (2012) 458.

[90] N. Chen, R. Frank, N. Asao, D.V. Louzguine, J.Q. Wang, X.Q. Xie, Y. Ishikawa, N. Hatakeyama, P. Sharma, Y.C. Lin, M. Esashi, Y. Yamamoto and A. Inoue, Formation and properties of Au-based nanograined metallic glasses, Acta Mater., 59 (2011) 6433. https://doi.org/10.1016/j.actamat.2011.07.007

[91] S.V. Ketov, X.T. Shi, G.Q. Xie, R. Kumashiro, A.Yu. Churyumov, A.I. Bazlov, N. Chen, Y. Ishikawa, N. Asao, H.K. Wu and D.V. Louzguine-Luzgin, Nanostructured Zr-Pd metallic glass thin film for biochemical applications, Scientific Reports, 5 (2015) 7799. https://doi.org/10.1038/srep07799

[92] T. Egami, Spectroscopic diffraction measurement with white x-rays. J. Mater. Sci. 13 (1978) 2587. https://doi.org/10.1007/BF02402745

[93]   N. Mattern, H. Hermann, S. Roth, J. Sakowski, P. Macht, P. Jovari and J. Jiang, Structural behavior of $Pd_{40}Cu_{30}Ni_{10}P_{20}$ bulk metallic glass below and above the glass transition, Appl. Phys. Lett. 82 (2003) 2589. https://doi.org/10.1063/1.1567457

[94]   A.R. Yavari, N. Nikolov, N. Nishiyama, T. Zhang, A. Inoue, J.L. Uriarte and G. Heunen, The glass transition of bulk metallic glasses studied by real time diffraction in transmission usinghigh energy synchrotron radiation. Mater. Sci. Eng. A 375–377 (2004) 709. https://doi.org/10.1016/j.msea.2003.10.224

[95]   A.R. Yavari, R Le Moulec, A. Inoue, N. Nishiyama, N. Lupu, E. Matsubara, W.J. Botta, G. Vaughan, M. Michiel and A. Kvick, Excess free volume in metallic glasses measured by X-ray diffraction, Acta Materialia 53 (2005) 1611. https://doi.org/10.1016/j.actamat.2004.12.011

[96]   Q.K. Jiang, Z.Y. Chang, X.D. Wang and J.Z. Jiang, Structures at glassy, supercooled liquid, and liquid states in La-based bulk metallic glasses, Metallurgical and Materials Transactions A 41A (2010) 1634. https://doi.org/10.1007/s11661-009-0041-9

[97]   M. Stoica, J. Das, G. Bednarčik, J. Wang, G. Vaughan, W.H. Wang and J. Eckert, Mechanical response of metallic glasses Insights from in-situ high energy X-ray diffraction, J. of Metals, 62 (2010) 76. https://doi.org/10.1007/s11837-010-0037-3

[98]   D.V. Louzguine, A.R. Yavari, K. Ota, G. Vaughan and A. Inoue, Synchrotron X-ray radiation diffraction studies of thermal expansion, free volume change and glass transition phenomenon in Cu-based glassy and nanocomposite alloys on heating, J Non-Cryst Sol 351 (2005) 1639. https://doi.org/10.1016/j.jnoncrysol.2005.04.054

[99]   D.V. Louzguine-Luzgin, A. Inoue, A.R. Yavari and G. Vaughan, Thermal expansion of a glassy alloy studied using a real-space pair distribution function, Applied Physics Letters 88 (2006) 121926. https://doi.org/10.1063/1.2187955

[100]   O. Mishima and H.E. Stanley, The relationship between liquid, supercooled and glassy water. Nature, 396 (1998) 329. https://doi.org/10.1038/24540

[101]   P.G. Debenedetti and F.H. Stillinger, Supercooled liquids and the glass transition, Nature, 410 (2001) 259. https://doi.org/10.1038/35065704

[102]   N. Jakse and A. Pasturel, Local order of liquid and supercooled zirconium by ab initio molecular dynamics, Phys. Rev. Lett., 91 (2003) 195501. https://doi.org/10.1103/PhysRevLett.91.195501

[103] H.B. Lou, X.D. Wang, Q.P. Cao, D.X. Zhang, J. Zhang, T.D. Hu, H.K. Mao and J.Z. Jiang, Negative expansions of interatomic distances in metallic melts, Proc. Natl. Acad. Sci., 110 (2013) 10068. https://doi.org/10.1073/pnas.1307967110

[104] J. Ding and E. Ma, Computational modeling sheds light on structural evolution in metallic glasses and supercooled liquids, NPJ Computational Materials 3 (2017) 9. https://doi.org/10.1038/s41524-017-0007-1

[105] L.H. Xiong, X.D. Wang, Q. Yu, H. Zhang, F. Zhang, Y. Sun, Q.P. Cao, H.L. Xie, T.Q. Xiao, D.X. Zhang, C.Z. Wang, K.M. Ho, Y. Ren, J.Z. Jiang, Temperature-dependent structure evolution in liquid gallium, Acta Materialia 128 (2017) 304. https://doi.org/10.1016/j.actamat.2017.02.038

[106] L. Battezzati and A.L. Greer, Thermodynamics of $Te_{80}Ge_{20-x}Pb_x$ glass-forming alloys, J. Mater. Res. 3 (1988) 570. https://doi.org/10.1557/JMR.1988.0570

[107] S.J. Cheng, X.F. Bian, J.-X. Zhang, X.B. Qin, Z.H. Wang, Correlation of viscosity and structural changes of indium melt, Materials Letters, 57 (2003) 4191. https://doi.org/10.1016/S0167-577X(03)00288-X

[108] K. Georgarakis, D.V. Louzguine-Luzgin, J. Antonowicz, G. Vaughan, A.R. Yavari, T. Egami and A. Inoue, Variations in atomic structural features of a supercooled Pd-Ni-Cu-P glass forming liquid during in situ vitrification, Acta Mater., 59 (2011) 708. https://doi.org/10.1016/j.actamat.2010.10.009

[109] D.V. Louzguine-Luzgin, R. Belosludov, A.R. Yavari, K. Georgarakis, G. Vaughan, Y. Kawazoe, T. Egami and A. Inoue, Structural basis for supercooled liquid fragility established by synchrotron-radiation method and computer simulation, J. Appl. Phys. 110 (2011) 043519. https://doi.org/10.1063/1.3624745

[110] G.J. Fan, J.F. Löffler, R.K. Wunderlich and H.-J. Fecht, Thermodynamics, enthalpy relaxation and fragility of the bulk metallic glass-forming liquid $Pd_{43}Ni_{10}Cu_{27}P_{20}$, Acta Materialia 52 (2004) 667. https://doi.org/10.1016/j.actamat.2003.10.003

[111] D.V. Louzguine-Luzgin, K. Georgarakis, A. Tsarkov, A. Solonin, V. Honkimaki, L. Hennet and A.R. Yavari, Structural changes in liquid Fe and Fe–B alloy on cooling, Journal of Molecular Liquids, 209 (2015) 233. https://doi.org/10.1016/j.molliq.2015.05.062

[112] N. Mattern, H. Hermann, S. Roth, J. Sakowski, M.P. Macht, P. Jovari and J. Jiang, Structural behavior of $Pd_{40}Cu_{30}Ni_{10}P_{20}$ bulk metallic glass below and above the glass transition, Appl. Phys. Lett. 82 (2003) 2589. https://doi.org/10.1063/1.1567457

[113] K. Georgarakis, D.V. Louzguine-Luzgin, J. Antonowicz, G. Vaughan, A.R. Yavari, T. Egami and A. Inoue, Variations in atomic structural features of a supercooled Pd–Ni–Cu–P glass forming liquid during in situ vitrification, Acta Materialia, 59 (2011) 708. https://doi.org/10.1016/j.actamat.2010.10.009

[114] D.V. Louzguine-Luzgin, A.I. Bazlov, A.Yu. Churyumov, K. Georgarakis and A.R. Yavari, Comparative analysis of the structure of palladium-based bulk metallic glasses prepared by treatment of melts with flux, Physics of the Solid State, 55 (2013) 1985. https://doi.org/10.1134/S1063783413100211

[115] A. Inoue, T. Negishi, H.M. Kimura, T. Zhang and A.R. Yavari, High packing density of Zr- and Pd-based bulk amorphous alloys, Mater. Trans. JIM 39 (1998) 318. https://doi.org/10.2320/matertrans1989.39.318

[116] D.V. Louzguine-Luzgin, A.R. Yavari, M. Fukuhara, K. Ota, G. Xie, G. Vaughan and A. Inoue, Free volume and elastic properties changes in Cu-Zr-Ti-Pd bulk glassy alloy on heating, J. Alloys and Comp. 431 (2007) 136. https://doi.org/10.1016/j.jallcom.2006.05.069

[117] D.S. Visthwanah, T.K. Ghosh, D.H.L. Prasad, N.K. Dutt and K.Y. Rani, Viscosity of liquids theory, estimation, experiment, and data, Springer, (2007) p. 643.

[118] R. Bohmer, K.L. Ngai, C.A. Angell, D.J. Plazek, Nonexponential relaxations in strong and fragile glass formers. J. Chem. Phys. 99 (1993) 4201. https://doi.org/10.1063/1.466117

[119] G.E.Abrosimova, A.S.Aronin, I.I. Zver'kova and Y.V.Kir'yanov, Phase trans-formations upon crystallization of amorphous Al-Ni-RE alloys, Phys. Met. Metallov., 94, (2002) 113.

[120] M.I. Ojovan, About activation energy of viscous flow of glasses and melts, Mater. Res. Soc. Symp. Proc. 15 (2015) 10. https://doi.org/10.1557/opl.2015.44

[121] D.V. Louzguine-Luzgin, Vitrification and devitrification processes in metallic glasses, Journal of Alloys and Compounds, 586 (2014) 2. https://doi.org/10.1016/j.jallcom.2012.09.057

[122] N.A. Mauro, M.L. Johnson, J.C. Bendert and K.F. Kelton, Structural evolution in Ni–Nb and Ni-Nb–Ta liquids and glasses — A measure of liquid fragility, Journal of Non-Crystalline Solids 362 (2013) 237. https://doi.org/10.1016/j.jnoncrysol.2012.11.022

[123] K.F. Kelton, Kinetic and structural fragility—a correlation between structures and dynamics in metallic liquids and glasses, Journal of Physics Condensed Matter 29 2 (2017) 023002.

[124] N.A. Mauro, M. Blodgett, M.L. Johnson, A.J. Vogt and K.F. Kelton, A structural signature of liquid fragility, Nat. Commun. 5 (2014) 4616. https://doi.org/10.1038/ncomms5616

[125] K.F. Kelton, G.W. Lee, A.K. Gangopadhyay, R.W. Hyers, T.J. Rathz, J.R. Rogers, M.B. Robinson and D.S. Robinson, A structural signature of liquid fragility, Phys. Rev. Lett. 90 (2003) 195504. https://doi.org/10.1103/PhysRevLett.90.195504

[126] Y.Q. Cheng, E. Ma and H. W. Sheng, Alloying strongly influences the structure, dynamics, and glass forming ability of metallic supercooled liquids, Appl. Phys. Lett. 93 (2008) 111913. https://doi.org/10.1063/1.2987727

[127] N. Jakse and A. Pasturel, Glass forming ability and short-range order in a binary bulk metallic glass by ab initio molecular dynamics, Applied Physics Letters 93 (2008) 113104. https://doi.org/10.1063/1.2976428

[128] R. Bhowmick, R. Raghavan, K. Chattopadhyay and U. Ramamurty, Plastic flow softening in a bulk metallic glass, Acta Mater. 54 (2006) 4221. https://doi.org/10.1016/j.actamat.2006.05.011

[129] F.O. Méar, B. Lenk, Y. Zhang and A.L. Greer, Structural relaxation in a heavily cold-worked metallic glass, Scripta Mater. 59 (2008) 1243. https://doi.org/10.1016/j.scriptamat.2008.08.023

[130] K.M. Flores, E. Sherer, A. Bharathula, H. Chen and Y.C. Jean, Sub-nanometer open volume regions in a bulk metallic glass investigated by positron annihilation, Acta Mater. 55 (2007) 3403. https://doi.org/10.1016/j.actamat.2007.01.040

[131] O. Haruyama, K. Kisara, A. Yamashita, K. Kogure, Y. Yokoyama and K. Sugiyama, Characterization of free volume in cold-rolled $Zr_{55}Cu_{30}Ni_5Al_{10}$ bulk metallic glasses, Acta Mater. 61 (2013) 3224. https://doi.org/10.1016/j.actamat.2013.02.010

[132] D.V. Louzguine-Luzgin, S.V. Ketov, Z. Wang, M.J. Miyama, A.A. Tsarkov and A.Yu. Churyumov, Plastic deformation studies of Zr-based bulk metallic glassy samples with a low aspect ratio, Mater. Sci. Eng. A, 616 (2014) 288. https://doi.org/10.1016/j.msea.2014.08.006

[133] A. Caron, A. Kawashima, H.-J. Fecht, D.V. Louzguine-Luzgin and A. Inoue, On the anelasticity and strain induced structural changes in a Zr-based bulk metallic glass, Appl. Phys. Lett. 99 (2011) 171907. https://doi.org/10.1063/1.3655999

[134] A.Yu. Churyumov, A.I. Bazlov, V.Yu. Zadorozhnyy, A.N. Solonin, A. Caron and D.V. Louzguine-Luzgin, Phase transformations in Zr-based bulk metallic glass cyclically loaded before plastic yielding, Mater. Sci. Eng. A 550 (2012) 358. https://doi.org/10.1016/j.msea.2012.04.087

[135] J. Ding, S. Patinet, M.L. Falk, Y.Q. Cheng and E. Ma, Soft spots and their structural signature in a metallic glass, PNAS 111 (2014) 14052. https://doi.org/10.1073/pnas.1412095111

[136] S.-J. Lee, B.-G. Yoo, J.-I. Jang and J.-C. Lee, Irreversible structural change induced by elastostatic stress imposed on an amorphous alloy and its influence on the mechanical properties, Metal. Mater. Int. 14 (2008) 9. https://doi.org/10.3365/met.mat.2008.02.009

[137] K.-W. Park, C.-M. Lee, M. Wakeda, Y. Shibutani, M.L. Falk and J.-C. Lee, Elastostatically induced structural disordering in amorphous alloys, Acta Mater. 56 (2008) 5440. https://doi.org/10.1016/j.actamat.2008.07.033

[138] H.B. Ke, P. Wen, H.L. Peng, W.H. Wang and A.L. Greer, Homogeneous deformation of metallic glass at room temperature reveals large dilatation, Scripta Mater. 64 (2011) 966. https://doi.org/10.1016/j.scriptamat.2011.01.047

[139] A.L. Greer and Y.H. Sun, Stored energy in metallic glasses due to strains within the elastic limit, Philos. Mag. 96 (2016) 1643. https://doi.org/10.1080/14786435.2016.1177231

[140] A.J. Cao, Y.Q. Cheng and E. Ma, Structural processes that initiate shear localization in metallic glass, Acta Mater. 57 (2009) 5146. https://doi.org/10.1016/j.actamat.2009.07.016

[141] Y. Ritter and K. Albe, Chemical and topological order in shear bands of $Cu_{64}Zr_{36}$ and $Cu_{36}Zr_{64}$ glasses, J. Appl. Phys. 111 (2012) 103527. https://doi.org/10.1063/1.4717748

[142] Y.C. Hu, P.F. Guan, Q. Wang, Y. Yang, H.Y. Bai and W.H. Wang, Pressure effects on structure and dynamics of metallic glass-forming liquid, The Journal of Chemical Physics, 146, (2017) 024507. https://doi.org/10.1063/1.4973919

[143] P.H. Poole, F. Sciortino, U. Essmann and H. Stanley, Phase behaviour of metastable water, Nature 360, (1992) 324. https://doi.org/10.1038/360324a0

[144] N. Giovambattista, T. Loerting, B.R. Lukanov and F.W. Starr, Interplay of the glass transition and the liquid-liquid phase transition in water, Scientific Reports 2 (2012) 390. https://doi.org/10.1038/srep00390

[145] O. Mishima, L.D. Calvert and E. Whalley, 'Melting ice' I at 77 K and 10 kbar: a new method of making amorphous solids, Nature, 310 (1984) 393. https://doi.org/10.1038/310393a0

[146] R.V. Belosludov, K.V. Gets, O.S. Subbotin, R.K. Zhdanov, Y.Y. Bozhko, V.R. Belosludov and J.I. Kudoh, Modeling the polymorphic transformations in amorphous solid ice, Journal of Alloys and Compounds 707, (2017) 108. https://doi.org/10.1016/j.jallcom.2016.12.197

[147] P.H. Poole, T. Grandea, F. Sciortinod, H.E. Stanley and C.A. Angel, Amorphous polymorphism, Computational Materials Science 4 (1995) 373. https://doi.org/10.1016/0927-0256(95)00044-9

[148] H.W. Sheng, H.Z. Liu, Y.Q. Cheng, J. Wen, P.L. Lee, W.K. Luo, S.D. Shastri and E. Ma. Polyamorphism in a metallic glass, Nature Materials 6 (2007) 192. https://doi.org/10.1038/nmat1839

[149] Q.S. Zeng, Y.C. Li, C.M. Feng, P. Liermann, M. Somayazulu, G.Y. Shen, H.-K. Mao, R. Yang, J. Liu, T.D. Hu and J.Z. Jian, Anomalous compression behavior in lanthanum/cerium-based metallic glass under high pressure, Proc. Nat'l. Acad. Sci. USA, 104 (2007) 13565. https://doi.org/10.1073/pnas.0705999104

[150] Q.-S. Zeng, Y. Ding, W.L. Mao, W. Yang, S.V. Sinogeikin, J. Shu, H.-K. Mao and J.Z. Jiang, Origin of pressure-induced polyamorphism in $Ce_{75}Al_{25}$ metallic glass, Phys. Rev. Lett., 104 (2010) 105702. https://doi.org/10.1103/PhysRevLett.104.105702

Chapter 4

Structural changes on heating

Glassy alloys are metastable at room temperature and devitrify/crystallize on heating above the temperature called T_x which, as well as T_g varies upon the heating rate used. Devitrification is a wider term which includes crystallization as well as the formation of the supercooled liquid on heating, however, both these terms are used in the literature. Glassy alloys may also crystallize even at room temperature but for the majority of glassy alloys the process is very slow. In general, amorphous alloys devitrify directly on heating while glassy alloys upon fast enough heating initially transform to a supercooled liquid which later crystallizes. This process is a subject of discussion in the present chapter.

Contents

4.1 Structural relaxation and rejuvenation

Glasses vitrified at different cooling rates have different T_g and properties. Upon heating, as the heating rate is usually lower than the cooling one from the melt, the structure of metallic glasses relaxes leading to their densification and decrease in the excess volume [1]. Structural relaxation of glasses below their glass transition is well-studied in general. Heat of structural relaxation can be readily measured below T_g (Fig. 4.1 (a)) starting at T_{br}. The glass transition on heating takes place in the temperature range between the conditional beginning (T_{gb}) and finish (T_{gf}) of glass transition temperatures (the area of change of C_p) as monitored in Fig. 4.1 (a) by the deflection of the DSC curve of a Zr-Cu-Fe-Al alloy on heating. Usually the structure relaxation produces a single broad exothermic peak. However, the relaxation process in the famous $Pd_{40}Cu_{30}Ni_{10}P_{20}$ alloy [2] clearly indicated two processes: a high and a low-temperature one [3]. Direct measurement of the isothermal relaxation of the volume and the enthalpy of the $Au_{50}Cu_{25.5}Ag_{7.5}Si_{17}$ bulk metallic glass was performed recently [4].

Relaxation processes are also studied by dynamic mechanical analysis (DMA). DSC (Fig. 4.1 (b)) and (DMA) (Fig. 4.1 (c)) of the $Zr_{65}Cu_5Ni_{10}Al_{7.5}Pd_{12.5}$ bulk metallic glassy sample indicate that the loss factor (tanδ) related to the viscoelastic component of the deformation exhibits a maximum at a characteristic temperature T_α of about 710 K [5]. This temperature is close to the glass transition temperature obtained by DSC. At higher temperature the loss factor shows a decrease associated to the formation of the quasicrystalline phase (Fig. 4.1(c)) [5]. However, tanδ increases again at the higher temperature region owing to the decreasing viscosity of the residual supercooled liquid phase until it starts to crystallize.

(a)

(b)

(c)

Fig.4.1. (a) DSC trace of a Zr-Cu-Fe-Al bulk metallic glass indicating the heat effect of structural relaxation and glass transition processes. The heating rate was 0.67 K/s. (b) DSC and (c) DMA traces of the $Zr_{65}Cu_5Ni_{10}Al_{7.5}Pd_{12.5}$ metallic glass. (b,c) Reproduced from [5] with permission of Elsevier.

The thermal expansion and volume changes due to structural relaxation during heat treatment in a glassy $Cu_{55}Hf_{25}Ti_{15}Pd_5$ alloy were studied in situ using synchrotron X-ray radiation diffractometry. Good correspondence of the thermal expansion coefficients was obtained independently by the thermo-mechanical length change measurements and synchrotron X-ray radiation diffractometry. Good correlation was also found for the glass transition temperature obtained by DSC and Synchrotron X-ray radiation diffractometry [6].

The slope of the shift in the main peak position in the reciprocal space gives the linear coefficient of thermal expansion $\beta=1.2\pm0.1*10^{-5}$ K^{-1} before the structural relaxation and $\beta=1.4\pm0.1*10^{-5}$ K^{-1} after structural relaxation as well as volume coefficient of thermal expansion $\alpha=3.6\pm0.3*10^{-5}$ K^{-1} before the structural relaxation and $\alpha=4.2\pm0.3*10^{-5}$ K^{-1} after the structural relaxation. At temperatures higher than 600 K, structural relaxation of the glass towards equilibrium causes deviation from linear expansion as the excess free volume anneals out in the temperature range of 600-750 K. In the supercooled liquid state $\beta = 4.2*10^{-5}$ K^{-1}, while $\alpha = 1.26*10^{-4}$ K^{-1}, the values common for liquid metals. The linear coefficient of thermal expansion was also measured by the thermal mechanical analysis to be 1.5 ± 0.1 10^{-5} K^{-1} corresponding to those obtained from the Synchrotron X-ray radiation diffraction data.

The thermal expansion of a glassy $Cu_{55}Hf_{25}Ti_{15}Pd_5$ alloy studied by using the reciprocal space functions was also verified using a real-space pair distribution function [7]. The plot the atomic distance value for peak maximum of the PDF(r) at given temperature T divided by room temperature value for the $Cu_{55}Hf_{25}Ti_{15}Pd_5$ glassy alloy indicated the thermal expansion of the sample due to an increase in the interatomic distances. The linear coefficient of thermal expansion, $\beta=1.6*10^{-5}$ K^{-1}, is close to that obtained from change in the position of the center of mass of the main diffraction maximum in the reciprocal space and it is very close to that measured by the thermo-mechanical analysis.

The $Cu_{55}Hf_{25}Ti_{15}Pd_5$ and $Cu_{55}Zr_{30}Ti_{10}Pd_5$ glassy alloys were found to exhibit a large amount of the excess volume of about 0.8 % [8] significantly larger than the volume changes upon the structure relaxation of conventional bulk glassy alloys of about 0.1-0.3 % [9,10]. The density values of the $Cu_{55}Zr_{30}Ti_{10}Pd_5$ bulk glassy sample were 7518 kg/m^3 in as-prepared state, 7580 kg/m^3 for the glass relaxed at 740 K, 120 s and 7592 kg/m^3 after precipitation of the nanoscale icosahedral phase (after annealing at 785 K, for 300 s).

Structural rejuvenation is an opposite process. It can be achieved by heating the glass above T_g and fast enough cooling thus restoring the initial glassy structure. However, some part of the incubation period for crystallization has already passed in the sample and its thermal stability will be damaged. Structural rejuvenation can also be done by ion irradiation [11], mechanical [12], thermal [13] or cryogenic cycling [14]. Cryogenic thermal cycling of metallic glasses induces rejuvenation, reaching the states of higher energy (Fig. 4.2). It originates from the heterogeneous dynamics in liquids leading to the heterogeneities in the glasses. Thermal cycling is a non-destructive, isotropic and controllable method which induces no macroscopic residual stresses.

Fig. 4.2. Differential scanning calorimetry traces of melt-spun ribbons of the $La_{55}Ni_{20}Al_{25}$ and bulk rods of the $La_{55}Ni_{10}Al_{35}$ metallic glasses indicating the increased heat of structural relaxation. Reproduced from [14] with permission of Nature Publishing Group.

The excess (free) volume in the $Zr_{50.7}Cu_{28}Ni_9Al_{12.3}$ bulk metallic glass was preserved and redistributed upon annealing by using a crystalline W coating. The annealing process was found to enhance both strength and plasticity of the BMG sample [15].

4.2 Phase separation

Some glassy alloys exhibit phase separation by spinodal or binodal mechanism [16] owing to solid- or liquid-state immiscibility among the constituent elements. Spinodal decomposition is characterized by nearly regular distribution of two phases and continuous variation of composition with time until equilibrium compositions are reached. Each phase parts are interconnected. The interface between phases being initially diffuse eventually sharpens with time. Opposite to the spinodal decomposition mechanism binodal separation takes place by nucleation and growth. Binodal separation is characterized by invariance of precipitating phase composition with time at constant temperature. Interface between the phases has same quality during growth. The growing particles are randomly distributed in the parent phase. The daughter phase has nearly spherical morphology.

4.2.1 Phase separation on cooling liquids

Some glassy alloys exhibit phase separation [16] either by spinodal or binodal mechanism owing to solid- or liquid-state immiscibility among the constituent elements. The solid-state immiscibility between Zr and Y caused spinodal phase separation in the Zr-Y-Al-Ni system during heating upon the glass transition [17]. Zr-rich/La-rich glassy phases were obtained in the melt-spun Zr-La-Al-Ni-Cu alloy [18]. The structure of multi-phase metallic glass strongly depends on the alloy composition as well as the processing conditions. The $Zr_{30}RE_{30}Al_{15}Ni_{25}$ and $Zr_{15}RE_{45}Al_{15}Ni_{15}$ (RE-rare earth metal) alloys also showed similar behaviour [19]. A globular phase in the $Zr_{30}Nd_{30}Al_{15}Ni_{25}$ alloy is Nd-rich (Fig. 4.3a) while the surrounding matrix phase is a Zr-rich one. The elongated Zr-rich phase with a nominal atomic composition of $Zr_{53.0}Nd_{9.2}Al_{13.9}Ni_{23.9}$ produces larger halo ring of SAED pattern, while the Nd-rich second glassy phase has the $Zr_{17.8}Nd_{43.4}Al_{16.1}Ni_{22.3}$ composition.

Surprisingly phase separation was also observed in the $Fe_{90}Zr_{10}$ metallic glass [20]. The observed evolution of the structure from metallic glass to the two-phase crystal structure could be the solute partitioning preceding eutectic crystallization. A nonuniform structure with two amorphous phases one depleted another enriched in yttrium was found in the Al-Ni-Y metallic glass [21].

Fig. 4.3. (a) Bright-field transmission electron micrograph showing the two-phase structure of the $Zr_{30}Nd_{30}Al_{15}Ni_{25}$ sample and SAED patterns taken from the corresponding areas. Reproduced from [19] with permission of Elsevier. (b) A residual globular glassy phase in the $Cu_{35}Zr_{45}Ag_{20}$ alloy annealed at 722 K for 1 ks. Reproduced from [24] with permission of Elsevier.

4.2.2 Phase separation on heating metallic glasses

Phase separation competes with crystallization. However, the driving force is usually smaller and it takes place at a lower temperature. Binodal phase separation was observed in the $Cu_{36}Zr_{48}Al_8Ag_8$ glassy alloy [22] in the glassy/supercooled liquid phase at long-term annealing within the supercooled liquid region prior to crystallization (Fig. 4.4). Dark spots in the HRTEM image were analyzed by using EDX mapping and found to be Ag-rich (Fig. 4.5). Nanocrystallization of Ag-rich particles takes place in these regions after separation at long-term annealing (Fig. 4.4). However, if the sample is heated quickly enough intensive eutectic-like crystallization of the studied alloy starts at a higher temperature.

Fig. 4.4 HRTEM image of the $Cu_{36}Zr_{48}Al_8Ag_8$ glassy alloy sample treated for 25 ks at 675 K.

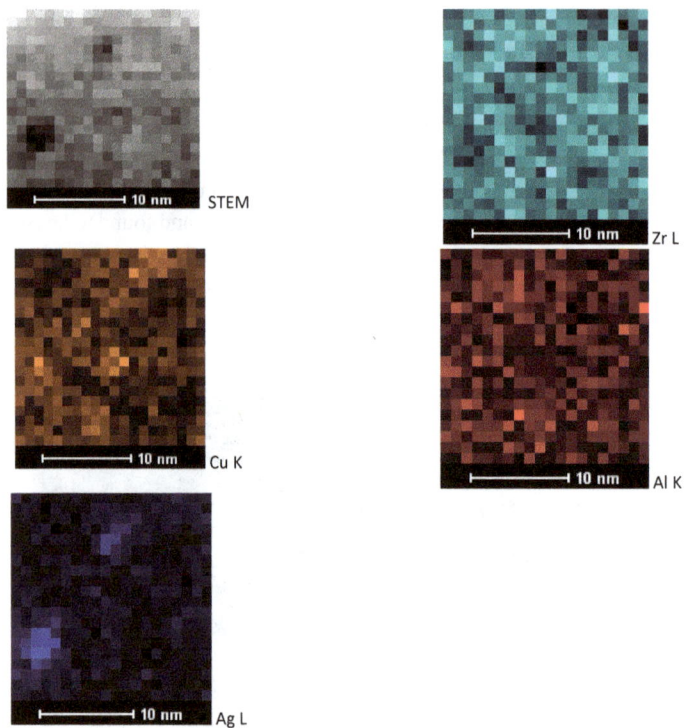

Fig. 4.5. STEM image of the $Cu_{36}Zr_{48}Al_8Ag_8$ glassy alloy sample annealed for 25 ks at 675 K as well as the corresponding EDX elemental maps as indicated. Courtesy of Dr. Hayasaka from the High-Voltage Microscopy Center of Tohoku University.

The $Zr_{62.5}Cu_{22.5}Al_{10}Fe_5$ alloy also undergoes nanoscale phase separation on heating prior to crystallization when kept at the supercooled liquid temperature owing to the repulsive Fe-Cu interaction (Fig. 4.6) [23].

Fig. 4.6. High-angle annular dark-field (HAADF) image of the $Zr_{62.5}Cu_{22.5}Al_{10}Fe_5$ sample annealed at 713 K for 300 s. Reproduced from [23] with permission of Taylor & Francis.

A small volume fraction (~10 vol.%) of a residual globular glassy phase was observed in the structure of the $Cu_{35}Zr_{45}Ag_{20}$ alloy (Fig. 4.3b) while no phase separation was found in the case of the as-solidified sample [24]. The sub-micrometer scale residual amorphous phase is depleted in Ag compared to the crystalline region. It contains only about 14 at.% Ag. This residual amorphous phase requires a higher temperature for crystallization. Such a phase separation in the Cu-Zr-Ag system can be expected according to a simple eutectic-type Cu-Ag phase diagram. It has been also observed in the $Cu_{43}Zr_{43}Al_7Ag_7$ [25] and $Ag_{20}Cu_{48}Zr_{32}$ [26] glassy alloys on a nanoscale while in the $Cu_{35}Zr_{45}Ag_{20}$ alloy it takes place on a submicron level.

4.3 Dissolution of a metastable crystalline phase in a supercooled liquid

The as-solidified $Cu_{50}Zr_{30}Ti_{10}Pd_{10}$ bulk glassy alloy was found to contain nanoparticles (Fig. 4.7 (a)) identified as CuZr cubic phase with a lattice parameter of 0.326 nm [27]. These particles nucleate but did not grow to the size more than 10 nm on cooling. However, dissolution of the CuZr nanoparticles was observed in the supercooled liquid (Fig 4.7 (b,c)). Dissolution of the CuZr nanoparticles in a supercooled liquid of the $Cu_{50}Zr_{30}Ti_{10}Pd_{10}$ alloy before devitrification shows their instability in the temperature range of 750-800 K. In accordance with the Cu-Zr phase diagram, the CuZr phase suffers eutectoid transformation at 988 K which is above the supercooled liquid region. Thus,

CuZr phase becomes thermodynamically unstable and dissolves in the supercooled liquid when atomic diffusivity is enhanced by temperature.

Fig. 4.7. $Cu_{50}Zr_{30}Ti_{10}Pd_{10}$, TEM. (a) bright-field image in as-solidified state. The insert – nanobeam diffraction pattern and its indexing according to CuZr phase. (b) bright-field and (c) dark-field images after heating up to T_g of 748 K and fast cooling. The insert in (b) – selected-area electron diffraction pattern. Reproduced from [27] with permission of Taylor & Francis.

4.4   Crystallization of glassy alloys.

4.4.1   General features.

Although it might be difficult to establish an intrinsic physical difference between amorphous and glassy alloys such a slightly arbitrary differentiation is useful. An alloy is "glassy" if it is produced by rapid solidification of a liquid. Many of metallic glasses transform to a supercooled liquid before devitrification at high enough heating rate. The formation of a supercooled liquid has a significant influence on the devitrification process in metallic glasses [28] owing to the change of the local atomic structure in the supercooled liquid region due to higher atomic mobility compared to that in the glassy phase.

In general, glassy alloys exhibiting the supercooled liquid region on heating have higher relative density (density of the glassy alloy compared to its crystalline state) and better glass-forming ability compared to amorphous alloys. Marginal glass-formers and growth controlled BMGs possess pre-existing nuclei or even nanoparticles in the amorphous matrix. Growth of these nuclei or particles starts at relatively low temperature without an incubation period (Fig. 2.14).

The following phase transformations take place during crystallization of the glassy alloys and supercooled liquids: polymorphous (a product phase has the same composition as the glassy phase), primary (a product phase has a composition different from that of the glassy phase), eutectic (two or more phases nucleate and grow conjointly). Crystal growth velocities in these processes can range from very low values in case of primary crystallization (of the order of an atomic layer per second and even less) [27] up to as high as ~100 m/s in case of pure metals [29] and chalcogenide glasses [29, 30]. Eutectic colonies exhibit intermediate growth rate values, for example ~100 nm/s in case of Pd-Ni-Si-P glassy alloys [31].

Below the liquidus temperature the liquid and glassy phases have a higher Gibbs free energy than the competing crystalline phase, and thus have a driving force for crystallization. However, there is an energy barrier. Nucleation of a crystalline phase can be homogeneous and heterogeneous. The classical homogeneous nucleation theory states that transformation of a glass or a supercooled liquid to a crystal produces the Gibbs free energy change ($\Delta G_v < 0$ per unit volume) which for a spherical nuclei is $(4/3)\pi r^3 \Delta G_v$. However, nucleation process leads to the creation of an interface with energy ($\Delta G_s > 0$ per unit interface) area which for a spherical nuclei is $4\pi r^2 \Delta G_s$. The energy change upon nucleation of a crystal with a radius r is

$$\Delta G = (4/3)\pi r^3 \Delta G_v + 4\pi r^2 \Delta G_s \qquad (4.1)$$

In case of a solid-solid glass→crystal transformation there is one more term connected to the energy increase due to elastic deformation ($\Delta G_e > 0$) because the parent and daughter phases usually have different unit volume. This energy term does not have an influence on the shape of the ΔG curve but changes ΔG_c and r_c, in some cases significantly. At small r influence of $4\pi r^2 \Delta G_s$ is significant while at larger r the $(4/3)\pi r^3 \Delta G_v$ term start to dominate (Fig. 4.8). There are two critical points in the plot where the first and second derivatives are equal to zero. The first one corresponds to a critical nuclei size r_c which equals to $-2\Delta G_s/\Delta G_v$. If we insert this value in eq (4.1) then the critical work will be

$\Delta G_c = (16\pi/3)\Delta G_s^3/\Delta G_v^2$. An inflection point where the second derivative is zero corresponds to $r_i = -\Delta G_s/\Delta G_v = r_c/2$. In the case of heterogeneous nucleation the energy barrier against nucleation can be significantly reduced enhancing nucleation.

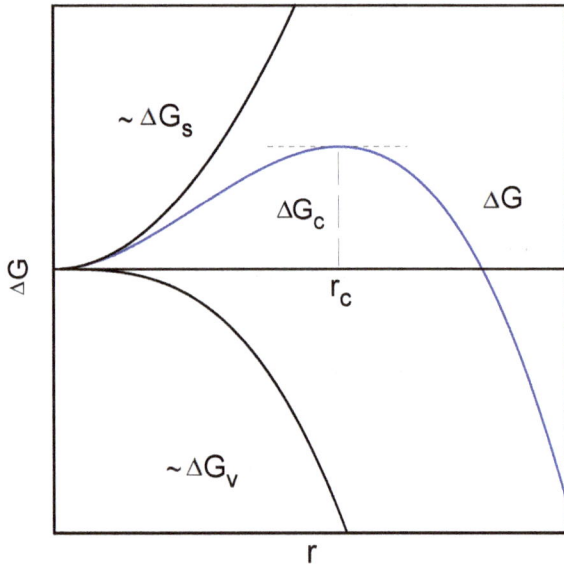

Fig. 4.8. Gibbs Free Energy difference between two phases as a function of the nucleus radius at constant temperature.

If crystallization occurs by nucleation and growth mechanism, then nucleation rate and growth rate have different temperature dependences. The number of critical-size nuclei per unit volume is proportional to $e^{-\Delta G_c/RT}$ while the frequency of atomic transfer through the interface is proportional to $e^{-Q_N/RT}$. Thus, the nucleation rate I is:

$$I = I_0 e^{-Q_N/RT} e^{-\Delta G_c/RT}, \tag{4.2}$$

where $Q_N \sim 1/\Delta G^2$ is the activation energy for transfer of atoms across the surface of a nucleus and ΔG_c as a function of supercooling (ΔT) is the energy required to form a critical nucleus. The growth rate (u) is proportional to the frequencies of the transition from liquid/glass v_{l-s} to solid and back v_{s-l}. $v_{l-s}=v\exp(-\Delta G/RT)$. Thus, in the case of interface controlled growth:

$$u=u_0 e^{-Q_g/RT}(1-e^{-\Delta G/RT}),$$ (4.3)

where Q_g is the activation energy for growth; ΔG is the Gibbs free energy difference. The nucleation and growth rate dependencies as a function of supercooling are schematically shown in Fig. 4.9.

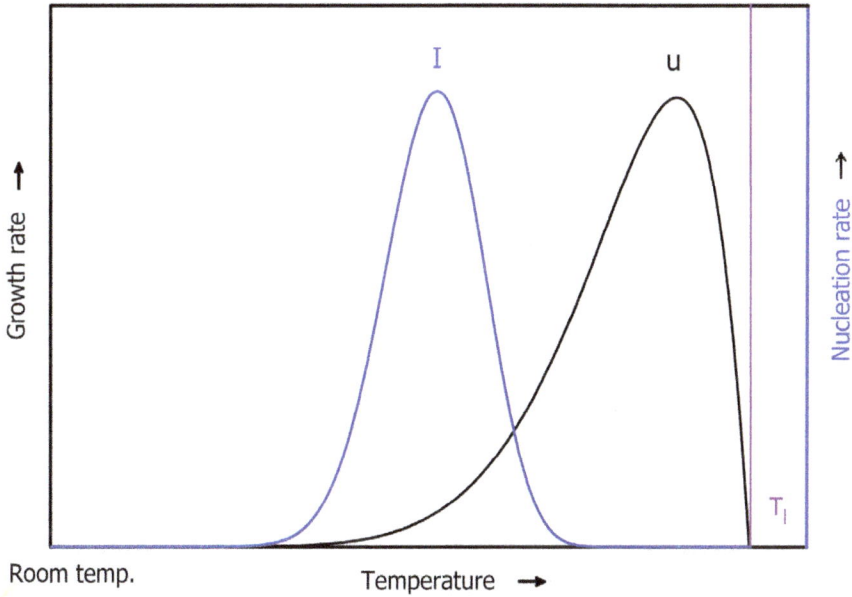

Fig. 4.9. Nucleation (I) and growth (u) rates as a function of supercooling.

As a result time-temperature-transformation (TTT) diagram can be plotted (Fig. 4.10).

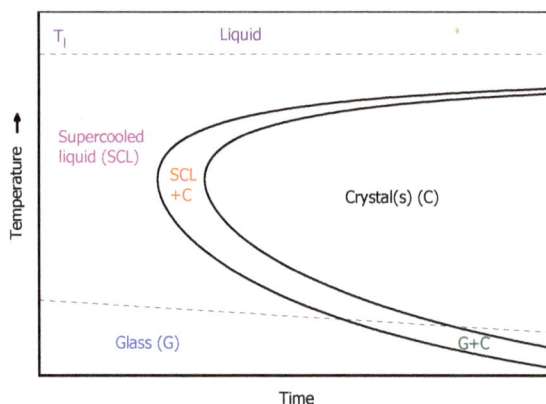

4.10. A schematic time-temperature-transformation (TTT) diagram.

Kinetics of the crystallization processes was studied substantially [32,33]. If I and u are time independent and the reaction is interface-controlled then kinetics of the formation of spherical crystals in the glassy alloys can be analyzed by Kolmogorov [34] -Johnson-Mehl [35] –Avrami [36] general exponential equation for the fraction transformed x:

$$x = 1 \ exp(- \pi / 3 * Iu^3 \ t^n) \tag{4.4}$$

Nevertheless, this equation is also used for other reactions. Depending on the Avrami exponent n different behaviour can be observed (Fig. 4.11). The n values close to 2.5 correspond to nucleation and diffusion controlled growth while n=4 corresponds to nucleation and interface controlled growth [37].

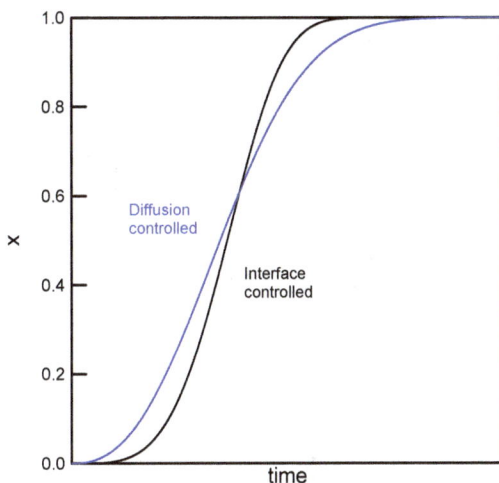

Fig. 4.11. Schematic representation of the fraction transformed (x) as a function of time in case of the interface and diffusion controlled growth. The kinetic constant, (related to the nucleation and growth rates) are not equal.

For example, the $Cu_{50}Zr_{30}Ti_{10}Ni_{10}$ alloy [38] exhibits a completion of the transition to a liquid state on heating at about 740 K (Fig. 4.12 (a)). Crystallization takes place by nucleation and growth of the $(Cu,Ni)_{10}(Zr,Ti)_7$ phase. The isothermal calorimetry traces (Fig. 4.12 (b)) taken within the supercooled liquid region, glass-transition region and from the glassy phase shown were used to create the Avrami plot (Fig. 4.12 (c)) and the TTT-diagram (Fig. 4.12 (d)).

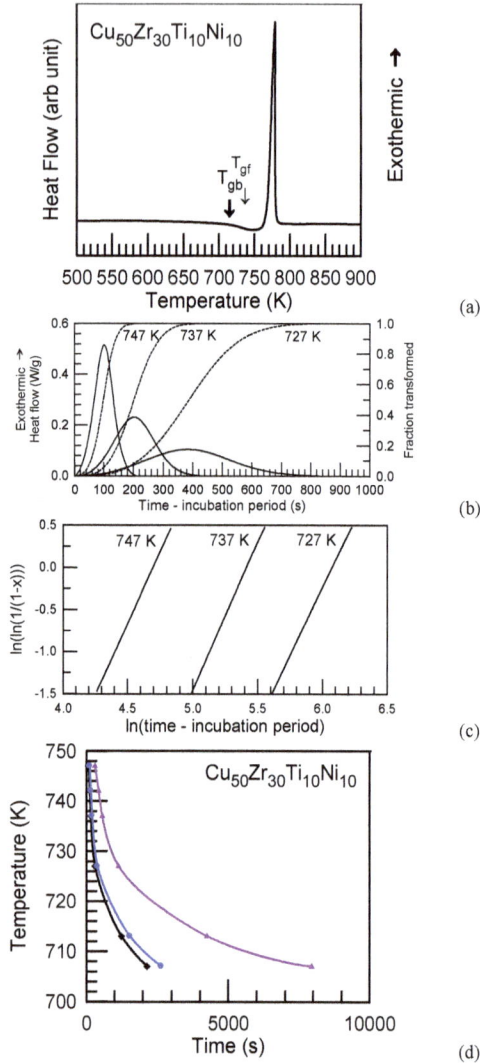

Fig. 4.12. DSC trace (a), isothermal calorimetry curves - solid lines and the fraction transformed - dotted lines (b), the Avrami plot (c) and a low-temperature part of the TTT diagram (diamond symbols and circles indicate the incubation period defined by using different methods and triangles indicate the completion of transformation). (d) of the $Cu_{50}Zr_{30}Ti_{10}Ni_{10}$ alloy. The data is partly taken from [39] https://www.scientific.net/JMNM.15-16.31.

The glass-formation, crystallization kinetics and structure changes in the $Cu_{55-x}Zr_{45}Ag_x$ alloys on heating were studied and compared with that of the binary $Cu_{55}Zr_{45}$ alloy [24, 39]. During the exothermic reactions the supercooled liquid crystallizes forming mainly oC68 $Cu_{10}Zr_7$ phase in the $Cu_{55}Zr_{45}$ alloy and a mixture of oC68 $(Cu,Ag)_{10}Zr_7$ and tP4 $(Ag,Cu)Zr$ solid solution phases in the Ag-bearing alloys. The crystallization started after a certain incubation period which indicates nucleation and growth transformation mechanism with a steady-state nucleation and 3-dimensional interface-controlled growth of nuclei (Fig. 4.13).

Fig. 4.13. The Avrami plots for the $Cu_{45}Zr_{45}Ag_{10}$ and $Cu_{35}Zr_{45}Ag_{20}$ glassy alloys constructed using the isothermal calorimetry. Reproduced from [24] with permission of Elsevier.

The XRD results indicate that the $(Cu,Ag)_{10}Zr_7$ phase is the main structure constituent in the $Cu_{45}Zr_{45}Ag_{10}$ alloy while the $Cu_{35}Zr_{45}Ag_{20}$ alloy contains a large fraction of $(Ag,Cu)Zr$ phase. A small volume fraction (~10 vol.%) of a residual globular glassy phase is observed in the structure of the $Cu_{35}Zr_{45}Ag_{20}$ alloy (Fig. 4.6) while no phase separation was found in the case of the as-solidified sample and the sample annealed at T_g. According to EDX analysis (Table 4.1) the residual amorphous phase is depleted in Ag compared to the crystalline region. It contains only about 14 at.% Ag.

Table 4.1. Chemical composition of the thermodynamic phases observed in the $Cu_{35}Zr_{45}Ag_{20}$ alloy annealed at 722 K for 1 ks. Reproduced from [24] with permission of Elsevier.

Phases\Elements	Cu, at.%	Zr, at.%	Ag, at.%
oC68 $(Cu,Ag)_{10}Zr_7$	34	43	23
tP4 (Ag,Cu)Zr	20	54	26
Residual amorphous	34	52	14

Small changes in composition often lead to drastic changes in the crystallization behavior of metallic glasses. Devitrification of the $Cu_{57.5}Zr_{30}Ti_{10}Ni_{2.5}$ glassy alloy starts from formation of the intermediate nanoscale intermetallic compound. The $Cu_{55}Zr_{30}Ti_{10}Ni_5$ and $Cu_{52.5}Zr_{30}Ti_{10}Ni_{7.5}$ glassy alloys devitrify forming directly the equilibrium oC68 $Cu_{10}Zr_7$ phase. Partial replacement of Zr by Ti and Cu by Ni in $Cu_{10}Zr_7$ caused a decrease of the lattice parameters. The $(Cu,Ni)_{10}(Zr,Ti)_7$ phase (Pearson symbol: oC68) is formed in the $Cu_{50}Zr_{30}Ti_{10}Ni_{10}$ glassy phase (Table 4.2, Fig. 4.14).

Table 4.2. Phase composition of the $Cu_{60-x}Zr_{30}Ti_{10}M_x$ glassy alloys after the initial, intermediate and final stages of the phase transformations corresponding to the DSC exothermic peaks. Reproduced from [39] with permission of Trans Tech Publications Ltd.

M element and its content	Primary stage	Intermediate	Fully crystallized
-	NIC* 1	NIC 2	$Cu_{10}Zr_7+Cu_3Ti_2+X^{**}$
2.5 at.% Ni	NIC 2	-	$Cu_{10}Zr_7+Cu_3Ti_2$
5 at.% Ni	$(Cu,Ni)_{10}(Zr,Ti)_7+NIC 3$	-	-***
7.5 at.% Ni	$(Cu,Ni)_{10}(Zr,Ti)_7+NIC 3$	-	-***
10 at.% Ni	$(Cu,Ni)_{10}(Zr,Ti)_7$	-	-***
5 at.% Co	NIC 2	NIC 2+X	$Cu_{10}Zr_7+X$

* NIC – nanoscale intermetallic compound; ** unidentified phase

*** at long enough annealing time the Cu_3Ti_2 phase starts to precipitate from $(Cu,Ni)_{10}(Zr,Ti)_7$

Fig. 4.14. Structure of the $Cu_{50}Zr_{30}Ti_{10}Ni_{10}$ alloy heat treated at 737 K for 0.9 ks, TEM. (a) Bright-field image. (b) and (c) SAED pattern and its indexing according to the Aba2 lattice. Zone axes are written in square brackets. Reproduced from [39] with permission of Trans Tech Publications Ltd.

The characteristic temperatures of the $Cu_{60-x}Zr_{30}Ti_{10}M_x$ glassy alloys are plotted in Fig. 4.15. Both T_x and T_g values at Ni content lower that 10 at.% change in an irregular way in consistent with the changes in the devitrification behavior observed (see Table 4.2).

Fig. 4.15. T_g, T_x and ΔT_x of the $Cu_{60-x}Zr_{30}Ti_{10}Ni_x$ glassy alloys as a function of Ni content. Reproduced from [39] with permission of Trans Tech Publications Ltd.

Alloys containing 5, 7.5, 10, 30 and 60 at.% Ni (see Fig. 4.12 (a)) exhibit a single strong DSC exothermic peak. As one can see in Fig. 4.15 small additions of Ni up to 10 at.% extend the supercooled liquid region which attains the maximum value of 58 K at 7.5 at.% Ni. The supercooled liquid region in the glassy alloys containing more than 3.75 at.% and less than 20 at.% Ni is about two times larger than that in the $Cu_{60}Zr_{30}Ti_{10}$ glassy alloy.

Highly negative mixing enthalpy values -49 kJ/mol and -35 kJ/mol, are found for Ni-Zr and Ni-Ti atomic pairs, respectively [40,41]. These values are larger than those for the Cu-Zr and Cu-Ti atomic pairs of -23 and -9 kJ/mol, respectively. Negative mixing enthalpy is favorable for glass-formation and may be responsible for the widening of ΔT_x in Ni-bearing alloys compared to $Cu_{60}Zr_{30}Ti_{10}$.

Three types of the metastable nanoscale intermetallic compounds (NIC) were obtained in the studied alloys on devitrification (see Table 4.2). NIC 1 having a cubic lattice was obtained in the $Cu_{60}Zr_{30}Ti_{10}$ alloy only. Formation of the primary NIC 2 was found in the $Cu_{57.5}Zr_{30}Ti_{10}Ni_{2.5}$ and $Cu_{55}Zr_{30}Ti_{10}Co_5$ alloys. NIC 3 was found in the $Cu_{55}Zr_{30}Ti_{10}Ni_5$ and $Cu_{52.5}Zr_{30}Ti_{10}Ni_{7.5}$ alloy. The $(Cu,Ni)_{10}(Zr,Ti)_7$ compound only precipitates in the $Cu_{60-10}Zr_{30}Ti_{10}Ni_{10-60}$ alloys (it does not contain Cu in the $Ni_{60}Zr_{30}Ti_{10}$ alloy) at the initial stage of devitrification.

The linearity of the Avrami plot of the $Cu_{50}Zr_{30}Ti_{10}Ni_{10}$ glassy alloy (Fig. 4.12(c)) indicates that this phase transformation obeys the kinetic law (4.4). The Arrhenius plot of the incubation time for devitrification as a function the testing temperature shows that the $Cu_{50}Zr_{30}Ti_{10}Ni_{10}$ glassy alloy has a high activation energy for nucleation (E_n) for the $(Cu,Ni)_{10}(Zr,Ti)_7$ of 340 kJ/mol (Fig 4.16).

$E_n = 340$ kJ/mol

Fig. 4.16. Arrhenius plot of the $Cu_{50}Zr_{30}Ti_{10}Ni_{10}$ glassy alloy. Reproduced from [39] with permission of Trans Tech Publications Ltd.

The time-temperature-transformation diagrams (4.12(c)) created in the isothermal mode or under continuous heating are useful for comparison of the thermal stabilities of different glasses against devitrification and for selection of the heat treatment regimes. Such diagrams have been created for various metallic glasses [42,43,44,45,46,47]. Comparison of the long-term thermal stabilities of different metallic glasses was performed using continuous heating transformation (CHT) diagrams (Fig. 4.17) [48]. CHT diagrams constructed by applying a corollary from the Kissinger analysis method using the DSC data at different heating rates allow estimating devitrification behavior of glassy alloys on a long-term scale. Early transition metals (ETMs): Ti-, Zr- and Hf-based metallic glasses were found to have a similar time-temperature gradient of the curves independently from their initial devitrification products (crystal or quasicrystal) and transformation mechanism. Metalloid-rich Si- and Ge-based glasses exhibit quite high long-term thermal stability and low average temperature gradient of the CHT curve. Late transition metals (LTMs): Cu- and Ni-based metallic glasses showed high thermal stability at a short time scale.

Fig. 4.17. Continuous heating transformation diagrams for different alloys numbered from 1 to 10. Corresponding alloy compositions are listed in Table 4.2. The curves correspond to T_p - peak temperature at which the transformation rate of the initial crystallization reaction attains the maximum. Reproduced from [48] with permission of Elsevier.

Table 4.3. Numbers (Nr.), compositions of metallic glasses, initial devitrification products (IDP), devitrification mechanism (M) and the apparent activation energy (E). Data from [48] with permission of Elsevier.

Nr	Composition	IDP[a]	M[b]	E, kJ/mol
1	$Hf_{60}Pd_{30}Ni_{10}$	C	PR	400
2	$Hf_{65}Pd_{17.5}Ni_{10}Al_{7.5}$	QC	PR	400
3	$Ni_{66}Hf_{20}Al_{14}$	C	CX	320
4	$Ti_{45}Ni_{20}Cu_{25}Sn_5Zr_5$	C	PR	320
5	$Si_{55}Al_{20}Fe_{10}Ni_5Cr_5Zr_5$	C	PM	340
6	$Ti_{50}Ni_{25}Cu_{25}$	C	PM	300
7	$Cu_{60}Zr_{30}Ti_{10}$	C	S, PM	265
8	$Ge_{68}Cr_{14}Al_{10}Ce_4Sm_4$	C	CX	360
9	$Zr_{55}Cu_{20}Ti_{15}Ni_{10}$	QC	PR	250
10	$Ge_{60}Ni_{35}La_5$	C	PM	240

[a] C-crystal, QC-quasicrystal

[b] PR-primary, PM-polymorphous, CX-complex, i.e., more than 2 phases simultaneously including eutectic-type transformation, S-possible spinodal

CHT diagrams also can be recalculated from the isothermal ones [49] using a method close to that used for steels [50]. The recalculation method is based on the calculation and integration of the fraction of the incubation period (fτ) passed on heating from room temperature at each step. The procedure is as follows: to sum all fτ=t_h/τ values starting from 298 K with the step of 1 K where t_h is the time used to heat the specimen for 1 K at a given heating rate β. Then the total heating time t_{ht} up to T_c is $(T_c-298)/\beta$. This method was proved to provide a good correspondence between the data at least in the mid-temperature range (4.18). The devitrification of metallic glasses is strongly temperature dependent since crystal nucleation and growth have different dependencies on temperature. However, as the incubation time representing an early stage of the reaction is mostly dominated by the nucleation process the present approach is useful.

Fig. 4.18. The TTT (curve 1) and CHT (curve 2) diagrams for the primary precipitation of a metastable phase in the $Cu_{60}Hf_{25}Ti_{15}$ glassy alloy. The dashed vertical line indicates a heating rate of 0.67 K/s. Primary transformation. Reproduced from [49] with permission of Elsevier.

4.4.2 Nanocrystallization.

Nanostructured materials can be defined as the substances having a very small grain size typically below 100 nm [51,52]. Invention of the age-hardenable Al-based alloys, with nanoscale Guinier-Preston zones, θ'' and θ' phases [53,54] was the first well known example of use of nanoparticles for practical purpose in physical metallurgy. A widely used method for the formation of nanomaterials is crystallization of the glassy alloys. Metallic glassy alloys crystallize at large supercooling when growth rate is relatively low while nucleation rate is usually high which with a small critical nucleus size lead to the formation of a nanostructure in various alloys. Some of the composites possess enhanced mechanical properties compared to single-phase glassy alloys.

Although there are some reports that, in particular cases, nanocrystalline structure can be obtained after eutectic [55] and polymorphous [56,57] crystallization of glassy alloys, the most common mechanism leading to formation of a nanostructure is primary crystallization. Nanostructured alloys are readily obtained on primary crystallization of glasses with a long-range diffusion controlled growth [58,59]. For example, the primary transformation reaction in the Al-rich amorphous alloys with high Al concentrations above 85 at.%, in many alloys, is related to the precipitation of a primary FCC-Al phase. Nanocrystallization was also observed in Fe-based glasses [60,61] and many others.

Different Al-RE-TM glasses show primary precipitation of the Al solid solution nanoparticles on heating [62] with extremely high nucleation rate exceeding 10^{20} m^{-3}s^{-1} [63,64].The formation of Al at the primary stage is quite typical for Al-based amorphous alloys and glass-formers, for example: Al-Y-Ni-Co [65], Al-Fe-Y, Al-Sm [66], Al-Ni-Nd [67] Al-Fe-Nd [68,69]. After reaching a certain size (typically few tens of nanometers) Al particles change their morphology from spherical to dendritic [70]. Pressure effects on crystallization of metallic glasses were also studied [71]. The neutron diffraction study of the Al-Y-Ni alloys indicated that Y-rich glasses are homogeneous while Ni-rich glasses have compositional inhomogeneity [72].

FCC Al lattice parameter measurements and atom probe ion field microscopy investigation [73] showed very low concentration of the alloying elements in nanocrystalline Al in accordance with the phase diagrams [74] of Al-RE and Al-TM systems. Segregation of the RE metal having low trace diffusivity in Al on the Al/amorphous phase interface is considered to be one of the most important reasons for the low growth rate of Al. By analogy with these results the extended X-ray absorption fine structure analysis of grain boundaries in the nanocrystalline $Fe_{85}Zr_7B_6Cu_2$ alloys showed very low Fe content in the grain boundaries between the BCC Fe nano grains. The primary crystallization requires a long-range diffusion that impedes crystal growth [75].

Another type of phase transformation in an amorphous solid leading to formation of a nanostructure is phase separation. For instance, it has been reported that in some Al–TM–RE glasses devitrification leading to formation of a nanostructure appears to be preceded by the amorphous phase separation [76,77].

In some alloys nanocrystallization only insignificantly affects flow ability of the supercooled liquid. The $Cu_{55}Zr_{30}Ti_{10}Co_5$ glassy sample exhibited a low viscosity at the temperatures above T_g till the end of the crystallization process owing to the equiaxed morphology of the nanoscale crystalline phase (Fig. 4.19). This alloy can be forged in the semi-solid state at low load and high speed, thus producing in-situ glassy/crystal composites [78].

Fig. 4.19. Volume fraction of crystalline phases (1) as a function of temperature calculated from the portion of heat release at each phase transformation divided by the total heat release and the viscosity of the $Cu_{55}Zr_{30}Ti_{10}Co_5$ samples (2). Reproduced from [78] with permission of Elsevier.

On heating the $Fe_{48}Cr_{15}Mo_{14}C_{15}B_6Y_2$ [79] and $Fe_{48}Cr_{15}Mo_{14}C_{15}B_6Tm_2$ [80] glassy alloys formed nanoscale particles of the χ-$Fe_{36}Cr_{12}Mo_{10}$ phase. Its structure is cI58 (α-Mn type) and it has a nominal chemical composition of $Fe_{62}Cr_{21}Mo_{17}$ [81]. Figure 4.20 shows X-ray diffraction patterns of the $Fe_{48}Cr_{15}Mo_{14}C_{15}B_6Tm_2$ alloy in the as-solidified and annealed state. One can see the formation of a very broad (330) peak related to the crystalline χ-$Fe_{36}Cr_{12}Mo_{10}$ phase after annealing for 420 s at 893 K. Precipitation of the nanoscale χ-$Fe_{36}Cr_{12}Mo_{10}$ phase was also observed in the $Fe_{48}Cr_{15}Mo_{14}C_{15}B_6Tm_2$ and $Fe_{48}Cr_{15}Mo_{14}C_{15}B_6Y_2$ alloys by TEM at an early crystallization stage [82]. Rather surprisingly this phase was also reported to form on crystallization in Cr- and Mo-free glassy alloys [83, 84, 85] though its size is significantly larger. After annealing $Fe_{48}Cr_{15}Mo_{14}C_{15}B_6Y_2$ at 973 K the (330) peak shows separation into two, owing to the formation of $Fe_{23}(B,C)_6$ phase (cF116 $Cr_{23}C_6$ type) and Fe_3Mo_3C (cF112 Fe_3W_3C type) likely by a quasi-peritectic reaction [86] (residual liquid+χ-$Fe_{36}Cr_{12}Mo_{10}$ phase $\rightarrow$ $Fe_{23}(B,C)_6$ phase + Fe_3Mo_3C).

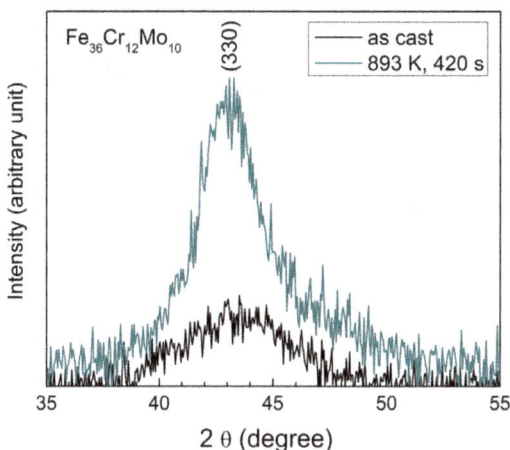

Fig. 4.20. X-ray diffraction patterns of the $Fe_{48}Cr_{15}Mo_{14}C_{15}B_6Tm_2$ alloy in as-solidified and annealed state.

The crystal size can be estimated from the main (330) X-ray diffraction peak of the primary nanocrystalline χ-$Fe_{36}Cr_{12}Mo_{10}$ phase of about 3 degrees of full width at half maximum (FWHM). According to the standard procedure after elimination of the instrumental broadening [87] and taking into account the effect of dispersed crystal blocks scattering only, direct application of the Scherrer formula suggests a coherent scattering crystallite size of about 3 nm diameter. Separating the broadening effects caused by small particle size and by inhomogeneous strain [88], and noting the crystal diameter of 10 nm provided by TEM observation, results in an estimate of the internal strain of 0.021. This is a very large value comparable to those obtained after intensive plastic deformation, for example ball milling [89]. This large inhomogeneous strain is likely to be related to the observed low growth rate of the particles compared to the relatively large particles of χ- phase found in the alloys containing no Cr, Mo and RE metals.

The large difference in the mass density between the metallic glassy and χ-$Fe_{36}Cr_{12}Mo_{10}$ phase, corresponding to large differences in the specific volume (about 4.2% and 6.5% for $Fe_{48}Cr_{15}Mo_{14}C_{15}B_6Tm_2$ and $Fe_{48}Cr_{15}Mo_{14}C_{15}B_6Y_2$ alloys, respectively), is likely to give a large hydrostatic tensile stress in the χ-$Fe_{36}Cr_{12}Mo_{10}$ phase. Such a stress may limit drastically the growth rate of this phase. On the other hand, these stresses in the crystals are not uniform. There is about one Y atom per unit cell of χ-$Fe_{36}Cr_{12}Mo_{10}$ phase of the

$Fe_{48}Cr_{15}Mo_{14}C_{15}B_6Y_2$ alloy. Taking into account the Avogadro number and volume of a spherical nanoparticle having 10 nm in diameter such a particle should contain $4.7 \cdot 10^4$ atoms total and about 960 of Y atoms. The number of Y atoms in a single nanoparticle is considered to be large enough to ensure its low growth rate.

Schematic transformation diagrams for the $Fe_{48}Cr_{15}Mo_{14}C_{15}B_6RE_2$ (Fe-RE2) and $Fe_{50}Cr_{15}Mo_{14}C_{15}B_6$ (Fe-RE0) alloys are shown in Fig. 4.21. The cooling and heating regimes correspond to: 1) ribbon samples with the glassy phase and pre-existing nuclei of χ-phase in the $Fe_{48}Cr_{15}Mo_{14}C_{15}B_6RE_2$ alloys; 2) bulk samples of 2.7 mm thickness (glassy in the case of $Fe_{48}Cr_{15}Mo_{14}C_{15}B_6RE_2$ and crystalline in the case of $Fe_{50}Cr_{15}Mo_{14}C_{15}B_6$); 3) bulk samples of 3.5 mm thickness (formation of nanoscale χ-phase in $Fe_{48}Cr_{15}Mo_{14}C_{15}B_6Y_2$); and 4) bulk samples of 4.2 mm thickness leading to the formation of crystalline phases in $Fe_{48}Cr_{15}Mo_{14}C_{15}B_6Y_2$. The heating paths 5 and 6 correspond to the DSC and isothermal calorimetry studies, respectively.

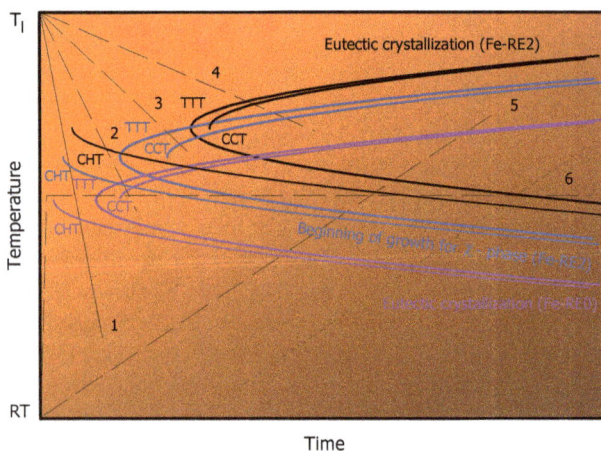

Fig. 4.21. Schematic time-temperature-transformation diagram for the $Fe_{48}Cr_{15}Mo_{14}C_{15}B_6RE_2$ (Fe-RE2) and $Fe_{50}Cr_{15}Mo_{14}C_{15}B_6$ (Fe-RE0) alloys. T_l and RT indicate liquidus and room temperature respectively. The numbered dashed lines indicate the cooling-heating pathways. TTT, CCT and CHT indicate time-temperature-transformation (isothermal), continuous cooling transformation and continuous heating transformation diagrams, respectively, marked according to the colors used. Reproduced from [79] with permission of Elsevier.

In contrast to the $Fe_{50}Cr_{15}Mo_{14}C_{15}B_6$ glass in which an incubation period is observed, crystallization of the $Fe_{48}Cr_{15}Mo_{14}C_{15}B_6RE_2$ (RE=Y or Tm) alloys starts immediately after heating. This lack of an incubation period was found to apply for annealing of both RE-bearing alloys below T_x in both bulk and ribbon states. For example, differential isothermal calorimetry of the $Fe_{48}Cr_{15}Mo_{14}C_{15}B_6Y_2$ alloy was carried out at 882 and 898 K and precipitation of the χ-$Fe_{36}Cr_{12}Mo_{10}$ phase was observed. Similar results were obtained for $Fe_{48}Cr_{15}Mo_{14}C_{15}B_6Tm_2$ while $Fe_{50}Cr_{15}Mo_{14}C_{15}B_6$ exhibited a clear nucleation and growth reaction with an incubation period and no χ-$Fe_{36}Cr_{12}Mo_{10}$ phase.

Thus, the $Fe_{50}Cr_{15}Mo_{14}C_{15}B_6$ glass is a typical BMG which GFA is limited by the rate of crystal nucleation, while GFA of the $Fe_{48}Cr_{15}Mo_{14}C_{15}B_6RE_2$ alloys is limited by the rate of crystal growth. The diffusion coefficients of Mo (D_{Mo}) at 873 and 903 K of $2.0*10^{-20}$ and $1.0*10^{-19}$ m^2/s, respectively, are very close to those for self diffusion of Fe (D_{Fe}) of $1.9*10^{-20}$ and $1.1*10^{-19}$ m^2/s in crystalline alloys [90]. Chromium also has a similar D value in this temperature region [91]. On the other hand, Y (and likely other RE metals) has much lower D value of about $4.2*10^{-23}$ m^2/s at 900 K [92] which corresponds to a diffusion length (L) (defined as $L = 2\sqrt{Dt}$ where t is the diffusion time) of 0.4 nm in 900 s. The self-diffusion coefficient of Fe in the $Fe_{91}Zr_9$ glassy alloy at 900 K is about $2*10^{-15}$ m^2/s [93] which is 4 orders in magnitude higher than that in α-Fe. In order to be able to diffuse 10 nm in 900 s, yttrium should have a diffusion coefficient of about 10^{-20} m^2/s which is a reasonable value compared to that in α-Fe. Slow diffusion of Y explains its homogeneous distribution in the partially transformed $Fe_{48}Cr_{15}Mo_{14}C_{15}B_6Y_2$ sample. Thus, the $Fe_{48}Cr_{15}Mo_{14}C_{15}B_6RE_2$ glassy alloys exhibit growth of pre-existing nuclei of the χ-$Fe_{36}Cr_{12}Mo_{10}$ phase, though these nuclei are too small to be clearly recognized even by high-resolution TEM. The χ-$Fe_{36}Cr_{12}Mo_{10}$ phase has the structural type of α-Mn in which local atomic structure is found to be close to that of the glassy phase and describes well the first maximum of the radial distribution function [94]. Y and Tm additions prevent the relatively fast eutectic crystallization observed in the $Fe_{48}Cr_{15}Mo_{14}C_{15}B_6$ alloy.

The structure of Fe-based soft magnetic alloys like: Finemet $Fe_{73.5}Cu_1Nb_3Si_{13.5}B_9$ [95] and Nanoperm $Fe_{84}Zr_{3.5}Nb_{3.5}B_8Cu_1$ [96] after annealing consists of bcc Fe nanocrystals below 20 nm in size finely dispersed in the amorphous matrix. Cu, Nb or Zr elements despite on their low contents are responsible for the formation of a nanostructure upon annealing. For example, devitrification of the $Fe_{73.5}Cu_1Nb_3Si_{13.5}B_9$ alloy, which initially has an isotropic amorphous structure, starts from the formation of Cu-enriched zones [97].

Atom probe field ion microscopy and high resolution transmission electron microscopy [98] indicated that Cu forms nano clusters in the $Fe_{73.5}Si_{13.5}B_9Nb_3Cu_1$ amorphous matrix which act as the sites for heterogeneous nucleation of the bcc Fe particles on devitrification [99]. X-ray absorption fine structure studies also showed that Cu clusters with near-FCC structure were present from very early stages of the crystallization process [100]. The density of the clusters estimated by three dimensional atom probe is in the order of 10^{24} m^{-3} at the average cluster size of about 2-3 nm [101].

Not only pure metals and their solid solutions but some intermetallic compound particles can also have a nanoscale size. For example, crystallization of the nanocrystal-forming Ti-based alloys, for example, the $Ti_{50}Ni_{20}Cu_{23}Sn_7$ alloy begins from the primary precipitation of a nanoscale equiaxed, almost spherical particles of cF96 Ti_2Ni solid solution phase (other alloying elements are partially dissolved in this phase) with a lattice parameter of 1.138 nm [102,103]. Formation of such a nanoscale cF96 phase having a large cubic unit cell was also observed in the Zr- and Hf-based alloys [104,105]. cF96 Ti_2Ni phase exhibits a diffusion-controlled growth.

Fig. 4.22. XRD patterns of the $Hf_{60}Co_{25}Al_{20}$ glassy alloy (a) in as-solidified state and annealed (b) at 907 K for 0.9 ks, (c) at 907 K for 2.7 ks and using DSC at 0.67 K/s up to 1100 K. 907 K corresponds to the first exothermic peak temperature. Reproduced from [106] with permission of Elsevier.

An extremely small size and low growth rate of cF96 crystals were observed in the $Hf_{55}Co_{25}Al_{20}$ glassy alloy [106]. This alloy undergoes a double-stage devitrification forming the primary cubic cF96 Hf_2Co (Fig. 4.22) phase from the supercooled liquid by the steady state nucleation and diffusion-controlled growth of nuclei followed by the subsequent devitrification of the residual glassy matrix forming Hf_2Al and an unidentified phase. The cF96 nanoscale particles having a very small size are hardly distinguishable from the glassy matrix (Fig. 4.23) likely owing to local structural similarity between cF96 Hf_2Co phase and the glassy matrix. They also have a very low average growth rate of the order of 10^{-11} m/s.

Fig. 4.23. HRTM image of the $Hf_{60}Co_{25}Al_{20}$ glassy alloy annealed at 907 K for 0.9 ks.

In many cases redistribution of the alloying elements on a short scale precedes crystallization like it was observed in the Mg-Ni-Mm and Mg-Ni-Y-Mm glassy alloys [107]. Local structure of the $Mg_{86}Ni_{10}Y_2Mm_2$ and $Mg_{82}Ni_{14}Y_2Mm_2$ metallic glasses containing MRO zones changes prior to formation of the crystalline phases. Splitting of the amorphous halo peak occurs after heating up to the first exothermic peak and indicates redistribution of the alloying elements in the amorphous matrix and formation of the Mg-enriched zones (Fig. 4.24). This process occurs without an incubation period and leads to embrittlement of the samples. At the same time, $Mg_{86}Ni_{10}Mm_4$ exhibits formation of the primary HCP Mg crystals.

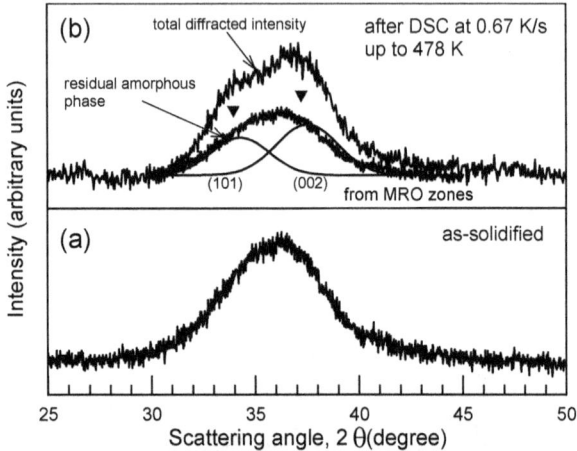

Fig. 4.24. X-ray diffraction patterns of the $Mg_{86}Ni_{10}Y_2Mm_2$ metallic glass. (a) as-solidified and (b) after DSC at 0.67 K/s up to 478 K. Triangles indicate positions of the (002) and (101) peaks of hcp Mg. Reproduced from [107] with permission of Taylor & Francis.

The amorphous Ge-Cr-Al-RE alloys also exhibit local order changes prior to nanoscale devitrification. For example, some structure changes occur in the $Ge_{70}Cr_{16}Al_{10}Nd_4$ alloy prior to devitrification. The position of the first peak of the radial distribution function (RDF) shifts to lower r-value after heating, and becomes closer to the values of pure amorphous (a-Ge) and crystalline Ge (c-Ge) which have tetrahedral configuration in the nearest neighbor region. The pronounced peak also appears at 0.41 nm in RDF for the heat treated sample. The ratio between the second and the first nearest neighbor distances (0.41 and 0.250 nm, respectively) is 1.64 which is almost identical with the value of 1.63 for a-Ge and c-Ge [108]. The results indicate that during heat treatment tetrahedral configuration of Ge atoms which is close to the configuration observed in pure a-Ge and c-Ge was formed. As a result primary Ge nanoparticles are also formed in many Ge-Al-Cr-RE alloys [109] on further heating. Anomalous X-ray scattering studied shows that Ge-Ge atomic pair in Ge-Al-Cr alloy has a different coordination compared to metal-metal and Ge-metal pairs [110].

The as-solidified microstructure of the rapidly solidified Ge-Al-Cr-RE alloys examined by TEM showed the presence of finely dispersed zones of about 1 nm in size [111] homogeneously distributed in the amorphous matrix [112]. These zones are responsible for the split of the first diffraction maximum in the XRD patterns in Fig. 4.25. Such a structure is typical for different rapidly solidified Ge-Al-Cr-RE and $Ge_{60}Ni_{35}La_5$ amorphous alloys [113]. Upon annealing nanoscale particles of cF8 Ge are formed (Fig. 4.25 and 4.26).

Fig. 4.25. XRD pattern of the ribbon sample of the $Ge_{55}Al_{30}Cr_{10}La_5$ alloy in as-solidified and annealed state as indicated (Cu Kα radiation). Reproduced from [111] with permission of Elsevier.

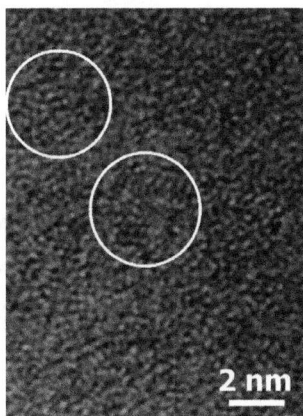

Fig. 4.26. High resolution TEM image of $Ge_{55}Al_{30}Cr_{10}La_5$ alloy in the annealed at 545 K for 20 min. Ge clusters are encircled.

4.4.3 Crystallization of metallic glasses above and below T_g

On heating glassy and amorphous $Al_{85}RE_8Ni_5Co_2$ alloys, namely, $Al_{85}Y_8Ni_5Co_2$, $Al_{85}Y_{8-x}Nd_xNi_5Co_2$ and $Al_{85}Gd_8Ni_5Co_2$ and others exhibit different crystallization behavior above and below glass-transition temperature (T_g) (Fig. 4.27 and Table. 4.4) [114]. The $Al_{85}RE_8Ni_5Co_2$ glassy alloys showed precipitation of the α-Al nanoparticles after continuous heating using DSC at high enough heating rates (0.67 K/s and higher) or isothermal annealing at the temperature above T_g [115]. At the same time, Y-, Gd- and Dy-bearing metallic glasses [116] as well as the $Al_{85}Y_4Nd_4Ni_5Co_2$ [117] showed simultaneous formation of the intermetallic compound(s) and α-Al nanoparticles, or primary formation of the intermetallic compound after annealing below T_g. For example, the $Al_{85}Y_8Ni_5Co_2$ alloy shows formation of an intermetallic compound conjointly with α-Al nanoparticles after isothermal annealing up to the completion of the primary phase transformation below T_g. After the completion of the primary phase transformation a lot of relatively small intermetallic compound particles below 50 nm size exist in the structure of the $Al_{85}Y_8Ni_5Co_2$ alloy [118]. The intermetallic compounds are metastable and have a multicomponent composition. The volume fraction of the intermetallic compound is higher than that of Al and the fraction of Al depends on the annealing temperature below T_g.

Table 4.4. Isothermal crystallizaiton of the $Al_{85}Y_4Nd_4Ni_5Co_2$ alloy in the vicinity of glass-transition region. T_g measured at 1.67 K/s is 561 K.

Transformation	Temperature, K	Crystalline Phases	n**
Primary,	568	nano-Al	2.6
diffusion-controlled growth	563	nano-Al+traces IMC*	3.1
Eutectic-like,	543	nano-Al+IMC	4
interface-controlled growth	538	nano-Al+IMC	4
	533	nano-Al+IMC	4

*IMC Intermetallic compound; ** the Avrami exponent

Fig. 4.27. (a) DSC curve of the $Al_{85}Ni_5Y_4Nd_4Co_2$ alloy obtained at a heating rate of 1.67 K/s, (b) Avrami plot for the isothermal calorimetry data at different temperatures above and below T_g of about 561 K. Reproduced from [114] with permission of Elsevier.

Below T_g most of the Al-based glassy alloys showed a single-stage transformation obeying the kinetic law (eq. 4.4) for the fraction transformed (x) as a function of time (t) with the Avrami exponent (n) close to 4 while n values on the primary crystallization of nanoscale α-Al in the $Al_{85}Dy_8Ni_5Co_2$, $Al_{85}Y_8Ni_5Co_2$, $Al_{85}Y_4Nd_4Ni_5Co_2$ and other metallic glasses above T_g varies not only with temperature but with time as well exhibiting non-steady state nucleation (Fig. 4.27 (b)). α-Al nanocrystals only are formed on rapid enough heating and annealing above T_g. The transition from the interface-controlled conjoint growth of α-Al and intermetallic compound (IM) to the diffusion-controlled growth of α-Al takes place within the glass-transition region. This behaviour was also observed in the $Al_{85}Ni_7Gd_8$ alloy [119].

The above-mentioned may indicate that the devitrification products in such alloys inherit the local structure (short-range order) of the glassy (amorphous) solid phase while the glasses devitrifying from the supercooled liquid state may inherit a short range order of the liquid [120]. The origin of such behaviour may also be connected with the difference in the structure, physical and thermal properties between the glassy phase and the supercooled liquid.

The crystallization of various Al-based glassy and amorphous alloys can be described by one of the three following types of mechanisms:

- If an alloy does not show T_g on heating prior to devitrification/crystallization and exhibits nucleation and growth transformation mechanism then it forms an intermetallic compound(s) (IM) or IM+nanoscale α-Al from the amorphous matrix.

- If an alloy does not show glass-transition on heating prior to devitrification and has pre-existing nuclei formed on cooling at high temperature then it forms nanoscale primary α -Al.

- If an alloy shows glass-transition on heating and exhibits the nucleation and growth transformation mechanism then it forms nanoscale fcc-Al above T_g and IM+α-Al or only IM below T_g.

These crystallization mechanisms can be associated with a new topological empirical criterion which was proposed for the design of multicomponent Al-rich amorphous alloys [121]. Such a criterion, is similar to the topological λ instability criterion [122], as the ratio of solute atoms (r_i) of concentration C_i to Al atomic radius (r_{Al}) defined by equation:

$$\lambda = \sum_{i=B}^{z} C_i \cdot \left| \frac{r_i^3}{r_{Al}^3} - 1 \right|$$

(4.5)

The λ parameter can successfully predict the thermal behavior of the Al-based alloys. It indicates compositions which exhibit the supercooled liquid region (λ>0.1), nanocrystalline behavior (λ<0.1) or alloys with an intermediate, nano-glassy behavior, where nanocrystallization is preceded by the supercooled liquid region (λ≈0.1). Following the concept of the effect of the electronegativity difference [123] on the width of the supercooled liquid region in Al-based metallic glasses the $Al_{85}Y_6Ni_5Co_2Zr_2$ and $Al_{84}Y_6Ni_4Co_2Sc_4$ glassy alloys with the largest supercooled liquid region known so far of about 50 K were created [124]. The $Al_{84}Y_6Ni_4Co_2Sc_4$ alloy has an empirical topological criterion λ=0.1 while the $Al_{85}Y_6Ni_5Co_2Zr_2$ one has λ=0.091 close to 0.1.

$Pd_{43}Ni_{10}Cu_{27}P_{20}$ [125] and $Zr_{41.2}Ti_{13.8}Cu_{12.5}Ni_{10}Be_{22.3}$ [126] BMG alloys also show some differences in the crystallization behavior above and below the TTT nose temperature and a transition from nucleation control to growth control has been suggested.

CHT diagrams of the amorphous alloys (Fig. 4.28) [127] which has no supercooled liquid prior to crystallization like the $Al_{84}Nd_8Ni_4Co_2$ one correspond well to the TTT diagrams while less correlation is found in the case of two other glassy alloys which transform to a supercooled liquid and have different crystallization behavior below and above T_g. This fact again is in line with the difference in their crystallization behavior above and below T_g.

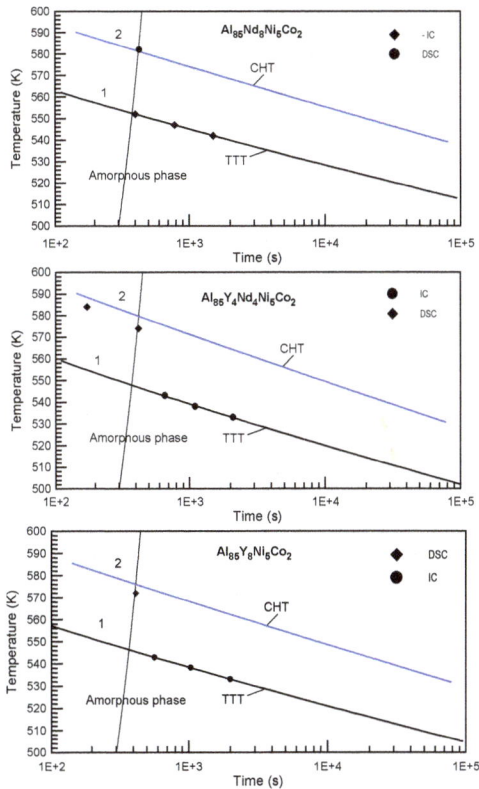

Fig. 4.28. The TTT (curve 1) and CHT (curve 2) diagrams for the initial phase transformations in the glassy alloys. The dashed vertical line indicates a heating rate of 0.67 K/s. Reproduced from [127] with permission of the Society of Glass Technology. http://www.ingentaconnect.com/content/sgt/ejgst/2009/00000050/00000001/art00006

145

Heat treatment of the $Al_{85}RE_8Ni_5Co_2$ glassy alloys (exhibiting glass-transition prior to crystallization) below the glass-transition temperature (T_g) leaded to formation of at least two phases one of which is Al, another – metastable intermetallic compound while these glassy alloys after continuous heating using DSC at high enough heating rate of 0.67 K/s and higher or annealing at the temperature above T_g showed precipitation of the α-Al nanoparticles only (Table 4.5). An exceptional case was in the case of the $Al_{85}Y_4Sm_4Ni_5Co_2$ alloy which showed formation of the only primary α-Al nanoparticles even at isothermal annealing below T_g. On the other hand amorphous alloys which crystallize directly and do not exhibit glass-transition on heating did not show any temperature dependence of their crystallization products. The alloys exhibiting no supercooled liquid region, for example La-, Mm- (Mischmetal) and Nd- bearing $Al_{85}RE_8Ni_5Co_2$ alloys ($Al_{85}La_8Ni_5Co_2$, $Al_{85}Nd_8Ni_5Co_2$, and $Al_{85}Mm_8Ni_5Co_2$), being annealed isothermally at various temperatures showed formation of the same crystallization products per alloy. It indicates an important role of the supercooled liquid in crystallization of the Al-based glasses.

Crystallization of a metallic glass forming a mixture of IM phase and Al looks similar to an irregular eutectic type. However, Al and IM phase in many cases do not have common interface and likely nucleate separately from the glassy phase. Below T_g most of the glassy alloys showed a single-stage transformation obeying the kinetic law (4.4) for the fraction transformed (x) as a function of time (t) with n close to 4 while the Avrami exponent of the primary phase transformation of the $Al_{85}Dy_8Ni_5Co_2$, $Al_{85}Y_8Ni_5Co_2$, $Al_{85}Y_4Nd_4Ni_5Co_2$ (Table 4.5) and other metallic glasses above T_g varies not only with temperature but with time as well exhibiting non-steady state nucleation.

Difference in the crystallization mechanism has been also observed in the $Al_{92}Sm_8$ amorphous alloy prepared by rapid solidification of the melt and by solid-state processing (cold rolling) at ambient temperature. No Al nanocrystals were formed in the samples produced by cold rolling while a large number density of Al was obtained by devitrification of the melt-spun samples [128].

Table 4.5. Compositions, glass transition (T_g) and crystallization temperature (T_x) of the $Al_{85}RE_8Ni_5Co_2$ glassy alloys crystallizing by nucleation and growth mechanism during continous heating in a DSC cell at high enough heating rate of 0.67 K/s.

Composition	T_g, K	T_x, K	Initial crystallization
$Al_{85}Y_8Ni_5Co_2$	544	573	Nanoscale Al
$Al_{85}La_8Ni_5Co_2$	-	582	$Al_4LaCo+X^*$
$Al_{85}Sm_8Ni_5Co_2$	559	574	Nanoscale Al
$Al_{85}Dy_8Ni_5Co_2$	549	574	Nanoscale Al
$Al_{85}Nd_8Ni_5Co_2$	-	578	X
$Al_{85}Mm_8Ni_5Co_2$	-	584	$Al_{11}La_3+Al_{11}Ce_3+X$
$Al_{85}Y_4Sm_4Ni_5Co_2$	545	568	Nanoscale Al
$Al_{85}Y_4Mm_4Ni_5Co_2$	555	581	Nanoscale Al
$Al_{85}Y_6Zr_2Ni_5Co_2$	567	597	Nanoscale Al

* X-unidentified phase(s)

4.4.4 Formation of in-situ glassy-nanocomposites upon rapid solidification

A nano-dispersed structure can be obtained directly from the melt upon rapid solidification at proper alloying. It is also found that the addition of Pd to Al-Y-Ni-Co alloys substituting Y caused disappearance of the supercooled liquid region as well as the formation of the highly dispersed primary FCC-Al nanoparticles about 3-7 nm in size homogeneously embedded in the glassy matrix upon solidification [129]. The value of the activation energy for the growth of α-Al nanocrystals of 146 kJ/mol obtained using the Kissinger analysis method is very close to the activation energy for self diffusion of pure Al. An extremely high density of precipitates of the order of 10^{24} m^{-3} is obtained (Fig. 4.29). This is the highest precipitation density observed so far in Al-based metallic glasses.

Fig. 4.29. The structure of the $Al_{85}Y_4Ni_5Co_2Pd_4$ glassy alloy in as-solidified state, TEM. (a) Dark-field image, (b) SAED and (c) nanobeam diffraction pattern. Reproduced from [129] with permission of Elsevier.

The broad X-ray diffraction peaks in the XRD pattern of the $Al_{85}Y_4Ni_5Co_2Pd_4$ glassy alloy (Fig. 4.30) were fitted with the Gaussian function. The resulted d-spacings for (111) and (200) of 0.2338 nm and 0.2030 nm, respectively, correspond very well to that of pure Al. The broad diffraction peak from the amorphous phase contributes to the intensity of (111).

Fig. 4.30. The X-ray diffraction pattern of the $Al_{85}Y_4Ni_5Co_2Pd_4$ as-rapidly solidified alloy and the peaks fitting with the Gaussian function. Reproduced from [129] with permission of Elsevier.

As indicated by the high-resolution TEM (HRTEM) (Fig. 4.31), in some areas where nuclei of α-Al nano-particles were located close enough to each other, the particles got in touch with each other upon growth and the crystal lattice of Al nanoparticles is heavily distorted.

Fig. 4.31. High-resolution TEM image of the $Al_{85}Y_4Ni_5Co_2Pd_4$ glassy alloy in as-solidified state.

As the diffraction peaks were significantly broadened it was possible to estimate the coherent scattering area ($D_{(hkl)}$) size (crystallite size). The resulted D_{hkl} values for two strong diffraction peaks (111) and (200) were 2.5 and 3.5 nm, respectively. The latter value is more reasonable as broad diffraction peak from the residual glassy phase contributes to the intensity of (111) peak. High degree of micro-strain within the α-Al lattice (see Fig. 4.31) is a reason for the slight discrepancy between the particles size calculated by using the XRD peaks and observed by TEM.

Compared to Al nanocrystals produced by the annealing of the glassy matrix which are usually found to be free of linear defects Al nanocrystals in the as-solidified Al-Y-Ni-Co-Pd sample contain microstrain and linear defects like dislocations. The source of the dislocations and the distortions of the crystalline lattice is impingement of the growing particles.

Also the microstructure of the $Si_{48}Al_{20}Fe_{10}Ge_7Ni_5Cr_5Zr_5$ alloy having an amorphous type X-ray diffraction pattern is inhomogeneous and contains Ge particles of less than 5 nm in size embedded in an amorphous matrix [130]. Ge having a lower mixing enthalpy with the other alloying elements than Si is rejected from the amorphous matrix and precipitates forming fine nanoparticles, though it has an unlimited solubility in crystalline Si. Similar results were obtained for Al-Si-Ge-TM alloys [131].

4.4.5 Crystallization upon deformation

Plastic deformation can cause nanocrystallization of a glassy phase in some alloys. For example, deformation of some Al-RE-TM amorphous alloys at room temperature causes precipitation of deformation-induced Al particles of 7-10 nm in diameter within the shear bands on bending [132] or nano-indentation [133]. Under the compressive testing, the nanoparticles were formed in shear bands while under tensile test they were found on the fracture surface only [134]. It was also shown that a high-energy ball milling also produced nanocrystalline particles from amorphous phase [135]. It has been suggested that a local temperature rise can play a role in mechanically-induced devitrification [136]. However, further studies on nano-indentation of glassy alloys at low strain rates also showed the formation of the nanoparticles when local rise of the temperature due to adiabatic heating is not supposed to take place [137]. This postulate has been confirmed by bending test at low temperature using dry ice [138].

At the same time some Al- and Zr-based glassy alloys do not show formation of nanoparticles on deformation [139]. The deformation-induced crystallization was studied in the glassy $Al_{85}Y_8Ni_5Co_2$ and $Al_{85}Gd_8Ni_5Co_2$ as well as amorphous $Al_{85}Nd_8Ni_5Co_2$ and $Al_{85}Mm_8Ni_5Co_2$ ribbon samples subjected to cold rolling with 33 % reduction in thickness [140]. The glassy $Al_{85}Y_8Ni_5Co_2$ and $Al_{85}Gd_8Ni_5Co_2$ alloys exhibiting the supercooled liquid region and the $Al_{85}Nd_8Ni_5Co_2$ and $Al_{85}Mm_8Ni_5Co_2$ alloys exhibiting just very beginning of the glass-transition phenomenon were chosen for the study. The studied alloys have close values of the crystallization temperatures and crystallize by the nucleation and growth reaction.

It is found that contrary to crystallization on heating, no intermetallic compounds but only a very small volume fraction of Al nano particles (<1 vol.%) was formed during the deformation-induced crystallization of the $Al_{85}Y_8Ni_5Co_2$ and $Al_{85}Nd_8Ni_5Co_2$ glassy alloys (Fig. 4.32), while the $Al_{85}Gd_8Ni_5Co_2$ and $Al_{85}Mm_8Ni_5Co_2$ samples remained fully disordered (Fig. 4.33). The most important factor responsible for the suppression of precipitation of the intermetallic compounds is believed to be the effect of pressure during cold rolling as FCC structure of Al is supposed to have a high atomic packing density of the FCC lattice. The applied pressure favors precipitation of the phases with high packing density. The effect of rising temperature within the shear bands also favors crystallization and enhances the diffusion-controlled growth. The higher growth rate of α-Al particles in the cold-rolled $Al_{85}Nd_8Ni_5Co_2$ amorphous alloy compared to the cold-rolled $Al_{85}Y_8Ni_5Co_2$ glassy alloy may indicate significantly higher deformation-induced diffusion rate of the alloying elements in the former alloy. As these alloys have a nucleation and growth crystallization behavior on heating they do not exhibit significant deformation induced crystallization compared to the Al-RE-Ni alloys studied earlier which can contain pre-existing nuclei and exhibit just growth behavior.

Fig. 4.32. The structure of the $Al_{85}Y_8Ni_5Co_2$ glassy alloy after cold-rolling, TEM. (a)-bright-field, (b)–dark-field image, (c) SAED pattern, (d) nanobeam diffraction pattern including the diffraction rings from a group of particles and a diffraction pattern from a single particle superimposed including their indexing according to cubic Al lattice. Dark-field image (b) was formed with the amorphous halo. The arrows indicate nanoscale α-Al particles. The insert in (a) is a high-resolution TEM image of the local region with high volume fraction of α-Al nanoparticles. Reproduced from [140] with permission of Elsevier.

It has been also shown that applied pressure enhances the precipitation of the α-Al from the amorphous phase in the $Al_{89}La_6Ni_5$ alloy by decrease in the crystallization temperature [141,142]. As these alloys have a nucleation and growth crystallization behavior on heating they do not exhibit significant deformation induced crystallization compared to the Al-RE-Ni alloys [143] studied earlier which can contain pre-existing nuclei and exhibit just growth behavior.

Fig. 4.33. The structure of the $Al_{85}Gd_8Ni_5Co_2$ glassy alloy after cold-rolling, TEM. (a)-bright-field, (b)–dark-field image, the insert SAED pattern. Reproduced from [140] with permission of Elsevier.

4.4.6 Formation of quasicrystals upon devitrification

An icosahedral phase - a quasicrystal having a long-range quasiperiodic translational order and an icosahedral orientational order (Fig. 4.34), but with no three-dimensional translational periodicity was initially discovered in Al-Mn alloys [144,145] and later in the other binary and ternary Al-TM-Cr [146] alloys. Thermodynamically stable icosahedral quasicrystals were found in binary Cd-Ca and Cd-Yb [147]; ternary Al-Li-Cu, Al-Cu-Fe, Al-Cu-Ru, etc…; as well as quaternary systems [148].

Fig. 4.34. HRTEM image of a single icosahedral quasicrystal formed in a bulk glassy $Zr_{65}Al_{7.5}Ni_{10}Pd_{17.5}$ alloy. The insert is the Fourier Transform image indicating 3-fold symmetry with irrational lattice spacing.

A dodecagonal quasicrystalline structure can be made of the perovskite barium titanate $BaTiO_3$, which forms long-range ordered dodecagonal films when grown on a Pt(111) substrate [149]. Oxide quasicrystals may possess useful optical and electrical properties which could be very important for engineering applications.

Different indexing methods of the diffraction patterns of quasicrystals are used [150, 151, 152]. The Al-TM base icosahedral structure is presumed [153] to consist of Mackay icosahedral clusters [154] while another type of icosahedral quasicrystals rather consists of Bergman-type clusters [155] and others [156]. The Mendeleev number is also used as a parameter for choosing the alloys forming a quasicrystalline phase [157]. The application of Pettifor maps [158] finds the middle path between purely empirical rules and detailed quantum mechanical calculations.

Synchrotron-beam X-ray diffraction has shed some light on the cluster structure of the $Cd_{5.7}Yb$ binary icosahedral phase [159]. A large difference in X-ray atomic scattering factors of Cd and Yb atoms and a series of approximants with close chemical composition in this system helped to produce some model of a real structure of this quasicrystal. The structure is described as a collection of the connected icosahedral YbCd clusters linked by ytterbium atoms located between the clusters.

Quasicrystals can be quasiperiodic in one, two or three dimensions: One-dimensional quasicrystals are quasiperiodic in one dimension [160]. Two-dimensional quasicrystals: octagonal quasicrystals [161] (Cr-Ni-Si, Mn-Fe-Si, etc.), decagonal quasicrystals (Al-Ni-Co, Al-Cu-Co, etc.) dodecagonal quasicrystals (Cr-Ni, V-Ni) are quasiperiodic in two dimensions [162, 163]. Three-dimensional quasicrystals: icosahedral quasicrystals (Al-Cu-Fe, Al-Li-Cu, etc.) are quasiperiodic in three dimensions. The symmetry of an icosahedral structure can be described by a six-dimensional space group.

Rational approximants are crystals similar to quasicrystals which exhibit translational periodicity [164, 165]. They have periodic structures with relatively large unit cells compared to ordinary crystals. Quasicrystalline-type short-range order can be found in metallic glassy alloys [166]. The existence of the chemical short range order clusters similar to those of a quasicrystal was found in the glassy state [167]. The icosahedral order exists in the liquid phase of Ti-Zr-Ni alloy [168]. The reduced supercooling (undercooling) before crystallization from the melt was found to be the smallest for quasicrystals, larger for crystal approximants and the largest for crystal phases [169]. A low energy barrier for nucleation of the icosahedral phase may explain the fact that only growth of the pre-existed icosahedral nuclei was observed in the $Zr_{55}Ni_{10}Al_{7.5}Cu_{7.5}Ti_{10}Ta_{10}$ alloy [170].

Fig. 4.35. The phase composition of the melt-spun $(Ti_xZr_yHf_z)_{80}Ni_{20}$ alloys. Reprinted from [171] with permission of Elsevier.

In the rapidly solidified $Ti_xZr_yHf_zNi_{20}$ system alloys the icosahedral phase is formed in the composition ranges close to the cI2 β solid solution phase and complex cF96 phase formation ranges (Fig. 4.35) [171].

Hf-based alloys have a higher tendency to form a cubic cF96 phase compared to Zr-based ones. The alloys in the systems in which an equilibrium Hf-based cF96 phase exists do not show the formation of the icosahedral phase from the amorphous matrix. The metastable cF96 phase and the icosahedral phase are formed by primary crystallization from the amorphous phase inheriting the structure of the icosahedral clusters [172]. cF96 phase nanoparticles, as well as those of the icosahedral phase, formed on crystallization were found to have similar equiaxed, close to spherical morphology.

4.4.7 Formation of nanoscale quasicrystals

Nanoscale quasicrystals are formed on devitrification in various metallic glassy alloys. The icosahedral phase was found in the devitrified Zr-Cu-Al, Zr-Al-Ni-Cu [173] and Zr-Ti-Ni-Cu-Al [174] glassy alloys containing an impurity of oxygen above about 1800 mass ppm, although no icosahedral phase is formed if oxygen content is lower than 1700 mass ppm. Later the nanoscale icosahedral phase was obtained in devitrified Zr-Al-Ni-Cu-Ag [175], Zr-Al-Ni-Cu-Pd [176], and various other alloys including binary Zr-Pd [177] and Zr-Pt [178,179] system alloys at low (typically about 800 mass ppm) oxygen content. It was also produced in the noble metal-free Zr–Cu–Ti–Ni [180]. The icosahedral phase in Zr-based alloys is often formed in cooperation with cF96 phase and with cI2 βZr solid solution phase [181].

A nanoscale icosahedral quasicrystalline phase has been also produced upon heating glassy Hf-based alloys such as $Hf_{65}Pd_{17.5}Ni_{10}Al_{7.5}$ [182], $Hf_{65}Al_{7.5}Ni_{10}Cu_{12.5}Pd_5$ [183], and $Hf_{65}Au_{17.5}(Ni,Cu)_{10}Al_{7.5}$ [184] alloys. According to the quasilattice constant value, the icosahedral phase in these alloys consists of 137-atom Bergmann rhombic triacontahedra. The activation energy for such a transformation was found to be 400 and 450 kJ/mol for Pd and Au-bearing alloys, respectively. No glassy phase was formed in the $Hf_{65}Ag_{17.5}Cu_{10}Al_{7.5}$ alloy. Also no icosahedral phase is formed during annealing of the glassy $Hf_{65}Pd_{17.5}Cu_{10}Al_{7.5}$ alloy.

Formation of the nanoscale (3-10 nm in size) Cu-based icosahedral phase was observed in the Cu-Zr-Ti-NM (NM-noble metal) alloys containing Pd [185] and Au [186] after heat treatment of the glassy phase. Although, due to the small particle size, the diffraction peaks of the icosahedral phase are broad, five strongest peaks were identified and indexed (Fig. 4.36 (a)). Calculation gives the quasilattice constants a_q of 0.495 and 0.496 nm for the $Cu_{55}Zr_{30}Ti_{10}Pd_5$ and $Cu_{55}Zr_{30}Ti_{10}Au_5$ alloys, respectively. At the next exothermic peak the icosahedral phase transforms to an intermediate intermetallic crystalline phase, which subsequently transforms to oC68 $(Cu,NM)_{10}(Zr,Ti)_7$ phase (solid solution of Ti and Pd or Au elements in $Cu_{10}Zr_7$) at higher temperature. This solid solution partially decomposes during long-time annealing forming other equilibrium phases.

Fig. 4.36. (a) X-ray diffraction patterns of the $Cu_{55}Zr_{30}Ti_{10}Pd_5$ and $Cu_{55}Zr_{30}Ti_{10}Au_5$ glassy alloys annealed for 600 s at 785 K and 420 s at 797 K, respectively. (b) DSC traces of these alloys. The heating rate 0.67 K/s. The data are used from [186,187] with permission of Elsevier.

Pd- and Au-bearing alloys have a wide supercooled liquid region (Fig. 4.36 (b)) while Ag and Pt-bearing alloys having narrow supercooled liquid do not form the icosahedral phase on heating [187]. The supercooled liquid region for $Cu_{55}Zr_{30}Ti_{10}Pd_5$ and $Cu_{55}Zr_{30}Ti_{10}Au_5$ reached about 50 K and 45 K, respectively (see Fig. 4.36 (b)). These values are very close to those of the $Cu_{55}Zr_{30}Ti_{10}Ni_5$ and $Cu_{55}Zr_{30}Ti_{10}Co_5$ alloys while the $Cu_{55}Zr_{30}Ti_{10}Ag_5$ and $Cu_{55}Zr_{30}Ti_{10}Pt_5$ glassy alloys exhibited narrower supercooled liquid regions.

The $Cu_{55}Zr_{30}Ti_{10}Pd_5$ glassy alloy shows a rather high Poisson's ratio exceeding 0.42 which is maintained after the structural relaxation and primary crystallization. The Young's and Shear moduli decrease upon primary crystallization while the bulk modulus exhibits a maximum after the structural relaxation (Fig. 4.37) [188].

Fig. 4.37. Some elastic moduli and the Poisson's ratio of the $Cu_{55}Zr_{30}Ti_{10}Pd_5$ alloy as a function of temperature. The studied materials were as-cast glass, glass relaxed at 740 K for 120 s, glass containing the nanoscale icosahedral phase (I), metastable intermetallic compound (MIMC) and stable intermetallic compound(s) (SIMC) formed at the first, second and third exothermic peaks of Fig. 4.35 (b), respectively. Reproduced from [188] with permission of Elsevier.

The structure of the $Cu_{55}Zr_{30}Ti_{10}Pd_5$ sample annealed for 0.6 ks at 785 K is shown in Fig. 4.38 (a,b). The average size of the quasicrystalline particles ranged from 3 to 10 nm. The sharp rings in the selected-area electron diffraction pattern correspond to the X-ray diffraction peaks. The nanobeam diffraction patterns shown in Fig. 4.38 (c,d) represent the five- and two-fold symmetry of the icosahedral phase, respectively. One should point out that some works impartially state that nanoscale icosahedral particles may belong to a high-order approximants. High symmetry of the observed nanobeam electron diffraction patterns indicated that the ordered phase in the $Cu_{55}Zr_{30}Ti_{10}Pd_5$ alloy has a high degree of the icosahedral symmetry, although, due to a small size it does not have all features of a quasicrystal.

Fig. 4.38. TEM images of the $Cu_{55}Zr_{30}Ti_{10}Pd_5$ alloy annealed for 0.6 ks at 785 K. (a) bright- and (b) dark-field images. (c) and (d) nanobeam diffraction pattern of five- and two-fold symmetry, respectively. (e) selected-area electron diffraction pattern.

Fig. 4.39 shows 3D atom probe (3DAP) elemental maps within the analyzed volume of approximately 4 nm × 4 nm × 130 nm of the $Cu_{55}Zr_{30}Ti_{10}Pd_5$ sample annealed at 785 K for 600 s [189]. Fig. 4.40 shows the atom probe concentration depth profiles of Cu, Zr, Ti, and Pd calculated from the volume of 2 nm × 2 nm × 130 nm that was selected from the analyzed volume in Fig. 4.39 and the concentration frequency distribution diagrams calculated from them. The data indicate that the concentration frequency diagrams of Cu, Zr and Ti show strong deviation from those expected from a homogeneous solid solution. This indicates that the nanoicosahedral phase in the $Cu_{55}Zr_{30}Ti_{10}Pd_5$ glassy alloy has slightly different composition from the composition of the glass as that of $Zr_{65}Al_{7.5}Ni_{10}Cu_{7.5}Ag_{10}$ [190]. Cu concentration is larger than 50 at.% in all regions indicating that the icosahedral phase in this alloy contains more than 50 at.% of Cu.

Fig. 4.39. 3D atom map for the $Cu_{55}Zr_{30}Ti_{10}Pd_5$ alloy sample after annealing at 785 K for 600 s. Reproduced from [189] with permission of the American Institute of Physics.

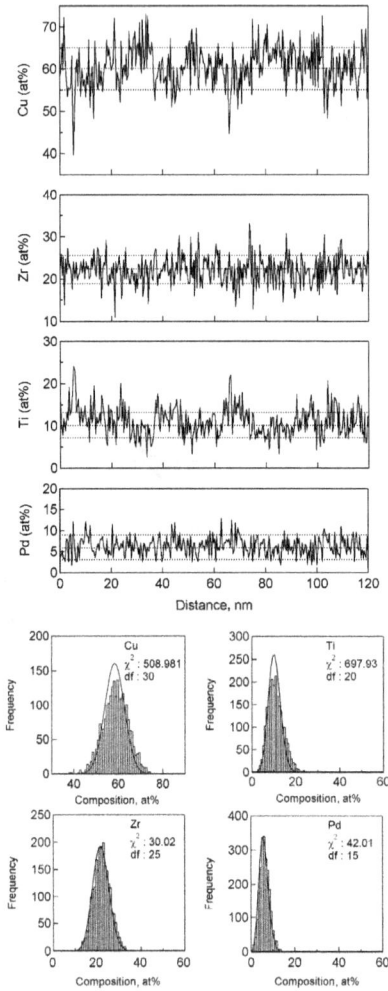

Fig. 4.40. 3D atom probe concentration depth profiles for different alloying elements obtained from the $Cu_{55}Zr_{30}Ti_{10}Pd_5$ sample after annealing at 785 K for 600 s and the concentration frequency distribution diagrams. Reproduced from [189] with permission of the American Institute of Physics.

3DAP results (Fig. 4.39 and 4.40) indicate that the composition of the icosahedral phase in the $Cu_{55}Zr_{30}Ti_{10}Pd_5$ alloy is close to that of the residual glassy matrix, though it is definitely different from that of the glassy matrix as the concentration fluctuation higher than the statistical error was detected. This is in line with the previously reported diffusion-controlled growth mechanism of the icosahedral phase. An alternative mechanism limiting growth rate of the icosahedral phase may be connected with incompatibility of 5-fold symmetry with 3-dimensional translational periodicity which may initiate internal elastic stresses. This idea is supported by the fact that the largest icosahedral phase particles in the $Cu_{55}Zr_{30}Ti_{10}Pd_5$ alloy hardly exceed 10 nm in size and transform to a metastable intermetallic compound of further heating.

The isothermal calorimetry trace for the $Cu_{55}Zr_{30}Ti_{10}Pd_5$ and $Cu_{55}Zr_{30}Ti_{10}Au_5$ alloys exhibited an exothermic peak after the certain incubation period which is required for nucleation of the icosahedral phase. Primary precipitation of the icosahedral phase obeys the kinetic law (eq. 4.5). The values of the Avrami exponent n=2.3 and 2.8 obtained for the $Cu_{55}Zr_{30}Ti_{10}Pd_5$ and $Cu_{55}Zr_{30}Ti_{10}Au_5$ alloys, respectively, are close to 2.5 which corresponds to 3-dimensional diffusion-controlled growth of nuclei at a constant nucleation rate.

Bulk glassy samples of 2 mm diameter were formed using the $Cu_{55}Zr_{30}Ti_{10}Pd_5$ ingots and $Cu_{55}Zr_{30}Ti_{10}Ag_5$ while the $Cu_{55}Zr_{30}Ti_{10}Au_5$ bulk alloy sample has had a mixed glassy and crystalline structure. No glassy phase was formed in the $Cu_{55}Zr_{30}Ti_{10}Pt_5$ bulk alloy sample whereas rapidly solidified ribbon samples with a glassy structure were obtained for all the $Cu_{55}Zr_{30}Ti_{10}NM_5$ alloys. Pd and Au additions significantly expand the supercooled liquid region of the $Cu_{60}Zr_{30}Ti_{10}$ glassy alloy.

In a later work the icosahedral phase was formed only in the $Cu_{55}Zr_{30}Ti_{10}Pd_5$ alloy while the $Cu_{55}Zr_{40}Pd_5$, $Cu_{55}Zr_{20}Ti_{20}Pd_5$, $Cu_{45}Zr_{40}Ti_{10}Pd_5$, $Cu_{60}Zr_{25}Ti_{10}Pd_5$ and $Cu_{65}Zr_{20}Ti_{10}Pd_5$ alloys did not form the icosahedral phase [191]. Notwithstanding on a limited homogeneity area of the icosahedral phase observed, according to the Avrami analysis this phase exhibited nucleation and 3-dimentional diffusion-controlled growth below and above the glass-transition region. The above mentioned also indicates (in addition to 3DAP results) that the composition of these two the glassy and icosahedral phases is not exactly equal.

Compared to the Zr-Cu-Pd alloys studied earlier the $Cu_{55}Zr_{40}Pd_5$ alloy exhibited formation of the equilibrium $(Cu,Pd)_{10}Zr_7$ phase from the supercooled liquid likely by polymorphous mechanism. This fact rules out possibility for the Cu-Zr-Ti-Pd icosahedral phase of being a Cu-rich part of the compositional homogeneity area of the Zr-Cu-Pd one. Thus, the filling sequence of the coordination shells in the Zr-Cu-Pd and Cu-Zr-Ti-Pd icosahedral phases must be different.

The $Zr_{60}Cu_{30-x}Co_xAl_{10}$ (x=0, 2.5, 5, 7.5, 10, 15, 20, 25, 30) metallic glasses were studied recently [192]. An ultrahigh grain number density (10^{25} m^{-3}) of a nanoscale metastable Co-rich icosahedral-like phase, with a size as small as about 1-2 nm, was obtained by partial crystallization of the $Zr_{60}Co_{30}Al_{10}$ metallic glass (Fig. 4.41). The nanoscale icosahedral phase remains stable over quite a large annealing temperature range and time. Nanoscale chemical and structural heterogeneity with Co rich areas were observed (Fig. 4.42). The interface energy, σ_{i-a} between the nano icosahedral phase and the amorphous matrix is about 14 mJ/m^2, indicating a low nucleation barrier. The calculated nucleation rate values are of the order of 10^{22}-10^{23} m^{-3}s^{-1}.

Fig. 4.41. (a) Dark-field TEM image, (b) HRTEM image of the $Zr_{60}Co_{30}Al_{10}$ alloy annealed at 823 K. The insert in (a) is the selected-area electron diffraction pattern; the inserts in (b) are the nanbeam diffraction patterns. (c) XRD pattern of the $Zr_{60}Co_{30}Al_{10}$ alloy heated to 823 K. Reproduced from [192] with permission of Elsevier.

Fig. 4.42. The HAADF image (a) and EDX mapping results (b-c) of the the $Zr_{60}Co_{30}Al_{10}$ alloy glassy ribbon up to the end of the first DSC crystallization peak. Reproduced from [192] with permission of Elsevier.

4.4.8 Crystallization of the amorphous alloys containing pre-existing nuclei and nanoparticles

As it has been mentioned in Chapter 2 the $Fe_{48}Cr_{15}Mo_{14}C_{15}B_6Tm_2$ and $Fe_{48}Cr_{15}Mo_{14}C_{15}B_6Y_2$ alloys are the example of growth limited BMGs. Fig. 2.14 showed the differential isothermal calorimetry (DIC) trace of $Fe_{48}Cr_{15}Mo_{14}C_{15}B_6Tm_2$ at 868 K which did not show an incubation period.

The $Al_{85}Y_8Ni_{5-3}Co_2Cu_{0-2}$ metallic glasses also exhibited a change from the nucleation and growth of a crystalline phase(s) in the $Al_{85}Y_8Ni_5Co_2$ one to growth of pre-existing nanoparticles (of 2-5 nm size) in $Al_{85}Y_8Ni_3Co_2Cu_2$ (Fig. 4.43) [193]. The addition of Cu also caused disappearance of the supercooled liquid region (Fig. 4.44). The heat released during primary exothermic heat effect (A) decreases from 42 J/g in Cu-free alloy up to 27 J/g and 23 J/g in the alloys containing 1 and 2 at.% Cu, respectively.

Fig. 4.43. TEM of the $Al_{85}Y_8Ni_3Co_2Cu_2$ metallic glass in as-solidified state. (a)-selected-area electron diffraction pattern, (b)- bright-field (c)- dark-field and (d) high-resolution image. Reproduced from [193] with permission of the Materials Research Society.

Fig. 4.44. DSC traces of the $Al_{85}Y_8Ni_{5-3}Co_2Cu_{0-2}$ metallic glasses obtained at 0.67 K/s. The arrows indicates increment of C_p related to glass-transition. Reproduced from [193] with permission of the Materials Research Society.

The melt-spun $(Fe_{0.75}Pt_{0.25})_{75-70}B_{25-30}$ ribbons were found to be X-ray amorphous [194]. These alloys were also found to possess good hard magnetic properties including a high intrinsic coercivity values up to 400 kA/m in the nano-crystallized state [195]. These alloys are promising to be applied as nanocomposite permanent magnets. Although only broad diffraction peaks are observed in the XRD pattern of the rapidly-solidified $(Fe_{0.75}Pt_{0.25})_{75}B_{25}$ alloy some nanoparticles were observed by high-resolution TEM (Fig. 4.45). These particles do not produce the detectable X-ray diffraction peaks due to their small size and limited volume fraction.

Fig. 4.45. High-resolution TEM image in as-solidified state. The insert: NBD pattern from the amorphous phase. Reproduced from [196] with permission of Elsevier.

No exothermic peak but just a variation of the baseline is observed during isothermal calorimetry trace taken at 647 K, which indicates absence of the incubation period (Fig. 4.46) [196].

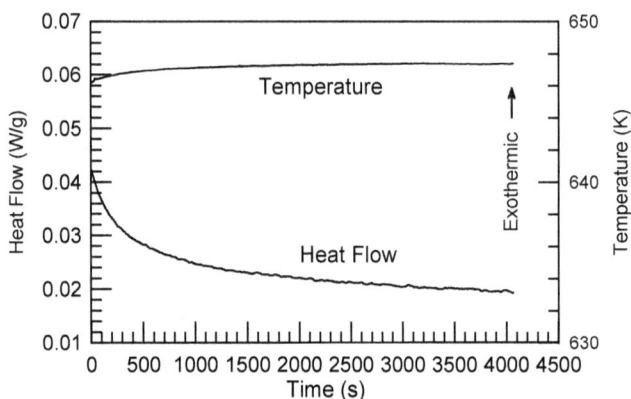

Fig. 4.46. DIC trace obtained at 647 K. Reproduced from [196] with permission of Elsevier.

Structure of the rapidly solidified $(Fe_{0.75}Pt_{0.25})_{75}B_{25}$ alloy contains a limited volume fraction of the pre-existing nuclei and nanoscale cubic cF4 Fe(Pt) solid solution particles embedded in the amorphous matrix. Nanoparticles of the cF4 Fe(Pt) phase start growing at the elevated temperature and then undergo ordering forming the tP4 FePt compound of about 15 nm in size which is followed by the formation of the tI12 Fe_2B phase from the residual amorphous matrix. Ordering of the cF4 phase takes place after the incubation period, existence of which allows anticipating nucleation and growth mechanism. This process takes place conjointly with further growth of the tP4 FePt particles in expense of the residual amorphous matrix. Formation of the tI12 Fe_2B phase from the residual amorphous matrix begins when the residual amorphous matrix is enriched in B up to a certain value. The observed low growth rate of tP4 FePt phase leading to the formation of a nanostructure is due to the diffusion-controlled growth mechanism and very low calculated diffusion coefficients of Fe and Pt in cF4 Fe at the temperature range of transformation (795-860 K).

It should be also mentioned that recently developed nanostructured metallic glasses exhibited high stability versus crystallization [197]. The Au-based nanoglass is kinetically ultrastable compared to its melt-spun ribbon counterpart. The nanoglass shows a 20 K higher glass transition temperature, a 32 K higher crystallization peak temperature, larger activation energy, slower crystal growth rate, and lower enthalpy compared to the ribbon sample consisting of the entire glassy phase.

4.5 Peritectic like reactions involving glassy phase

Such phase transformation involving a glassy/amorphous/supercooled liquid phase was initially observed in the rapidly-solidified $Ge_{60}Al_{25}La_{15}$ alloy in which an amorphous and a crystalline phase produce another crystalline phase [198]. Another rapidly solidified $Zr_{65}Ni_{10}Al_{7.5}Cu_{7.5}Ti_5Nb_5$ alloy has a mixed structure containing glassy and a submicron β-Zr solid solution phase (Fig. 4.47). The glassy+bcc β-Zr solid solution structure transforms to a mixture consisting of residual glassy and icosahedral phases after the completion of the first exothermic reaction, and this reaction is also a single-type reaction [199]. A common feature of these phase transformation is that they are both diffusion-controlled. The reactions described above look, in general, similar to the peritectic one though absence of the diffusion transfer through the product phase, lack of a common interface and absence of an inherited structure differentiate such type of reactions from a classical type peritectic reaction. A reaction "glassy+nanoscale I-phase → cF96 Hf_2Fe phase" observed in the melt-spun $Ti_{40}Zr_{20}Hf_{20}Fe_{20}$ alloy is closer to typical peritectic one [200].

Fig. 4.47. X-ray diffraction patterns of the $Zr_{65}Ni_{10}Al_{7.5}Cu_{7.5}Ti_5Nb_5$ amorphous alloy: (a) in as solidified state and (b) annealed for 300 s at 723 K.

References

[1]   A.L. Greer, Atomic transport and structural relaxation in metallic glasses, Journal of Non-Crystalline Solids 61-62 (PART 2) (1984) 737. https://doi.org/10.1016/0022-3093(84)90633-1

[2]   A. Inoue, N. Nishiyama and T. Matsuda, Preparation of bulk glassy $Pd_{40}Ni_{10}Cu_{30}P_{20}$ alloy of 40 mm in diameter by water quenching, Mater. Trans. JIM, 37 (1996) 181. https://doi.org/10.2320/matertrans1989.37.181

[3]   F.O. Méar, B. Lenk, Y. Zhang and A.L. Greer, Structural relaxation in a heavily cold-worked metallic glass Scripta Materialia 59 (2008) 1243. https://doi.org/10.1016/j.scriptamat.2008.08.023

[4]   J. Bünz and G. Wilde, Direct measurement of the kinetics of volume and enthalpy relaxation of an Au-based bulk metallic glass, Journal of Applied Physics 114 (2013) 223503. https://doi.org/10.1063/1.4843716

[5]   J.M. Pelletier, D.V. Louzguine-Luzgin, S. Li and A. Inoue, Elastic and viscoelastic properties of glassy, quasicrystalline and crystalline phases in $Zr_{65}Cu_5Ni_{10}Al_{7.5}Pd_{12.5}$ alloys, Acta Materialia 59 (2011) 2797. https://doi.org/10.1016/j.actamat.2011.01.018

[6]   D.V. Louzguine, A.R. Yavari, K. Ota, G. Vaughan and A. Inoue, Synchrotron X-ray radiation diffraction studies of thermal expansion, free volume change and glass transition phenomenon in Cu-based glassy and nanocomposite alloys on heating, Journal of Non-Crystalline Solids, 351 (2005) 1639. https://doi.org/10.1016/j.jnoncrysol.2005.04.054

[7]   D.V. Louzguine-Luzgin, A. Inoue, A.R. Yavari and G. Vaughan, Thermal expansion of a glassy alloy studied using a real-space pair distribution function, Appl. Phys. Lett. 88 (2006) 121926. https://doi.org/10.1063/1.2187955

[8]   D.V. Louzguine-Luzgin, A.R. Yavari, M. Fukuhara, K. Ota, G. Xie, G. Vaughan and A. Inoue, Free volume and elastic properties changes in Cu-Zr-Ti-Pd bulk glassy alloy on heating, Journal of Alloys and Compounds 431 (2007) 136. https://doi.org/10.1016/j.jallcom.2006.05.069

[9]   O. Haruyama, M. Kohda, N. Nishiyama and T. Egami, Volume relaxation in ternary bulk Pd- and Pt-P based metallic glasses, Journal of Physics: Conference Series 144 (2009) 012050. https://doi.org/10.1088/1742-6596/144/1/012050

[10]   A. Ishii, F. Hori, A. Iwase, Y. Fukumoto, Y. Yokoyama and T.J. Konno, Relaxation of free Volume in $Zr_{50}Cu_{40}Al_{10}$ bulk metallic glasses studied by positron

annihilation measurements, Materials Transactions, 49 (2008) 1975.
https://doi.org/10.2320/matertrans.MAW200826

[11] D.J. Magagnosc, G. Kumar, J. Schroers, P. Felfer, J.M. Cairney and D.S. Gianola, Effect of ion irradiation on tensile ductility, strength and fictive temperature in metallic glass nanowires, Acta Mater. 74, (2014) 165. https://doi.org/10.1016/j.actamat.2014.04.002

[12] D.V. Louzguine-Luzgin, V.Yu. Zadorozhnyy, S.V. Ketov, Z. Wang, A.A. Tsarkov and A.L. Greer, On room-temperature quasi-elastic mechanical behaviour of bulk metallic glasses, Acta Materialia, 129 (2017) 343. https://doi.org/10.1016/j.actamat.2017.02.049

[13] J. Saida, R. Yamada, M. Wakeda and S. Ogata, Thermal rejuvenation in metallic glasses, Science and Technology of Advanced Materials, 18, (2017) 152. https://doi.org/10.1080/14686996.2017.1280369

[14] S.V. Ketov, Y.H. Sun, S. Nachum, Z. Lu, A. Checchi, A.R. Beraldin, H.Y. Bai, W.H. Wang, D.V. Louzguine-Luzgin, M.A. Carpenter and A.L. Greer, Rejuvenation of metallic glasses by non-affine thermal strain, Nature, 524 (2015) 200. https://doi.org/10.1038/nature14674

[15] Z.Q. Chen, L. Huang, F. Wang, P.Huang, T.J. Lu, K.W. Xu, Suppression of annealing-induced embrittlement in bulk metallic glass by surface crystalline coating, Materials and Design, 109 (2016) 179. https://doi.org/10.1016/j.matdes.2016.07.069

[16] J.W. Cahn and R.J. Charles, The initial stages of phase separation in glasses. Physics and Chemistry of Glasses, 6 (1965) 181.

[17] A. Inoue, S. Chen and T. Masumoto, Zr-Y base amorphous alloys with two glass transitions and two supercooled liquid regions, Mater. Sci. Eng. A, 179/180 (1994) 346. https://doi.org/10.1016/0921-5093(94)90224-0

[18] A. Kündig, M. Ohnuma, D. Ping, T. Ohkubo and K. Hono, In situ formed two-phase metallic glass with surface fractal microstructure, Acta Mater. 52 (2004) 2441. https://doi.org/10.1016/j.actamat.2004.01.036

[19] T. Wada, D.V. Louzguine-Luzgin and A. Inoue, Preparation of Zr-based metallic glass nanowires and nanoparticles by selective etching, Scripta Materialia, 57 (2007) 901. https://doi.org/10.1016/j.scriptamat.2007.07.032

[20]   G.E. Abrosimova and A.S. Aronin, Evolution of the amorphous-phase structure in metal–metal type metallic glasses, J. Synch. Investig. 9 (2015) 887. https://doi.org/10.1134/S1027451015050031

[21]   G.E. Abrosimova and A.S. Aronin, Two-phase amorphous alloys of Al-Ni-Y system, J. Synch. Investig. 9 (2015) 134.https://doi.org/10.1134/S102745101501022X

[22]   D.V. Louzguine-Luzgin, T. Wada, H. Kato, J. Perepezko and A. Inoue, In situ phase separation and flow behavior in the glass transition region, Intermetallics, 18 (2010) 1235. https://doi.org/10.1016/j.intermet.2010.03.018

[23]   D.V. Louzguine-Luzgin, G. Xie, Q. Zhang and A. Inoue, Effect of Fe on the glass-forming ability, structure and devitrification behavior of Zr-Cu-Al bulk glass-forming alloys, Philosophical Magazine, 90 (2010) 1955. https://doi.org/10.1080/14786430903571495

[24]   D.V. Louzguine-Luzgin, G. Xie, W. Zhang and A. Inoue, Devitrification behavior and glass-forming ability of Cu–Zr–Ag alloys, Materials Science and Engineering: A 465 (2007) 146. https://doi.org/10.1016/j.msea.2007.02.039

[25]   J.C. Oh, T. Ohkubo, Y.C. Kim, E. Fleury and K. Hono, Phase separation in $Cu_{43}Zr_{43}Al_7Ag_7$ bulk metallic glass, Scripta Mater. 53 (2005) 165. https://doi.org/10.1016/j.scriptamat.2005.03.046

[26]   A.A. Kundig, M. Ohnuma, T. Ohkubo, T. Abe and K. Hono, Glass formation and phase separation in the Ag–Cu–Zr system, Scripta Materialia, 55 (2006) 449. https://doi.org/10.1016/j.scriptamat.2006.05.012

[27]   D.V. Louzguine, A.R. Yavari and A. Inoue, Devitrification behaviour of Cu-Zr-Ti-Pd bulk glassy alloys, Philosophical Magazine, 83 (2003) 2989. https://doi.org/10.1080/1478643031000151394

[28]   D.V. Louzguine and A. Inoue, Influence of a supercooled liquid on devitrification of Cu-, Hf- and Ni- based metallic glasses, Mater. Sci. and Eng. A 375-377 (2004) 346. https://doi.org/10.1016/j.msea.2003.10.140

[29]   J. Orava, A.L. Greer, B. Gholipour, D. W. Hewak and C.E. Smith, characterization of supercooled liquid $Ge_2Sb_2Te_5$ and its crystallization by ultrafast-heating calorimetry, Nat. Mater. 11 (2012) 279. https://doi.org/10.1038/nmat3275

[30]   P. Zalden, A. von Hoegen, P. Landreman, M. Wuttig and A.M. Lindenberg, How supercooled liquid phase-change materials crystallize: snapshots after femtosecond

optical excitation, Chem. Mater., 27 (2015) 5641.
https://doi.org/10.1021/acs.chemmater.5b02011

[31]   N. Chen, L. Gu, G.Q. Xie, D.V. Louzguine-Luzgin, A.R. Yavari, G. Vaughan, S.D. Imhoff, J.H. Perepezko, T. Abe and A. Inoue, Flux-induced structural modification and phase transformations in a $Pd_{40}Ni_{40}Si_4P_{16}$ bulk-glassy alloy, Acta Materialia, 58 (2010) 5886. https://doi.org/10.1016/j.actamat.2010.07.003

[32]   J.H. Perepezko, R.J. Hebert, W.S. Tong, J. Hamann, H.R. Rösner and G. Wilde, Nanocrystallization reactions in amorphous aluminum alloys, Mater. Trans. 44 (2003) 1982. https://doi.org/10.2320/matertrans.44.1982

[33]   J.H. Perepezko, R.J. Hebert and W.S. Tong, Amorphization and nanostructure synthesis in Al alloys, Intermetallics. 10 (2002) 1079. https://doi.org/10.1016/S0966-9795(02)00144-9

[34]   A.N. Kolmogorov, A statistical theory for the recrystallisation of metals, Isz. Akad. Nauk. USSR Ser. Matem. 3 (1937) 355.

[35]   M.W.A. Johnson and K.F. Mehl, Reaction kinetics in processes of nucleation and growth, Transactions of the American Institute of Mining and Metallurgical Eng. Trans. Am. Inst. Mining. Met. Eng. 135 (1939) 416.

[36]   M. Avrami, Granulation, phase change, and microstructure kinetics of phase change. III J. Chem. Phys. 9 (1941) 177. https://doi.org/10.1063/1.1750872

[37]   J.W. Christian, The theory of transformations in metals and alloys, Pergamon Press Ltd., Oxford, (1975) 369.

[38]   D.V. Louzguine and A. Inoue, Effect of Ni on stabilization of the supercooled liquid and devitrification of Cu-Zr-Ti bulk glassy alloys, J. Non-Cryst. Sol., 325 (2003) 187. https://doi.org/10.4028/www.scientific.net/JMNM.15-16.31

[39]   D.V. Louzguine and A. Inoue, Influence of Ni and Co additions on supercooled liquid region, devitrification behaviour and mechanical properties of Cu-Zr-Ti bulk metallic glass, Journal of Metastable & Nanocrystalline Materials, 15-16, (2003) 31.

[40]   F.R. De Boer, R. Boom, W.C.M. Mattens, A.R. Miedema and A.K. Niessen: Cohesion in metals: transition metal alloys (Cohesion and structure), Elsevier Science Publishers, North-Holland, (1988) 31.

[41]   A. Takeuchi and A. Inoue, Calculations of mixing enthalpy and mismatch entropy for ternary amorphous alloys, Mater. Trans., JIM 41 (2000) 1372.

[42]   A. Peker and W.L. Johnson, Time-temperature-transformation diagram of a highly processable metallic glass, Mater. Sci. Eng. A., 179-180 (1994) 173. https://doi.org/10.1016/0921-5093(94)90187-2

[43]   N. Nishiyama and A. Inoue, Supercooling investigation and critical cooling rate for glass formation in Pd–Cu–Ni–P alloy, Acta Mater. 47 (1999) 1487. https://doi.org/10.1016/S1359-6454(99)00030-0

[44]   J. Loeffler, J. Schroers and W.L. Johnson, Tme-temperature-transformation diagram and microstructures of bulk glass forming $Pd_{40}Cu_{30}Ni_{10}P_{20}$, Appl. Phys. Lett. 77 (2000) 681. https://doi.org/10.1063/1.127084

[45]   R. Janlewing, and U. Köster, Nucleation in crystallization of Zr–Cu–Ni–Al metallic glasses Mater. Sci. Eng. A, 304-306 (2001) 833. https://doi.org/10.1016/S0921-5093(00)01574-4

[46]   J.H. Kim, S.G. Kim and A. Inoue, In situ observation of solidification behavior in undercooled Pd-Cu-Ni-P alloy by using a confocal scanning laser microscope, Acta Mater. 49 (2001) 615. https://doi.org/10.1016/S1359-6454(00)00353-0

[47]   J. Schroers and W.L. Johnson, Dispersion technique studies on $Pd_{43}Ni_{10}Cu_{27}P_{20}$, Mater. Sci. Eng. A 375-377 (2004) 781. https://doi.org/10.1016/j.msea.2003.10.176

[48]   D.V. Louzguine and A. Inoue, Comparison of the long-term thermal stability of various metallic glasses under continuous heating, Scripta Materialia 47 (2002) 887. https://doi.org/10.1016/S1359-6462(02)00268-3

[49]   D.V. Louzguine-Luzgin and A. Inoue, Relation between time–temperature transformation and continuous heating transformation diagrams of metallic glassy alloys, Physica B: Condensed Matter, 358 (2005) 174. https://doi.org/10.1016/j.physb.2005.01.141

[50]   R.A. Grange and J.M. Kiefer, Transformation of austenite on continuous cooling and its realation to transformation at constant temperature, Trans. ASM 29 (1941) 85.

[51]   H. Gleiter, Nanocrystalline materials, Progress in Materials Science 33 (1989) 223. https://doi.org/10.1016/0079-6425(89)90001-7

[52]   A.L. Greer, Metallic glasses, Science 267 (1995) 1947. https://doi.org/10.1126/science.267.5206.1947

[53]   A. Guinier, Structure of age-hardened aluminium-copper alloys, Nature 142 (1938) 569. https://doi.org/10.1038/142569b0

[54]   G.D. Preston, Structure of age-hardened aluminium-copper alloys, Nature 142 (1938) 570. https://doi.org/10.1038/142570a0

[55]   L. Battezzati, C. Antonione, G. Riontino, F. Marino and H.R. Sinning, The crystallization of a $Pd_{60}Ti_{20}Si_{20}$ metallic glass, Acta Metall. Mater. 39 (1991) 2107. https://doi.org/10.1016/0956-7151(91)90181-Y

[56]   W.N. Myung, L. Battezzati, M. Baricco, K. Oaki, A. Inoue, and T. Masumoto, Kinetic and thermodynamic aspects of crystallization in Cu-Ti-Ni and Cu-Ti-Al metallic glasses, Mater. Sci. Eng. A 179-180 (1994) 371. https://doi.org/10.1016/0921-5093(94)90229-1

[57]   U. Koster and J. Meinhardt, Crystallization of highly undercooled metallic melts and metallic glasses around the glass transition temperature ,Mater. Sci. Eng. A 178 (1994) 271. https://doi.org/10.1016/0921-5093(94)90553-3

[58]   A.R. Yavari and D. Negri, Effect of concentration gradients on nanostructure development during primary crystallization of soft-magnetic iron-based amorphous alloys and its modelling Nanostructured Materials 8 (1997) 969. https://doi.org/10.1016/S0965-9773(98)00047-6

[59]   D.V. Louzguine-Luzgin, Aluminum-base amorphous and nanocrystalline materials, Metal Science and Heat Treatment, 53 (2012) 472. https://doi.org/10.1007/s11041-012-9417-3

[60]   A.R. Yavari and O. Drbohlav, Thermodynamics and kinetics of nanostructure formation in soft-magnetic nanocrystalline alloys, Mater. Trans. JIM 36 (1995) 896. https://doi.org/10.2320/matertrans1989.36.896

[61]   A. Mitra, H.Y. Kim, D.V. Louzguine, N. Nishiyama, B. Shen and A. Inoue, Structure and magnetic properties of amorphous and nanocrystalline $Fe_{40}Co_{40}Cu_{0.5}Zr_9Al_2Si_4B_{4.5}$ alloys, Journal of Magnetism and Magnetic Materials 278 (2004) 299. https://doi.org/10.1016/j.jmmm.2003.12.1315

[62]   M. Gogebakan, P.J. Warren and B. Cantor, Crystallization behaviour of amorphous $Al_{85}Y_{11}Ni_4$ alloy, Mater. Sci. Eng., A 226-228 (1997) 168. https://doi.org/10.1016/S0921-5093(96)10611-0

[63]   J.C. Foley, D.R. Allen and J.H. Perepezko, Analysis of nanocrystal development in Al-Y-Fe and Al-Sm glasses, Scripta Mater. 35 (1996) 655. https://doi.org/10.1016/1359-6462(96)00196-0

[64] M. Calin, A. Rudiger and U. Koester, Primary crystallization of Al-based metallic glasses, J. Metastable and Nanocryst. Mater., 8 (2000) 359. https://doi.org/10.4028/www.scientific.net/JMNM.8.359

[65] N. Bassim, C.S. Kiminami and M.J. Kaufman, Phases formed during crystallization of amorphous $Al_{84}Y_9Ni_5Co_2$ alloy, J. Non-Cryst. Solids 273 (2000) 271. https://doi.org/10.1016/S0022-3093(00)00135-6

[66] R.I. Wu, G. Wilde and J.H. Perepezko, Glass formation and primary nanocrystallization in Al-base metallic glasses, Materials Science and Engineering A301 (2001) 12. https://doi.org/10.1016/S0921-5093(00)01390-3

[67] A. Inoue, Amorphous, nanoquasicrystalline and nanocrystalline alloys in Al-based systems, Prog. Mater. Sci. 43 (1998) 365. https://doi.org/10.1016/S0079-6425(98)00005-X

[68] J.H. Paik, F.W.J. Botta and A.R. Yavari, Al-based nanostructures obtained from amorphous precursors, Mater. Sci. Forum, 225–227 (1996) 305. https://doi.org/10.4028/www.scientific.net/MSF.225-227.305

[69] A.R. Yavari, W.J.Botta, C.A.D. Rodrigues, C. Cardoso and R.Z. Valiev, Nanostructured bulk $Al_{90}Fe_5Nd_5$ prepared by cold consolidation of gas atomised powder using severe plastic deformation, Scr. Mater., 46 (2002) 711. https://doi.org/10.1016/S1359-6462(02)00057-X

[70] M.C. Gao and G.J. Shiflet, Devitrification phase transformations in amorphous $Al_{85}Ni_7Gd_8$ alloy, Intermetallics 10 (2002) 1131. https://doi.org/10.1016/S0966-9795(02)00139-5

[71] Y.X. Zhuang, J.Z. Jiang, T.J. Zhou, H. Rasmussen, L. Gerward, M. Mezouar, W. Crichton and A. Inoue, Crystallization of $Pd_{40}Cu_{30}Ni_{10}P_{20}$ bulk glass under pressure, Appl. Phys. Lett. 77 (2000) 4133. https://doi.org/10.1063/1.1332409

[72] Z. Altounian, S. Saini, J. Mainville and R. Bellissent, A neutron diffraction study of Al-rich glasses, Physica B: Condensed Matter, 241-243 (1997) 915. https://doi.org/10.1016/S0921-4526(97)00752-7

[73] K. Hono, Y. Zhang, A.P. Tsai, A. Inoue and T. Sakurai, Solute partitioning in partially crystallized Al-Ni-Ce(-Cu) metallic glasses, Scripta Mater. 32 (1995) 191. https://doi.org/10.1016/S0956-716X(99)80035-1

[74] T.B. Massalski, Binary alloy phase diagrams, ASM International, Materials Park, Ohio (1990) 30.

[75]   A.R. Yavari and D. Negri, Effect of concentration gradients on nanostructure development during primary crystallization of soft-magnetic iron-based amorphous alloys and its modeling, Nanostr. Mater. 8 (1997) 969. https://doi.org/10.1016/S0965-9773(98)00047-6

[76]   A.K. Gangopadhyay, T.K. Croat and K.F. Kelton, The effect of phase separation on subsequent crystallization in A1-Gd-Ni, Acta Mater. 48 (2000) 4035. https://doi.org/10.1016/S1359-6454(00)00196-8

[77]   K.F. Kelton, T.K. Croat, A.K. Gangopadhyay, L.Q. Xing, A.L. Greer, M. Weyland, X. Li and K. Rajan, Mechanisms for nanocrystal formation in metallic glasses, J. Non-Cryst. Sol. 317 (2003) 71.

[78]   S.V. Ketov, A. Inoue, H. Kato and D.V. Louzguine-Luzgin, Viscous flow of $Cu_{55}Zr_{30}Ti_{10}Co_5$ bulk metallic glass in glass-transition and semi-solid regions, Scripta Materialia 68 (2013) 219. https://doi.org/10.1016/j.scriptamat.2012.10.037

[79]   D.V. Louzguine-Luzgin, A.I. Bazlov, S.V. Ketov, A.L. Greer and A. Inoue, Crystal growth limitation as a critical factor for formation of Fe-based bulk metallic glasses, Acta Materialia, 82 (2015) 396. https://doi.org/10.1016/j.actamat.2014.09.025

[80]   D.V. Louzguine-Luzgin, A.I. Bazlov, S.V. Ketov and A. Inoue, Crystallization behavior of Fe- and Co-based bulk metallic glasses and their glass-forming ability, Materials Chemistry and Physics, 162 (2015) 197. https://doi.org/10.1016/j.matchemphys.2015.05.058

[81]   J.S. Kasper, The ordering of atoms in the chi-phase of the ironchromium-molybdenum system, Acta Metallurgica, 2 (1954) 456. https://doi.org/10.1016/0001-6160(54)90066-8

[82]   A. Hirata, Y. Hirotsu, K. Amiya and A. Inoue, Crystallization process and glass stability of an $Fe_{48}Cr_{15}Mo_{14}C_{15}B_6Tm_2$ bulk metallic glass, Physical Review B 78 (2008) 144205. https://doi.org/10.1103/PhysRevB.78.144205

[83]   I.V. Lyasotskii, N.B. Dyakonova, E.N. Vlasova, D.L. Dyakonov and M. Yu. Yazvitskii, Metastable and quasiperiodic phases in rapidly quenched Fe–B–Si–Nb(Cu) alloys, Phys. Stat. Sol. A, 203 (2006) 259. https://doi.org/10.1002/pssa.200521126

[84]   I.V. Lyasotsky, N.B. Dyakonova and D.L. Dyakonov, Structure transformations in multicomponent amorphous alloys in connection with grain boundary segregations and temper embrittlement of steels, Journal of ASTM International, 9 (2012) 104153. https://doi.org/10.1520/JAI104153

[85]   I.V. Lyasotsky, N.B. Dyakonova and D.L. Dyakonov, Metastable primary precipitation phases in multicomponent glass forming Fe-base alloys with metalloids, Journal of Alloys and Compounds 586 (2014) 20. https://doi.org/10.1016/j.jallcom.2013.03.112

[86]   G. Sha, K.A.Q. O'Reilly, B. Cantor, J.M. Titchmarsh and R.G. Hamerton, Quasi-peritectic solidification reactions in 6xxx series wrought Al alloys, Acta Mater., 51 (2003) 1883. https://doi.org/10.1016/S1359-6454(02)00595-5

[87]   H.H. Tian and M.M. Atzmon, Comparison of X-ray analysis methods used to determine the grain size and strain in nanocrystalline materials. Philos. Mag. A, 79 (1999) 1769. https://doi.org/10.1080/01418619908210391

[88]   E.J. Mittemeijer and U. Welzel, The "state of the art" of the diffraction analysis of crystallite size and lattice strain. Z. Kristallogr., 223 (2008) 552. https://doi.org/10.1524/zkri.2008.1213

[89]   E. Purushotham and N. Krishna, X-ray determination of crystallite size and effect of lattice strain on Debye–Waller factors of platinum nano powders, Bull. Mater. Sci., 36 (2013) 973. https://doi.org/10.1007/s12034-013-0553-1

[90]   H. Nitta, K. Miura and Y. Iijima, Self-diffusion in iron-based Fe–Mo alloys, Acta Mater., 54 (2006) 2833. https://doi.org/10.1016/j.actamat.2006.02.020

[91]   K.Takasawa, Y. Yamazaki, S. Takaki, K. Abiko and Y. Iijima, Diffusion of Cr and Fe in a high-purity- Fe-50 mass%Cr-8 mass%W alloy, Mater. Trans., 43 (2002) 178. https://doi.org/10.2320/matertrans.43.178

[92]   C. Hin, B.D. Wirth and J.B. Neaton, Formation of Y_2O_3 nanoclusters in nanostructured ferritic alloys during isothermal and anisothermal heat treatment: A kinetic Monte Carlo study, Phys. Rev. B, 80 (2009) 134118. https://doi.org/10.1103/PhysRevB.80.134118

[93]   F. Faupel, W. Frank, M.-P. Macht, H. Mehrer, V. Naundorf, K. Rätzke, H.R. Schober, S.K. Sharma and H. Teichler, Diffusion in metallic glasses and supercooled melts, Rev. Mod. Phys., 75 (2003) 237. https://doi.org/10.1103/RevModPhys.75.237

[94]   D.V.Louzguine-Luzgin, A.R. Yavari, G. Vaughan and A. Inoue, Clustered crystalline structures as glassy phase approximants, Intermetallics, 17 (2009) 477. https://doi.org/10.1016/j.intermet.2008.12.008

[95] H. Yoshizawa, K. Yamauchi, T. Yamane and H. Sugihara, Common mode choke cores using the new Fe - based alloys composed of ultrafine grain structure, J. Appl. Phys. 64 (1988) 6047. https://doi.org/10.1063/1.342150

[96] K. Suzuki, N. Kataoka, A. Inoue, A. Makino and T. Masumoto, High saturation magnetization and soft magnetic properties of BCC Fe-Zr-B alloys with ultrafine grain structure, Mater. Trans., JIM 31 (1990) 743.

[97] K. Hono, Nanoscale microstructural analysis of metallic materials by atom probe field ion microscopy, Progress in Materials Science 47 (2002) 621. https://doi.org/10.1016/S0079-6425(01)00007-X

[98] K. Hono, K. Hiraga, Q. Wang, A. Inoue and T. Sakurai, The microstructure evolution of a $Fe_{73.5}Si_{13.5}B_9Nb_3Cu_1$ nanocrystalline soft magnetic material, Acta Metall. et Mater. 40 (1992) 2137. https://doi.org/10.1016/0956-7151(92)90131-W

[99] W.J. Botta F., D. Negri and A.R. Yavari, Crystallization of Fe-based amorphous alloys, J. Non-Cryst. Sol. 247 (1999) 19.

[100] J.D. Ayers, V.G. Harris, J.A. Sprague, W.T. Elam and H.N. Jones, On the formation of nanocrystals in the soft magnetic alloy $Fe_{73.5}Nb_3Cu_1Si_{13.5}B_9$, Acta Materialia 46 (1998) 1861. https://doi.org/10.1016/S1359-6454(97)00436-9

[101] M. Ohnuma, K. Hono, H. Onodera, J.S. Pedersen and S. Linderoth, Cu clustering stage before the crystallization in Fe–Si–B–Nb–Cu amorphous alloys, Nanostructured Materials 12 (1999) 693. https://doi.org/10.1016/S0965-9773(99)00219-6

[102] D.V. Louzguine and A. Inoue, Nanocrystallization of Ti-Ni-Cu-Sn amorphous alloy, Scripta Materialia 43 (2000) 371. https://doi.org/10.1016/S1359-6462(00)00425-5

[103] G. He, J. Eckert and W. Loser, Stability, phase transformation and deformation behavior of Ti- base mettalic glass and composite, Acta Materialia 51 (2003) 1621. https://doi.org/10.1016/S1359-6454(02)00563-3

[104] Z. Altounian, E. Batalla, J.O. Strom-Olsen and J.L. Walter, The influence of oxygen and other impurities on the crystallization of $NiZr_2$ and related metallic glasses, J. Appl Phys 61 (1987) 149. https://doi.org/10.1063/1.338847

[105] L.Q. Xing, J. Eckert, W. Loser, L. Schultz and D.M. Herlach, Crystallization behaviour and nanocrystalline microstructure evolution of a $Zr_{57}Cu_{20}Al_{10}Ni_8Ti_5$ bulk amorphous alloy, Phil. Mag. A 79 (1999) 1095. https://doi.org/10.1080/01418619908210349

[106] D.V. Louzguine, H. Kato, H.S. Kim and A. Inoue, Formation of 2–5 nm size pre-precipitates of cF96 phase in a Hf-Co-Al glassy alloy, J. Alloys and Comp. 359 (2003) 198. https://doi.org/10.1016/S0925-8388(03)00292-5

[107] D.V. Louzguine, L.V. Louzguina and A. Inoue, Multistage devitrification of Mg-Ni-Mm and Mg-Ni-Y-Mm metallic glasses (Mm=mischmetal), Philosophical Magazine 83 (2003) 203. https://doi.org/10.1080/0141861021000032687

[108] D.V. Louzguine, A. Inoue, M. Saito and Y. Waseda, Structural relaxation in Ge-Cr-Al-Nd amorphous alloy, Scripta Materialia 42 (2000) 289. https://doi.org/10.1016/S1359-6462(99)00348-6

[109] D.V. Louzguine and A. Inoue, Formation of nanocrystalline nuclei in the amorphous phase of $Ge_{55}Al_{30}Cr_{10}Y_5$ alloy, Materials Letters 39 (1999) 211. https://doi.org/10.1016/S0167-577X(99)00008-7

[110] D.V. Louzguine, M. Saito, Y. Waseda and A. Inoue, Structural study of amorphous $Ge_{50}Al_{40}Cr_{10}$ alloy, Journal of the Physical Society of Japan, 68 (1999) 2298. https://doi.org/10.1143/JPSJ.68.2298

[111] D.V. Louzguine and A. Inoue, Ge-Al-Cr-La amorphous alloys containing crystalline-like zones, NanoStructured Materials 11 (1999) 115. https://doi.org/10.1016/S0965-9773(99)00024-0

[112] D.V. Louzguine and A. Inoue, Influence of rare earth metals (RE) on formation range and structure of amorphous phase in Ge-Al-Cr-RE system, Mater. Trans., JIM 40 (1999) 485.

[113] D.V. Louzguine, A. Takeuchi and A. Inoue, Structure and crystallization behavior of Al-free Ge-based amorphous alloys produced by rapid solidification of the melt, Journal of Non-Crystalline Solids 289 (2001) 196. https://doi.org/10.1016/S0022-3093(01)00702-5

[114] D.V. Louzguine and A. Inoue, Crystallization behaviour of Al-based metallic glasses below and above the glass-transition temperature, Journal of Non-Crystalline Solids, 311 (2002) 281. https://doi.org/10.1016/S0022-3093(02)01375-3

[115] K. Pekala, P. Jaskiewicz, J. Latuch and A. Kokoszkiewicz, Kinetics of crystallization processes in Al−Y−Ni−Co studied by electrical resistivity, J. Non-Cryst. Solids 211 (1997) 72. https://doi.org/10.1016/S0022-3093(96)00627-8

[116] D.V. Louzguine and A. Inoue, Comparative analysis of crystallization of $Al_{85}RE_8Ni_5Co_2$ (RE-Rare Earth Metals) metallic glasses with and without

supercooled liquid region, Proceedings of the ISMANAM 2001 International Symposium on Metastable, Mechanically Alloyed and Nanocrystalline Materials, June 24-29, 2001, Ann Arbor Michigan, USA. Materials Science Forum, 386-388 (2002) 117.

[117] D.V. Louzguine and A. Inoue, Strong influence of supercooled liquid on crystallization of the $Al_{85}Ni_5Y_4Nd_4Co_2$ metallic glass, Appl. Phys. Lett. 78 (2001) 3061. https://doi.org/10.1063/1.1371795

[118] D.V. Louzguine and A. Inoue, Influence of a supercooled liquid on crystallization behaviour of Al-Y-Ni-Co metallic glass, Materials Letters, 54 (2002) 75. https://doi.org/10.1016/S0167-577X(01)00542-0

[119] F.Q. Guo, S.J Poon and G.J. Shiflet, in Supercooled liquids, glass transition, and bulk metallic glasses, edited by, T. Egami, A.L. Greer, A. Inoue, S. Ranganathan, MRS Symposium Proceedings 754 (2003) CC11.6.

[120] D.V. Louzguine and A. Inoue, Influence of a supercooled liquid on devitrification of Cu-, Hf- and Ni- based metallic glasses, Mater. Sci. and Eng. A, 375-377 (2004) 346. https://doi.org/10.1016/j.msea.2003.10.140

[121] R. D. Sá Lisboa, C. Bolfarini, W. J. Botta F. and C. S. Kiminami, Topological instability as a criterion for design and selection of aluminum-based glass-former alloys, Appl. Phys. Lett., 86 (2005) 211904. http://dx.doi.org/10.1063/1.1931047

[122] T. Egami and Y. Waseda: Atomic size effect on the formability of metallic glasses, J. Non-Crystall. Sol. 64 (1984) 113.

[123] D.V. Louzguine and A. Inoue, Electronegativity of the constituent rare-earth metals as a factor stabilizing the supercooled liquid region in Al-based metallic glasses, Appl. Phys. Lett., 79 (2001) 3410. https://doi.org/10.1063/1.1420781

[124] D.V. Louzguine-Luzgin, A. Inoue and W.J. Botta, Reduced electronegativity difference as a factor leading to the formation of Al-based glassy alloys with a large supercooled liquid region of 50 K, Appl. Phys. Lett. 88 (2006) 011911. https://doi.org/10.1063/1.2159420

[125] J. Schroers, Y. Wu, R. Busch and W.L. Johnson, Transition from nucleation controlled to growth controlled crystallization in $Pd_{43}Ni_{10}Cu_{27}P_{20}$ melts, Acta Mater., 49 (2001) 2773. https://doi.org/10.1016/S1359-6454(01)00159-8

[126] W.L. Johnson, G. Kaltenboeck, M.D. Demetriou, J.P. Schramm, X. Liu, K. Samwer, C.P. Kim and D.C. Hofmann, Beating crystallization in glass-forming

metals by millisecond heating and processing. Science, 332 (2011) 828. https://doi.org/10.1126/science.1201362

[127] D.V. Louzguine-Luzgin and A. Inoue, The outline of glass transition phenomenon derived from the viewpoint of devitrification process, Physics and Chemistry of Glasses - European Journal of Glass Science and Technology Part B, 50, (2009) 27.

[128] G. Wilde, H. Sieber and J.H. Perepezko, Glass formation in Al-rich Al–Sm alloys during solid state processing at ambient temperature, J. Non-Cryst. Solids, 250-252 (1999) 621. https://doi.org/10.1016/S0022-3093(99)00147-7

[129] D.V. Louzguine-Luzgin and A. Inoue, Structure and transformation behaviour of a rapidly solidified Al–Y–Ni–Co–Pd alloy, J. Alloys and Comp. 399 (2005) 78. https://doi.org/10.1016/j.jallcom.2005.02.018

[130] D.V. Louzguine and A. Inoue, Effect of Ge addition to $Si_{50-55}Al_{25-20}Fe_{10}Ni_5Cr_5Zr_5$ alloys obtained by melt spinning, Proc. 4-th Int. Conf. IUMRS-ICA-97 (16-18 Sept. 1997, Mat. Res. Soc. Chiba, Tokyo) the special issue of the Journal NanoStructured Materials, 8 (1997) 1007. https://doi.org/10.1016/S0965-9773(98)00043-9

[131] D.V. Louzguine and A. Inoue, Precipitation of nanogranular Ge particles in rapidly solidified Al-Si-Fe-Cr-Ge alloys, Materials Transactions, JIM, 39, (1998) 504.

[132] H. Chen, Y. He, G.J. Shiflet and S.J. Poon, Deformation-induced nanocrystal formation in shear bands of amorphous alloys, Nature (London) 367 (1994) 541. https://doi.org/10.1038/367541a0

[133] W.H. Jiang, F.E. Pinkerton and M. Atzmon, Effect of strain rate on the formation of nanocrystallites in an Al-based amorphous alloy during nanoindentation, J. Appl. Phys. 93 (2003) 9287. https://doi.org/10.1063/1.1571234

[134] W.H. Jiang and M. Atzmon, The effect of compression and tension on shear-band structure and nanocrystallization in amorphous $Al_{90}Fe_5Gd_5$: a high-resolution transmission electron microscopy study, Acta Mater. 51 (2003) 4095. https://doi.org/10.1016/S1359-6454(03)00229-5

[135] Y. He, G.J. Shiflet and S.J. Poon, Ball milling-induced nanocrystal formation in aluminum-based metallic glasses, Acta Metall. Mater. 43 (1995) 83. https://doi.org/10.1016/0956-7151(95)90264-3

[136] A.A. Csontos and G.J. Shiflet, Formation and chemistry of nanocrystalline phases formed during deformation in aluminum-rich metallic glasses, Nanostruct. Mater. 9 (1997) 281. https://doi.org/10.1016/S0965-9773(97)90068-4

[137] J.J. Kim, Y. Choi, S. Suresh and A.S. Argon, Nanocrystallization during nanoindentation of a bulk amorphous metal alloy at room temperature, Science 295 (2002) 654.

[138] W.H. Jiang, F.E. Pinkerton and M. Atzmon, Deformation-induced nanocrystallization in an Al-based amorphous alloy at a subambient temperature, Scr. Mater. 48 (2003) 1195. https://doi.org/10.1016/S1359-6462(02)00568-7

[139] J. Li, Z.L. Wang and T.C. Hufnagel, Characterization of nanometer-scale defects in metallic glasses by quantitative high-resolution transmission electron microscopy, Phys. Rev. B 65 (2002) 144. https://doi.org/10.1103/PhysRevB.65.144201

[140] D.V. Louzguine-Luzgin and A. Inoue, Comparative study of the effect of cold rolling on the structure of Al-RE-Ni-Co (RE = rare-earth metals) amorphous and glassy alloys, Journal of Non-Crystalline Solids 352 (2006) 3903. https://doi.org/10.1016/j.jnoncrysol.2006.06.022

[141] F. Ye and K. Lu, Pressure effect on crystallization kinetics of an Al–La–Ni amorphous alloy, Acta Materialia, 47 (1999) 2449. https://doi.org/10.1016/S1359-6454(99)00104-4

[142] Y.X. Zhuang, J.Z. Jiang, T.J. Zhou, H. Rasmussen, L. Gerward, M. Mezouar, W. Crichton and A. Inoue, Pressure effects on $Al_{89}La_6Ni_5$ amorphous alloy crystallization, Appl. Phys. Lett. 77 (2000) 4133. https://doi.org/10.1063/1.1332409

[143] H. Chen, Y. He, G.J. Shiflet and S.J. Poon, Deformation-induced nanocrystal formation in shear bands of amorphous alloys, Nature 367 (1994) 541. https://doi.org/10.1038/367541a0

[144] D. Shechtman, I.A. Blech, D. Gratias and J.W. Cahn, Metallic phase with long-range orientational order and no translational symmetry, Phys. Rev. Lett. 53 (1984) 1951. https://doi.org/10.1103/PhysRevLett.53.1951

[145] D. Levine and R.J. Steinhardt, Quasicrystals: a new class of ordered structures, Phys. Rew. Lett. 53 (1984) 2477. https://doi.org/10.1103/PhysRevLett.53.2477

[146] A. Inoue, H.M. Kimura and T. Masumoto, Formation, thermal stability and electrical resistivity of quasicrystalline phase in rapidly quenched Al-Cr alloys, J. Mater. Sci. 22 (1987) 1758. https://doi.org/10.1007/BF01132404

[147] A.P. Tsai, J.Q. Guo, E. Abe, H. Takakura and T.J. Sato, A stable binary quasicrystal, Nature 408 (2000) 537. https://doi.org/10.1038/35046202

[148] D.V. Louzguine-Luzgin and A. Inoue, Formation and properties of quasicrystals, Annual Review of Materials Research, 38 (2008) 403. https://doi.org/10.1146/annurev.matsci.38.060407.130318

[149] S. Förster, K. Meinel, R. Hammer and W. Widdra, Quasicrystalline structure formation in a classical crystalline thin-film system, Nature 502 (2013) 215. https://doi.org/10.1038/nature12514

[150] V. Elser, Indexing problems in quasicrystal diffraction, Phys. Rev. B 32 (1985) 4892. https://doi.org/10.1103/PhysRevB.32.4892

[151] P.A. Bancel, P.A. Heiney, P.W. Stephens, A.I. Goldman and P.M. Horn, Structure of rapidly quenched Al-Mn, Phys Rev Lett 54 (1985) 2422. https://doi.org/10.1103/PhysRevLett.54.2422

[152] J.W. Cahn, D. Shechtman and D. Gratias, Metallic phase with long-range orientational order and no translational symmetry, J Mater. Res. 1 (1986) 13. https://doi.org/10.1557/JMR.1986.0013

[153] C.L. Henley and V. Elser, Quasicrystal structure of (Al, Zn)$_{49}$Mg$_{32}$, Philos. Mag. B 53 (1986) L59. https://doi.org/10.1080/13642818608240638

[154] A.L. Mackay, Dense non-crystallographic packing of equal spheres, Acta Crystallogr. 15 (1962) 916. https://doi.org/10.1107/S0365110X6200239X

[155] G. Bergman, J.L.T. Waugh and L. Pauling, The crystal structure of the metallic phase Mg$_{32}$(Al, Zn)$_{49}$, Acta Crystalogr. 10 (1957) 254. https://doi.org/10.1107/S0365110X57000808

[156] S. Ranganathan and A. Inoue, An application of Pettifor structure maps for the identification of pseudo-binary quasicrystalline intermetallics, Acta Mater., 54 (2006) 3647. https://doi.org/10.1016/j.actamat.2006.01.041

[157] H.S. Jeevan and S. Ranganathan, A new basis for the classification of quasicrystals, J. Non-Cryst. Solids 334 (2004) 184. https://doi.org/10.1016/j.jnoncrysol.2003.11.035

[158] D.G. Pettifor, A chemical scale for crystal-structure maps, Solid State Commun. 51 (1984) 31. https://doi.org/10.1016/0038-1098(84)90765-8

[159] H. Takakura, C.P. Gómez, A. Yamamoto, M. de Boissieu and A.P. Tsai, Atomic structure of the binary icosahedral Yb–Cd quasicrystal, Nature Mater. 6 (2007) 58. https://doi.org/10.1038/nmat1799

[160] R. Merlin, K. Bajema, R. Clarke, F.Y. Juang and P.K. Bhattacharya, Quasiperiodic GaAs-AlAs heterostructures, Phys. Rev. Lett. 55 (1985) 1768. https://doi.org/10.1103/PhysRevLett.55.1768

[161] N. Wang, H. Chen and K.H. Kuo, Two-dimensional quasicrystal with eightfold rotational symmetry, Phys. Rev. Lett. 59 (1987) 1010. https://doi.org/10.1103/PhysRevLett.59.1010

[162] L. Bendersky, Quasicrystal with one-dimensional translational symmetry and a tenfold rotation axis, Phys. Rev. Lett. 55 (1985) 1461. https://doi.org/10.1103/PhysRevLett.55.1461

[163] S. Ranganathan, K. Chattopadhyay, A. Singh and K.F. Kelton, Decagonal quasicrystals, Progr. Mater. Sci. 41 (1997) 195. https://doi.org/10.1016/S0079-6425(97)00028-5

[164] A.K. Srivastava and S. Ranganathan, Quasicrystals, crystals and multiple twins in rapidly solidified AlCrSi, AlMnSi and AlMnCrSi alloys, Acta. Mater. 44 (1996) 2935. https://doi.org/10.1016/1359-6454(95)00353-3

[165] W. Steurer, Approximants — the key to the structure of quasicrystals, Mat. Res. Soc. Symp. 643 (2001). https://doi.org/10.1557/PROC-643-K3.2

[166] W.W. Warren Jr, H.S. Chen and J.J. Hauser, NMR spin-echo spectra of icosahedral quasicrystals, Phys. Rev. 32 (1985) 7614. https://doi.org/10.1103/PhysRevB.32.7614

[167] E. Matsubara, M. Sakurai, T. Nakamura, M. Imafuku, S. Sato, J. Saida and A. Inoue, Environmental structural studies in amorphous and quasicrystalline $Zr_{70}Al_6Ni_{10}Pt_{14}$ alloys, Scripta Mater. 44 (2001) 2297. https://doi.org/10.1016/S1359-6462(01)00895-8

[168] K. F. Kelton, G.W. Lee, A.K. Gangopadhyay, R.W. Hyers, T.J. Rathz, J.R. Rogers, M.B. Robinson and D.S. Robinson, First X-Ray Scattering studies on electrostatically levitated metallic liquids: demonstrated influence of local icosahedral order on the nucleation barrier, Phys. Rev. Lett. 90 (2003) 195504. https://doi.org/10.1103/PhysRevLett.90.195504

[169] K.F. Kelton, Crystallization of liquids and glasses to quasicrystals, J. Non-Cryst. Sol. 334-335 (2004) 253. https://doi.org/10.1016/j.jnoncrysol.2003.11.052

[170] Lj. Ouyang, D.V. Louzguine, H.M. Kimura, T. Ohsuna, S. Ranganathan and A. Inoue, Influence of icosahedral clusters on crystallization of

$Zr_{55}Ni_{10}Al_{7.5}Cu_{7.5}Ti_{10}Ta_{10}$ glassy alloy, Journal of Metastable and Nanocrystalline Materials 18 (2003) 37. https://doi.org/10.4028/www.scientific.net/JMNM.18.37

[171] N. Chen, D.V. Louzguine, S. Ranganathan and A. Inoue, Formation ranges of icosahedral, amorphous and crystalline phases in rapidly solidified Ti–Zr–Hf–Ni alloys, Acta Materialia 53 (2005) 759. https://doi.org/10.1016/j.actamat.2004.10.027

[172] D.V. Louzguine and A. Inoue, Nanoscale cF96 cubic versus icosahedral phase in devitrified Hf-based metallic glasses, Annales de Chimie - Science des Matériaux 27 (2002) 91. https://doi.org/10.1016/S0151-9107(02)80049-6

[173] U. Koster, J. Meinhardt, S. Roos and H. Liebertz, Formation of quasicrystals in bulk glass forming Zr–Cu–Ni–Al alloys, Appl. Phys. Lett. 69 (1996) 179. https://doi.org/10.1063/1.117364

[174] L.Q. Xing, J. Eckert, W. Loser and L. Schultz, Effect of cooling rate on the precipitation of quasicrystals from the Zr–Cu–Al–Ni–Ti amorphous alloy, Appl. Phys. Lett. 73 (1998) 2110. https://doi.org/10.1063/1.122394

[175] M. W. Chen, T. Zhang, A. Inoue, A. Sakai and T. Sakurai, Quasicrystals in a partially devitrified $Zr_{65}Al_{7.5}Ni_{10}Cu_{12.5}Ag_5$ bulk metallic glass, Appl. Phys. Lett. 75 (1999) 1697. https://doi.org/10.1063/1.124793

[176] A. Inoue, J. Saida, M. Matsushita and T. Sakurai, Formation of an icosahedral quasicrystalline phase in $Zr_{65}Al_{7.5}Ni_{10}M_{17.5}$ (M=Pd, Au or Pt) alloys, Mater. Trans. JIM 41 (2000) 362. https://doi.org/10.2320/matertrans1989.41.362

[177] J. Saida, M. Matsushita and A. Inoue, Transformation kinetics of nanoicosahedral phase from a supercooled liquid region in $Zr_{70}Pd_{30}$ binary glassy alloy, J. Appl. Phys. 88 (2000) 6081. https://doi.org/10.1063/1.1322377

[178] B.S. Murty, D.H. Ping, M. Ohnuma and K. Hono, Nanoquasicrystalline phase formation in binary Zr–Pd and Zr–Pt alloys, Acta Mater. 49 (2001) 3453. https://doi.org/10.1016/S1359-6454(01)00254-3

[179] J. Saida, M. Matsushita and A. Inoue, Nano icosahedral phase in Zr–Pd and Zr–Pt binary alloys, Journal of Alloys and Compounds 342 (2002) 18. https://doi.org/10.1016/S0925-8388(02)00117-2

[180] D.V. Louzguine and A. Inoue, Formation of a nanoquasicrystalline phase in Zr-Cu-Ti-Ni metallic glass, Applied Physics Letters 78 (2001) 1841. https://doi.org/10.1063/1.1358362

[181] X.Y. Yang, M.J. Kramer, E.A. Rozhkova and D.J. Sordelet, Coincident lattice sites between cubic β-Zr(Pt) and an isochemical icosahedral phase in rapidly solidified $Zr_{80}Pt_{20}$ alloys, Scripta Materialia 49 (2003) 885. https://doi.org/10.1016/S1359-6462(03)00440-8

[182] D.V. Louzguine, M.S. Ko and A. Inoue, Nanoquasicrystalline phase produced by devitrification of Hf-Pd-Ni-Al metallic glass, Appl. Phys. Lett. 76 (2000) 3424. https://doi.org/10.1063/1.126667

[183] C. Li, J. Saida, M. Matsushita and A. Inoue, Precipitation of icosahedral quasicrystalline phase in $Hf_{65}Al_{7.5}Ni_{10}Cu_{12.5}Pd_5$ metallic glass. Appl. Phys. Lett. 77 (2000) 528. https://doi.org/10.1063/1.127033

[184] D.V. Louzguine, M.S. Ko and A. Inoue, Nanoscale icosahedral phase produced by devitrification of Hf-Au-Ni-Al and Hf-Au-Cu-Al metallic glasses, Scripta Mater. 44 (2001) 637. https://doi.org/10.1016/S1359-6462(00)00606-0

[185] D.V. Louzguine and A. Inoue, Nanoparticles with icosahedral symmetry in Cu-based bulk glass former induced by Pd addition, Scripta Mater. 48 (2003) 1325. https://doi.org/10.1016/S1359-6462(03)00018-6

[186] D.V. Louzguine and A. Inoue, Gold as an alloying element promoting formation of a nanoicosahedral phase in a Cu-based alloy, J. Alloys Comp. 361 (2003) 153. https://doi.org/10.1016/S0925-8388(03)00409-2

[187] D.V. Louzguine, H. Kato and A. Inoue, Investigation of mechanical properties and devitrification of Cu-based bulk glass formers alloyed with noble metals, Sci. Tech. Adv. Mater. 4 (2003) 327. https://doi.org/10.1016/j.stam.2003.09.003

[188] D.V. Louzguine-Luzgin, A.R. Yavari, M. Fukuhara, K. Ota, G. Xie, G. Vaughan, and A. Inoue, Free volume and elastic properties changes in Cu-Zr-Ti-Pd bulk glassy alloy on heating, Journal of Alloys and Compounds 431, (2007) 136. https://doi.org/10.1016/j.jallcom.2006.05.069

[189] D.V. Louzguine-Luzgin, A. Inoue, D. Nagahama and K. Hono, Composition and structure of Cu-based nanoicosahedral phase in Cu-Zr-Ti-Pd alloy, Applied Physics Letters 87 (2005) 211918. https://doi.org/10.1063/1.2135143

[190] J.Z. Jiang, A.R. Rasmussen, C.H. Jensen, Y. Lin and P.L. Hansen, Change of quasilattice constant during amorphous-to-quasicrystalline phase transformation in $Zr_{65}Al_{7.5}Ni_{10}Cu_{7.5}Ag_{10}$ metallic glass, Appl. Phys. Lett. 80 (2002) 2090. https://doi.org/10.1063/1.1463207

[191] D.V. Louzguine-Luzgin and A.Yu. Churyumov, Dual-phase glassy/nanoscale icosahedral phase materials in Cu–Zr–Ti–Pd system alloys, Materials Characterization 96 (2014) 6. https://doi.org/10.1016/j.matchar.2014.07.014

[192] Z. Wang, S.V. Ketov, C.L. Chen, Y. Shen, Y. Ikuhara, A.A. Tsarkov, D.V. Louzguine-Luzgin and J.H. Perepezko, Nucleation and thermal stability of an icosahedral nanophase during the early crystallization stage in Zr-Co-Cu-Al metallic glasses, Acta Materialia, 132 (2017) 298. https://doi.org/10.1016/j.actamat.2017.04.044

[193] D.V. Louzguine and A. Inoue, Investigation of structure and properties of the Al-Y-Ni-Co-Cu metallic glasses, Journal of Materials Research, 17, (2002) 1014. https://doi.org/10.1557/JMR.2002.0149

[194] A. Inoue and W. Zhang, Nanocrystalline Fe-Pt-B base hard magnets with high coercive force obtained from amorphous precursor, J. Appl. Phys. 97 (2005) 10H308. https://doi.org/10.1063/1.1854252

[195] W. Zhang, D.V. Louzguine and A. Inoue, Synthesis and magnetic properties of Fe-Pt-B nanocomposite permanent magnets with low Pt concentrations, Applied Physics Letters 85 (2004) 4998. https://doi.org/10.1063/1.1824172

[196] D.V. Louzguine-Luzgin, W. Zhang and A. Inoue, Nanoscale precipitates and phase transformations in a rapidly-solidified Fe-Pt-B amorphous alloy, Journal of Alloys and Compounds 402 (2005) 78. https://doi.org/10.1016/j.jallcom.2005.03.089

[197] J.Q. Wang, N. Chen, P. Liu, Z. Wang, D.V. Louzguine-Luzgin, M.W. Chen and J.H. Perepezko, The ultrastable kinetic behavior of an Au-based nanoglass, Acta Materialia 79 (2014) 30. https://doi.org/10.1016/j.actamat.2014.07.015

[198] D.V. Louzguine and A. Inoue, The structure and phase transformation behaviour of rapidly solidified alloys in the Ge-Al-La system, Mater. Res. Bull. 34 (1999) 1991. https://doi.org/10.1016/S0025-5408(99)00200-7

[199] D.V. Louzguine, Lj. Ouyang, H.M. Kimura and A. Inoue, Transformation from glassy + beta-Zr to glassy +icosahedral structure in Zr-based alloy, Scripta Mater. 50 (2004) 973. https://doi.org/10.1016/j.scriptamat.2004.01.005

[200] N. Chen, D.V. Louzguine-Luzgin, S. Ranganathan and A. Inoue, Glassy and icosahedral phases in rapidly solidified Ti-Zr-Hf-(Fe, Co or Ni) alloys, J. Non-Cryst. Sol. 351 (2005) 2547. https://doi.org/10.1016/j.jnoncrysol.2005.06.041

Chapter 5

Mechanical properties and deformation behavior

Owing to the absence of a crystalline lattice and dislocations, a unique deformation mechanism is realized in bulk glassy alloys, which thus, exhibit high strength, high hardness, good wear resistance and large elastic deformation. In the present chapter the mechanical properties and deformation behavior of BMGs at different temperatures are discussed.

Contents

5.1   Mechanical properties of bulk metallic glasses and their deformation behavior at room temperature

5.1.1   Elastic properties

The moduli of elasticity define how a material responds to external loading until plastic deformation begins. The bulk modulus (K) is related to the hydrostatic pressure required to change the average interatomic distances in a body. The Young's modulus (E) is a physical value characterizing the properties of the material to resist tensile, compressive or bending elastic deformation. The shear modulus (G) characterizes the ability of a material to resist shear deformation. The moduli of elasticity and the Poisson's ratio (v) are mutually related. For example, in a homogeneous isotropic material (like a BMG at a long enough length scale) the shear modulus is related to the Young's modulus by the Poisson's ratio:

$$G = \frac{E}{2(1+v)} \tag{5.1}$$

The elastic properties of bulk metallic glasses are shown in Table 5.1. θ_D represents the Debue temperature.

Table 5.1. Elastic properties of bulk metallic glasses from Ref. [1].

Alloys	K (GPa)	G (GPa)	E (GPa)	v	θ_D (K)
$Zr_{55}Al_{10}Ni_{10}Cu_{15}Be_{10}$	111.6	34.4	93.7	0.36	297
$Zr_{65}Al_{10}Ni_{10}Cu_{15}$	112.0	35.9	97.3	0.36	293
$Zr_{61.88}Al_{10}Ni_{10.12}Cu_{18}$	108.3	29.1	80.1	0.38	263
$Zr_{62.325}Cu_{17.55}Ni_{10.125}Al_{10}$	108.3	29.9	82.0	0.37	266
$Zr_{61}Cu_{18.3}Ni_{12.8}Al_{7.9}$	101.4	28.9	79.1	0.37	260
$Zr_{62}Al_{10}Ni_{12.6}Cu_{15.4}$	109.0	28.9	79.7	0.38	262
$Pd_{40}Ni_{10}Cu_{30}P_{20}$	172.6	35.5	99.8	0.40	279
$Pd_{40}Ni_{40}P_{20}$	175.0	36.5	105.0	0.40	292
$Pt_{57.5}Cu_{14.7}Ni_{5.3}P_{22.5}$	198.7	33.3	94.8	0.42	206
$Fe_{48}Cr_{15}Mo_{14}C_{15}B_6Er_2$	191.9	85.8	224.0	0.31	489
$Fe_{41}Co_7Cr_{15}Mo_{14}C_{15}B_6Y_2$	192.5	84.1	220.2	0.31	488
$Mg_{60}Cu_{25}Gd_{15}$	46.6	19.9	52.2	0.31	261
$Mg_{55}Cu_{25}Ag_{10}Gd_{10}$	53.5	23.6	61.7	0.31	274
$Mg_{58.5}Cu_{30.5}Y_{11}$	49.4	20.4	53.9	0.32	293
$Ca_{50}Mg_{20}Cu_{30}$	29.0	12.6	33.2	0.31	–
$Sr_{50}Mg_{20}Zn_{20}Cu_{10}$	17.1	9.0	22.9	0.28	169
$Sr_{60}Mg_{20}Zn_{15}Cu_5$	15.3	7.8	19.9	0.28	157
$Ce_{60}Al_{15}Ni_{15}Cu_{10}$	36.6	18.7	48.0	0.28	146
$Ce_{65}Al_{10}Ni_{10}Cu_{10}Nb_5$	29.8	11.6	30.9	0.33	145
$Nd_{60}Al_{10}Fe_{20}Co_{10}$	46.5	20.7	54.1	0.31	159
$Nd_{60}Al_{10}Ni_{10}Cu_{20}$	42.8	13.5	36.5	0.36	189
$Er_{55}Al_{25}Co_{20}$	60.7	27.1	70.7	0.31	229
$La_{66}Al_{14}Cu_{10}Ni_{10}$	34.9	13.4	35.7	0.33	161

As one can see in Fig. 5.1 although BMGs have reduced E modulus values compared to crystal their stress-strain curves exhibit good linearity. A typical coefficient of determination for the linear fit R^2 can be as high as 0.9999.

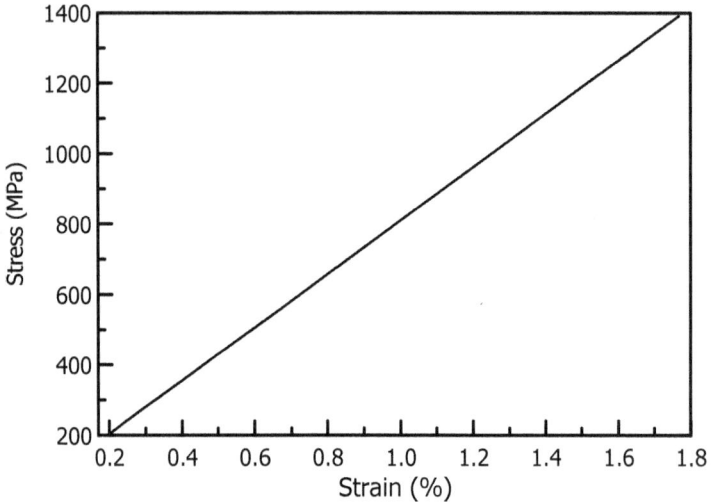

Fig. 5.1. A stress-strain loading curve for $Zr_{61}Cu_{27}Fe_2Al_{10}$ bulk sample up to 1400 MPa.

These elastic constants are found to be sensitive probes for evaluation of the structure changes and analyzing phase transformations in glassy alloys. The elastic moduli of the $Zr_{65}Ni_{10}Cu_5Al_{7.5}Pd_{12.5}$ alloy were studied in the glassy, quasicrystalline and crystalline states [2]. A relation between the structure and elasticity of the glassy alloy was established. The high Poisson's ratio, large values of K, λ and K/G ratio were found in the as-cast state (Fig. 5.2). The high Poisson's ratio of this alloy (partly possibly owing to low: 110 mass ppm (0.06 at.%) oxygen content) correlates with large compressive plasticity of Zr-based BMGs containing Pd [3]. The reduction of K and λ on heating is associated with the formation of an icosahedral phase.

Fig. 5.2. Elastic parameters as a function of the annealing temperature (and time at 1000 K the last points). Reproduced from [2] with permission of Elsevier.

5.1.2   Mechanical strength, hardness and plasticity

Owing to the absence of crystalline lattice and dislocations, a unique deformation mechanism is realized in bulk metallic glassy alloys [4], which thus, exhibit high strength, high hardness, good wear resistance [5] and large elastic deformation [6]. For example, the $(Al_{0.84}Y_{0.09}Ni_{0.05}Co_{0.02})_{95}Sc_5$ glassy alloy has an ultra-high tensile fracture strength exceeding 1500 MPa which surpasses those for all other Al-based fully crystalline and amorphous alloys reported up to date [7]. 1.5-1.8 GPa strength values are typical for Zr- and Pd-based alloys. Cu-Zr-based alloys become stronger with increase in Cu content which is related to the smaller excess volume in Cu-rich alloys [8]. For example, compressive yield strength values for Cu-Zr-Ti glassy alloys are higher than those for Zr-based bulk glassy alloys [9]. The $Cu_{57.5}Zr_{30}Ti_{10}Ni_{2.5}$ bulk glassy alloy showed an ultimate compressive strength of 2.2 MPa [10]. Similar values were found for the Cu–Zr–Ti–(Pd, Ag, Pt and Au) bulk glass-formers [11]. The highest strength value among Cu-based alloys (2.3 GPa) was found for a Cu-Zr-Ti-Co bulk metallic glass [12]. Ferrous-metals based alloys have much higher strength. For example, the Fe–(Co,Ni)–(Zr,Nb,Ta)–(Mo,W)–B system bulk glassy alloys exhibit a high compressive strength of 3.8 GPa and good corrosion resistance [13].

The (Fe, Co)-Cr-Mo-C-B-Tm system glassy alloys prepared in a cylindrical form with a diameter up to 18 mm demonstrate an excellent GFA and high strength exceeding 4 GPa [14]. The $Co_{43}Fe_{20}Ta_{5.5}B_{31.5}$ glassy alloy exhibited ultrahigh fracture strength exceeding 5 GPa, high Young's modulus of 270 GPa and high specific strength [15].

Mechanical properties of bulk metallic glasses in terms of the yield strength (σ_y), Vickers microhardness (HV) as well as the Young's modulus (E) are summarized in Table 5.2 (compressive) and Table 5.3 (tensile). A correlation is found between HV and σ_y as expected for isotropic solids like BMGs. One can note that the Zr-Cu-Fe-Al alloys with good plasticity showed HV/σ_y values [16] lower than 3 (Fig. 3a). Such alloys are marked with diamonds in Fig. 3a. It is interesting to note that this ratio calculated using the tensile strength of Zr-Cu-Ni-Al alloys (Table 5.3) tested in tension is also lower than 3. It can be connected with high purity of Zr used for the samples preparation in that work [17] which might have caused reduction in hardness. Also, the majority of these alloys were reported to exhibit high Charpy impact fracture toughness exceeding 100 kJ/m². Formation of a more open structure with a larger excess volume can also be a reason for higher plasticity of these glasses [18]. Also there is a correlation between strength and the Young's modulus (Fig. 3b).

Table 5.2. Composition and compressive mechanical properties of bulk glassy alloys. The values of σ_y are rounded to the nearest tens.

Nr	Content, at. %	σ_y (MPa)	E (GPa)	HV (kgf/mm²)	HV (MPa)	HV/σ_y	Ref.
1	$Cu_{60}Hf_{25}Ti_{15}$	2010	124				19
2	$Cu_{60}Hf_{25}Ti_{15}$	1920	120				20
3	$Cu_{47.5}Zr_{47.5}Al_5$	1550	87				21
4	$Cu_{46}Zr_{46}Al_8$	1900		580	5690	3.0	22
5	$Cu_{48}Zr_{48}Al_4$	1200	103				23
6	$Cu_{47}Zr_{47}Al_6$	1730		580	5690	3.3	24
7	$Cu_{60}Zr_{30}Ti_{10}$	1790	114				25
8	$La_{45}Al_{45}Ni_{10}$	1080	52	330	3237	3.0	26
9	$La_{45}Al_{35}Ni_{20}$	1000	46	305	2992	3.0	26

10	$La_{50}Al_{35}Ni_{15}$	950	41	290	2845	3.0	26
11	$La_{50}Al_{30}Ni_{20}$	930	41	285	2796	3.0	26
12	$La_{55}Al_{25}Ni_{20}$	740	34	225	2207	3.0	24
13	$Mg_{80}Cu_{10}Y_{10}$	630		220	2158	3.4	27
14	$Zr_{70}Al_{10}Ni_{20}$	1410	61	432	4238	3.0	28
15	$Zr_{65}Al_{10}Ni_{25}$	1580	65	484	4748	3.0	28
16	$Zr_{65}Al_{15}Ni_{20}$	1610	71	494	4846	3.0	28
17	$Zr_{60}Al_{15}Ni_{25}$	1640	73	502	4925	3.0	28
19	$Zr_{60}Ni_{20}Al_{20}$	1790	78	549	5386	3.0	29
20	$Zr_{70}Ni_{20}Al_{10}$	1410	61	432	4238	3.0	30
21	$Zr_{65}Ni_{25}Al_{10}$	1520	65	484	4748	3.1	30
22	$Zr_{65}Ni_{20}Al_{15}$	1610		494	4846	3.0	30
23	$Zr_{60}Ni_{25}Al_{15}$	1640		502	4925	3.0	30
24	$Cu_{56.4}Hf_{23.5}Ti_{14.1}Ta_6$	2125	104				31
25	$Cu_{58.8}Zr_{29.4}Ti_{9.8}Y_2$	1780	115				32
26	$Pd_{79}Cu_6Si_{10}P_5$	1475	82				33
27	$Zr_{55}Al_{20}Co_{20}Cu_5$	2000	92				34
28	$Zr_{60}Cu_{30}Al_{10}$	1710	93	465	4562	2.7	35
29	$Zr_{60}Cu_{25}Fe_5Al_{10}$	1650	92	479	4699	2.9	35, 36
30	$Zr_{60}Cu_{22.5}Fe_{7.5}Al_{10}$	1720	100	485	4758	2.8	35
31	$Zr_{60}Cu_{20}Fe_{10}Al_{10}$	1710	104	492	4827	2.8	35
32	$Zr_{60}Cu_{17.5}Fe_{12.5}Al_{10}$	1730	95	498	4885	2.8	35
33	$Zr_{62.5}Cu_{22.5}Fe_5Al_{10}$	1580	88	460	4513	2.8	36
34	$Zr_{65}Cu_{20}Fe_5Al_{10}$	1590	91	451	4424	2.8	36
35	$Zr_{67.5}Cu_{17.5}Fe_5Al_{10}$	1500	91	434	4258	2.8	36

Table 5.3. Tensile mechanical properties of the cast Zr-Cu-Ni-Al bulk glassy alloys: σ_t is tensile strength, HV is Vickers microhardness, and E is the Young's modulus [17]. The values of σ_y are rounded to the nearest tens.

Nr	Alloy composition	σ_t (MPa)	E (GPa)	HV (kgf/mm^2)	HV (MPa)	HV/σ_t
1	$Zr_{46}Cu_{34}Ni_8Al_{12}$	1780	111	562	5513	3.1
2	$Zr_{48}Cu_{28}Ni_{12}Al_{12}$	1910	102	530	5199	2.7
3	$Zr_{48}Cu_{30}Ni_{10}Al_{12}$	1980	92	528	5180	2.6
4	$Zr_{48}Cu_{32}Ni_8Al_{12}$	2100	102	527	5170	2.5
5	$Zr_{48}Cu_{34}Ni_6Al_{12}$	1900	94	529	5189	2.7
6	$Zr_{50}Cu_{26}Ni_{12}Al_{12}$	1880	88	498	4885	2.6
7	$Zr_{50}Cu_{28}Ni_{10}Al_{12}$	1990	92	517	5072	2.5
8	$Zr_{50}Cu_{30}Ni_8Al_{12}$	1820	92	526	5160	2.8
9	$Zr_{50}Cu_{32}Ni_6Al_{12}$	1880	92	521	5111	2.7
10	$Zr_{50}Cu_{34}Ni_4Al_{12}$	1910	91	517	5072	2.7
11	$Zr_{52}Cu_{26}Ni_{10}Al_{12}$	1960	89	509	4993	2.5
12	$Zr_{52}Cu_{28}Ni_8Al_{12}$	1798	94	512	5023	2.8
13	$Zr_{52}Cu_{30}Ni_6Al_{12}$	1820	93	506	4964	2.7
14	$Zr_{52}Cu_{32}Ni_4Al_{12}$	1780	88	501	4915	2.8
15	$Zr_{48}Cu_{32}Ni_{10}Al_{10}$	1890	94	513	5033	2.7
16	$Zr_{50}Cu_{28}Ni_{12}Al_{10}$	1810	89	518	5082	2.8
17	$Zr_{50}Cu_{32}Ni_8Al_{10}$	1800	87	508	4983	2.8
18	$Zr_{50}Cu_{30}Ni_{10}Al_{10}$	1960	92	509	4993	2.5
19	$Zr_{52}Cu_{32}Ni_6Al_{10}$	1890	86	490	4807	2.5
20	$Zr_{52}Cu_{28}Ni_{10}Al_{10}$	1890	88	498	4885	2.6
21	$Zr_{52}Cu_{30}Ni_8Al_{10}$	1860	89	498	4885	2.6
22	$Zr_{54}Cu_{28}Ni_6Al_{10}$	1780	83	477	4679	2.6

23	$Zr_{54}Cu_{28}Ni_8Al_{10}$	1570	89	485	4758	3.0
24	$Zr_{50}Cu_{28}Ni_{14}Al_8$	1800	86	509	4993	2.8
25	$Zr_{50}Cu_{32}Ni_{10}Al_8$	1960	96	504	4944	2.5
26	$Zr_{50}Cu_{34}Ni_8Al_8$	1890	98	503	4934	2.6
27	$Zr_{52}Cu_{28}Ni_{12}Al_8$	1900	92	490	4807	2.5
28	$Zr_{52}Cu_{30}Ni_{10}Al_8$	1830	84	490	4807	2.6
29	$Zr_{52}Cu_{32}Ni_8Al_8$	1800	95	492	4827	2.7
30	$Zr_{52}Cu_{34}Ni_6Al_8$	1860	92	488	4787	2.6
31	$Zr_{54}Cu_{30}Ni_8Al_8$	1820	97	469	4601	2.5
32	$Zr_{54}Cu_{32}Ni_6Al_8$	1810	97	475	4660	2.6

Deformation behavior of bulk metallic glassy alloys has been under intense investigations since the last century [37,38]. Although, the majority of bulk metallic glasses tested at ambient temperature fracture shortly after yielding, considerable plasticity was observed in several specific bulk metallic glassy alloys [39,40], particularly, in the Zr-Cu-Fe-Al system alloys [41] (Fig. 4), Zr-rich alloys of the Zr-Cu-Ni-Al system [17], Zr-Al-Ni-Pd [42] and Pt-Cu-Ni-P [43] system alloys. Moreover, good room temperature tensile ductility of Zr-rich metallic glasses is found in small samples [44,45].

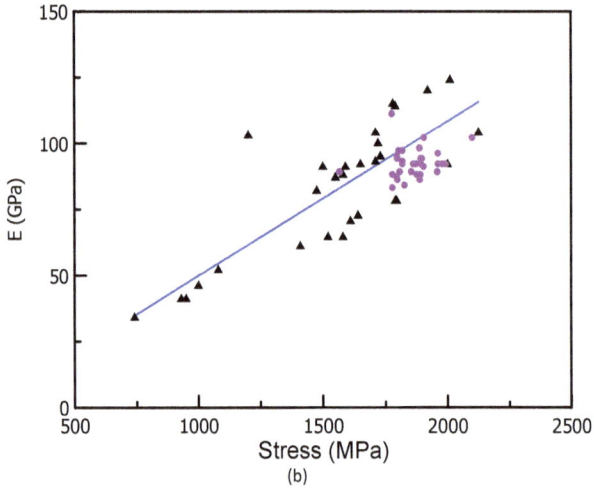

Fig. 5.3. Correlation between the mechanical properties of bulk metallic glasses. (a) HV/σ_y versus σ_y. Here σ_y denotes the yield strength. (b) E as a function of σ_y Compressive (triangles) and σ_t tensile (circles).

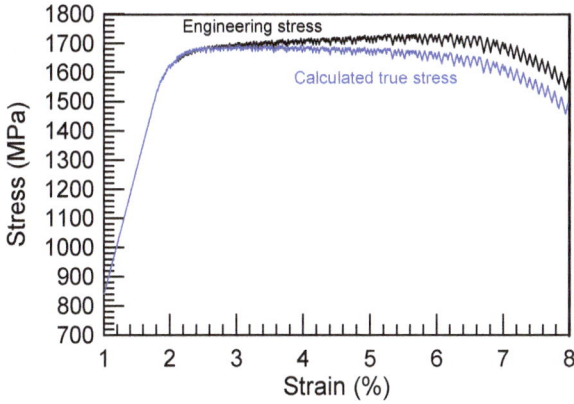

Fig. 5.4. Stress-strain curves of a Zr-Cu-Fe-Al BMG alloy. Note that if the sample is not bended upon compression the calculated true stress is nearly constant before a dominant shear band is formed. Starting at about 7% strain there is an apparent decrease owing to the reduction of the cross-section and formation of cracks in stick-slip mode.

Ductility of the $Pd_{40}Ni_{40}Si_xP_{20-x}$ [46,47] BMGs is drastically improved by application of the fluxing technique with B_2O_3 [47]. It is likely because fluxing removes possible inclusions acting as stress concentrations. Zr-Cu-Fe-Al bulk glassy alloys exhibit phase separation on heating prior to crystallization [48] which also may be responsible for their higher room-temperature ductility compared to the Zr-Cu-Al alloys [41]. In other alloys the improved plasticity can be related to the Poisson's ratio or the elastic moduli ratio [49], in-situ nano-crystallization [50] or glassy phase separation [51]. An empirical rule of the brittle-ductile transition in BMGs was established based on the Poisson's ratio [52]. There is an abrupt brittle–ductile threshold around the Poisson's ratio $v_0 \approx 0.31$. BMGs with the Poisson's ratio higher than v_0 demonstrate the plastic deformation. However, there are some exceptions. For example, the Poisson's ratio of the $Au_{49}Cu_{26.9}Ag_{5.5}Pd_{2.3}Si_{16.3}$ BMG $v= 0.406$ is larger then v_0 but this alloy does not exhibit any plastic deformation at room temperature [53].

Further investigations on this subject showed that embrittlement of the Mg- and La-based alloys is caused by the inclusions of crystalline oxide particles acting as stress concentrators [54]. This is because oxygen is insoluble in solid Mg and La, while it dissolves very well in Zr and Ti [55]. It explains reduced plasticity of these metallic glasses.

Table 5.4. Mechanical properties of the $Zr_{63-x}Cu_{24}Al_xNi_{10}Co_3$ alloys obtained by compression testing. $\sigma_{0.2}$ is yield strength. e_p is the engineering plastic strain and its maximum value is shown instead of the average one owing to large data scattering. The data is taken from [59] with permission of Elsevier.

Al content	Sample Nr	$\sigma_{0.2}$ (MPa)	e_p (%)
10 at. %	1	1775	10
	2	1780	2.6
	3	1790	3.6
	4	Failure*	Failure*
	5	Failure*	Failure*
	6	1780	4.4 (stopped**)
	Average	1781±10	Max: 10
13 at. %	1	1850	12.6
	2	1900	2.5
	3	Failure*	Failure*
	4	1900	1.9
	5	1920	3 (stopped**)
	Average	1893±48	Max: 12.6
15 at. %	1	Failure*	Failure*
	2	Failure*	Failure*
	3	1900	Failure*
	4	1815	5.5
	5	1920	0.4
	Average	1878±138	Max: 5.5

* Failure means that the sample was broken before reaching $\sigma_{0.2}$ at much lower stress (owing to the internal defects). ** The measurement was stopped to check the shape of the deformed sample.

Like the Hadfield steel in which locking of dislocations takes place [56] BMGs exhibit serrated flow (Fig. 5.4) with regular stress drops typical for ductile bulk metallic glassy alloys [57,58] and negative strain rate sensitivity. Contrary to crystalline alloys one can see general absence of strain hardening except for the initial transition period from the elastic to plastic deformation mode. The average stress drop magnitude increases with strain, reaching a critical value at fracture.

One should mention that the fracture strength of BMGs varies significantly depending on the existence of defects (likely pores and surface imperfections) which cause premature fracture, for example, as found for the $Zr_{63-x}Cu_{24}Al_xNi_{10}Co_3$ alloys at 15 at. % Al (Table. 5.4) [59]. It explains large scattering in the values of the percentage strain from sample to sample [60].

Unlike the prediction by the weighted rule of mixing for the elastic properties of the $Zr_{63-x}Cu_{24}Al_xNi_{10}Co_3$ BMGs, the measured moduli G and E show stiffening while increasing the aluminum content likely owing to partly covalent bonding. The Poisson's ratio exhibits a clear maximum $v = 0.32$ at the aluminum concentration $c_{Al} = 13$ at.%. The enhanced shear softness at $c_{Al} = 13$ at.% was also accompanied by an enhancement of the ductility in terms of the maximal plastic strain ε_p, during a quasi-static compression test, and in terms of creep deformation at constant load and room temperature. According to the density functional theory calculations the presence of an Al atom in the Cu–Zr clusters introduces new well localized states at low energies that are characterized by covalent-like bond promoting a higher mechanical strength [61]. Al atoms alter significantly the electronic structure and significant charge transfer was found to take place upon mechanical deformation.

5.1.3 Plastic deformation mechanism

Room-temperature plastic deformation of ductile crystalline solids generally proceeds via dislocation activity within the slip bands or by the formation of twins producing elongation. Due to work-hardening the plastic flow in ductile crystalline alloys is delocalized because the strained regions become more resistant to further deformation. On the other hand the inhomogeneous plastic flow of glassy alloys at relatively low homologous temperature (at least several tens degrees Kelvin below T_g) occurs by propagation of shear bands [62], which are 10-20 nm thick [63,64] (shear bands observed in crystalline alloys are much thicker) and make steps on the surface up to several micrometers in height (Fig. 5.5) [65]. Nevertheless, even in crystalline materials deformation by shear band starts to dominate over dislocation slip and twinning after certain strain or at high strain rate [66].

A strongly localized shear deformation at room temperature [67,68] without strain hardening limits practical application of BMGs. However, in some BMGs formation of multiple shear bands solves this problem [36,46]. A certain band once formed propagates within the sample and can gradually reach the other surface of the sample, leaving either a relatively flat or curved interface between two pieces of the sample (see in Fig. 5.5 marked with arrows) further exhibiting the slip-stick process [69] and concentration of the subsequent deformation activity in this band. In some alloys the onset of the shear band activity was observed well below the yield strength in the strain-stress curves [70].

Fig. 5.5. Shear bands in the $Zr_{61.88}Cu_{18}Ni_{10.12}Al_{10}$ BMG with large compressive plasticity, SEM. The mechanical test was stopped after the formation of a dominant shear band (marked with the arrows) and stick-slip behavior leading to large shear offsets on the surface.

As the localized shear deformation is a dominant plastic-deformation mode at room temperature, tensile ductility of metallic glasses is not found except for few special cases in thin sections of hypoeutectic alloys [71] tested at a relatively high strain rate [44,72]. Tensile deformation behavior of Zr-based glassy thin foils has been also studied recently in-situ in TEM and the foils were found to be more ductile than larger samples [73,74]. Hypoeutectic Zr-based alloys were found to be much more ductile compared to hypereutectic ones owing to a larger excess (free) volume [75]. Large plasticity at room temperature was achieved in the Zr-Cu-Ni-Al system alloys with appropriate elastic moduli combination [76]. Deformation of BMGs obeys the Mohr-Coulomb yield criterion [77]. Metallic glasses as well as granular materials like powders, ball, sands exhibit dilatation on shear localization.

The free volume model [78] explains deformation in BMGs as a series of diffusion-like local atomic jumps into vacant sites in the regions of large excess volume. The shear transformation zones model [79] and the atomic-jump free volume model [38] represent the alternative approach. If the atoms are used as the particles in a granular model of a metallic glass, a shear-band thicknesses of 2.5–3.1 nm would be expected according to the Goldschmidt atomic diameters. However as the structure of a metallic glasses is an efficient packing of clusters the outer diameter of the clusters (approximately three times the diameter of the atoms of the base element in the glasses) suggest the shear-band thicknesses shall be close to the observed thickness of 10-20 nm [80].

Formation of a shear band by shear deformation localization as modeled by molecular dynamics (MD) [81] is shown in Fig. 5.6 as the von-Mises shear strain invariant η_s which is defined in terms of the relative displacement of the particle configuration as a function of strain (e) from the initial state. Despite on the rather small sample size compared to the experimental one, one can observe a heterogeneous distribution of the higher η_s regions even in the early stage of 0.2 strain. There is also a clear tendency for strain localization in the shear bands after 0.5 strain as marked with arrows.

Fig. 5.6. Sliced view (at x=0 plane) of a set of local shear strain maps and a surface view at strain (e)=0.5 obtained in molecular dynamics simulation of the sample deformed to 0.7 strain at an initial strain rate of $1 \cdot 10^{-3}$. Reproduced from [81] with permission of Elsevier.

Figure 5.7 shows the results of finite element modeling of plastic strain of the non-deformable sample matrix separated by the shear bands. Strain is localized at the intersections of the shear bands while local stress maxima are found in the non-deformed glassy matrix, especially close to the loading punches.

(a)   (b)

(c)   (d)

Fig. 5.7. Finite element modeling of plastic deformation of the non-deformable sample matrix separated by the shear bands. (a,c) strain (dimensionless) and (b,d) stress distribution (the scale in MPa) up to 0.03 and 0.07 strain, respectively. Reproduced from [81] with permission of Elsevier.

A dislocation is clearly distinct from the crystal lattice until it disappears on the surface of the sample or at a grain boundary while a shear transformation zone, if it exists, is difficult to identify and may be not distinct after shear [82], though it possesses a larger volume. However, when the spatial displacement distribution was measured around the shear band tip terminated in a deformed bulk metallic glass using a digital image correlation technique good agreement was found between the experimentally observed and theoretically predicted displacement fields for mixed dislocations [83]. Here the shear step on the surface of a BMG sample is considered as a result of propagation of a mixed dislocation (Fig. 5.8).

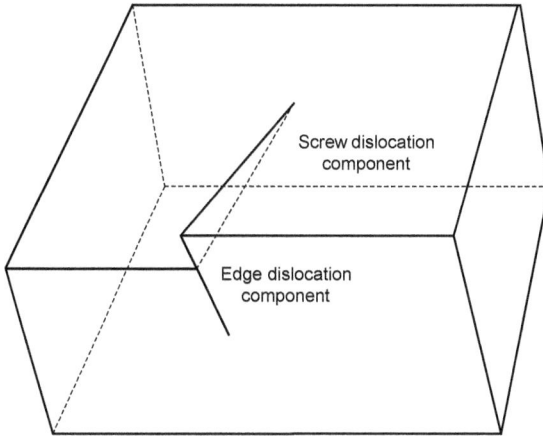

Fig. 5.8. Schematic representation of a shear band in metallic glass having a screw and edge component.

Shear transformation zones are more likely to be found in the areas with a higher structural disorder [8], larger excess volume [84], or areas of high atomic-level stresses [85]. These regions can therefore be treated as fertile sites for shear transformations in metallic glasses [86]. The formation of a shear transformation zone is likely in the areas of a higher local stress near a stress concentrator (flaw or surface notch) leading to its structural disordering [87].

Recently, Greer et al. [88] published a seminal work on shear bands in metallic glasses. A shear band could be initiated from the percolation of the shear transformation zones. Nucleation of the shear band includes: its creation by structural rejuvenation and sliding along the rejuvenated plane. Heterogeneous nucleation of shear bands indicates that the propagating shear band could be divided into the zones that include shear-rejuvenated glass near the shear band tip, glue zone, and liquid or near-liquid tail [89]. Activation of the shear transformation zones along the band makes the structure more disordered [90].

Nucleation kinetics of shear bands in metallic glasses was studied by nanoindentation [91]. The strain rate dependence suggested that the nucleation of shear bands exhibits thermally activated behavior. The activation energy is found to be close to that of the slow-β relaxation in metallic glasses.

The experimental and computational methods indicate that the internal structure of metallic glasses plays important role in their mechanical behavior. Processing history of a bulk metallic glassy alloy, i.e. annealing and cooling rate, leads to a change in the chemical and topological local order, and thus to alteration of the mechanical behavior [92,93].

Computer simulations [94] were performed in order to investigate both the structural effect, in terms of the configurational potential energy and the fraction of full icosahedral clusters and fragmented ones, and the compositional effect in Cu-Zr metallic glasses on their elastic and plastic properties. While the bulk modulus K remains independent on structural change, the shear modulus G and thus the Poisson's ratio v (or G/K-ratio) widely change upon the cooling rate during sample preparation: G increases at slower cooling, whereas v decreases. These changes were accompanied by an increase of the potential energy and an increase of the fraction of full icosahedral clusters at the cost of the fragmented ones. The liquid-like clusters are prone to favor the delocalization of shear transformation.

Molecular dynamics (MD) simulation also indicated structural disordering of the icosahedra into less-shear-resistant, fragmented clusters that are prone to structural changes [95]. In the $Cu_{64}Zr_{36}$ glass, the total number of icosahedral units inside the shear band decreased, the number of icosahedra in the $Cu_{36}Zr_{64}$ glass did not change [96]. Both topological and chemical short range order in Cu-Zr metallic glasses are affected inside the shear bands, while the surrounding matrix is slightly affected [96]. The tensile behavior of the $Ni_{60}Nb_{40}$ metallic glass has been investigated by using *ab-initio* simulation with a large cell containing 1024 atoms. With increasing tensile strain, the local atomic strain was found to become more heterogeneous [97].

The excess (free) volume is created inside the shear bands. Nanocrystallization is suggested to take place within the shear bands if nanoscale voids are not formed [98]. The formation of nanovoids reduces the excess volume. Independent of the alloy composition, plastic deformation inside the shear bands was found to increase the fraction of Cu-Cu and Zr-Zr bonds and decrease the fraction of unlike Cu-Zr bonds. On the other hand the chemical composition of a shear band of a Cu-Zr metallic glass is found to be the same as of the non-deformed sample [99]. The authors also showed that the deformation can vary from homogeneous flow to localized flow depending on the excess volume content.

Formation of nanovoids was observed in the shear bands [100] and described with a revised classical nucleation theory [101]. Molecular dynamics simulations illustrated the initial stages of cavitation in the $Pd_{82}Si_{18}$ and $Cu_{46}Zr_{54}$ metallic glasses on deformation [102].

The atomic structure of the $Pd_{53}Ni_{34}Cu_{13}$ metallic glass was studied by molecular dynamics simulation at different temperatures and no anisotropy was found in the initial state. After deformation the anisotropy was revealed by significant differences in the partial pair distribution functions. The anisotropy was found to be constrained into the shear plane. Residual anisotropy was found at a low temperature, while close to the glass transition the anisotropy induced by the initial shear process was removed by the subsequent recovery process [103].

The shear band activity was monitored by a high-speed camera [104] and by thermography [105]. In-situ monitoring of the deformation behavior of bulk metallic glasses performed using an infrared camera [106] showed a weak rise of temperature during a shear band operation. The instant melting of BMG specimens at the moment of the failure was captured by thermography while temperature rise in the shear band was much lower.

Shear softening can be a result of local heating or driven by stresses [107,108]. The temperature in the central area of a shear band could be as high as several thousand Kelvin depending on the shear band propagation speed [109]. At the same time much lower temperature rise inside a shear band was also suggested [110,111]. It was shown that upon four-point bending of the Zr-based BMG coated by sputtered tin creation and propagation of shear bands caused the local temperature rise in the BMG and the appropriate calculations of the temperature rise were provided [107].

Nevertheless, regardless of the source of strain localization at high enough strain rate within the shear band the central shear zone heats up due to friction forces [112] and its viscosity can drop several orders in magnitude further increasing the shear speed in the band. The band propagation can be stopped likely owing to the decrease in axial stress waiting for the next round of the increase in stress. At the critical shear offset sudden fracture takes place owing to the low viscosity reduced by the acting friction force heat release. Also, no coating (with sputtered tin, Rose metal and indium) melting was observed later near individual shear bands. Melting occurred only in some places near the fracture surface, near microcracks and near places of shear band concentrations where total heat release was large enough to initiate melting (Fig 5.9). It was suggested that the shear localization and banding is a stress driven process and the temperature rise that was observed in the shear bands is due to the friction forces in the thin liquid layer within the shear bands.

(a)   (b)

Fig. 5.9. SEM images the areas near the fracture surface of the bulk sample coated by indium (a) and Rose metal (b) after the compression test. Reproduced from [112] with permission of Nature Publishing Group.

By using the free volume model it was shown that a sharp increase of the excess/free volume at yielding is mainly responsible for the low viscosity of a propagating shear band [113]. MD simulation showed the excess volume production on the deformation process in metallic glasses [114] as a result of the reformation of the atomic bonds during the deformation process. The volume change is an important factor in the mechanical response of metallic glasses subjected to mechanical deformation. It is suggested that the adiabatic heating may not be necessary as a mechanism for shear band formation and at large strains, the excess volume change could reach or even exceed that of the supercooled liquid at the glass transition.

Exceptionally high diffusion coefficient in the shear band also indicates high excess (free) volume concentration [115]. Generation of a large fraction of the excess volume during deformation of a metallic glassy sample leading to enormous softening was observed by nanoindentation in terms of hardness [116], by using MD simulation [114] and by an atomic force microscope in a tapping mode in terms of the reduction of the Young's modulus [117].

The idea of generation and redistribution of the excess volume in a shear band is supported by the work where a large decrease in the hardness within a zone up to 0.1 mm and even more was observed near the dominant shear band [116]. Thickness of the softened zone (centered on the shear band) increases with increasing the shear offset and total strain. Measurements of dimension change perpendicular to the shear band suggested dilatation within the band of about 1 %. DSC exhibited high enthalpy of relaxation in that area which is reduced by annealing.

Fig. 5.10. The Young's modulus distribution image (a) and the normalized logE values picked from chosen areas of the polished sample near the shear band core (initial offset of the shear band was about 2 μm) as a function of the coordinate X (b). The normalized logE of the region far from the shear band edge (more than 40 μm away) (c). The normalized logE measured near the shear band core after the annealing (d). Reproduced from [117] with permission of Elsevier.

The elastic modulus variation in the surface regions of the deformed $Zr_{60}Cu_{20}Co_{10}Al_{10}$ BMG near and far from a shear band with a shear step of about 2 μm was measured using the atomic force microscope quantitative nanomechanical mapping method (Fig. 5.10). It was found that the Young's modulus decreases gradually towards the shear band's central region on the micrometer scale. Maximum decrease in the Young's modulus near the shear band's central region is about 70%. Again this effect is attributed to the generation and migration of the excess volume during the shear band operation [117].

In order to measure the velocity of shear bands the $Pd_{40}Ni_{40}P_{20}$ bulk metallic glass was tested in compression with a piezoelectric load cell. The data indicate that in case of simultaneous sliding of the entire sample the displacement rate is approximately 2 mm/s while if the displacement occurs as a localized propagating front, the velocity of the front is approximately 2.8 m/s [118]. The velocity of shear bands measured with high-speed video cameras and the operation time was found to be close to their temporal resolution [119]. Digital image correlation procedure was applied to the images to determine the displacement adjacent to the shear band as a function of time and position across the width of the sample [120].

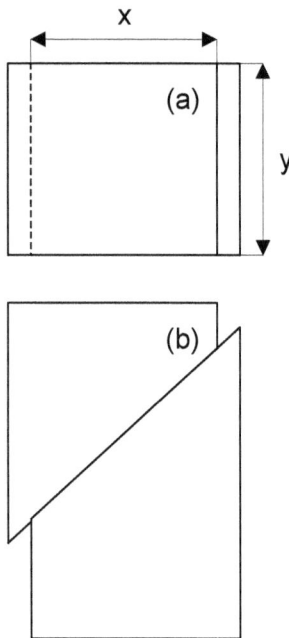

Fig. 5.11. Reduction of the effective cross-section area of the rectangular sample with a dominant shear band upon uniaxial compression (top view (a) and side view (b)) propagating by the stick-slip process.

It was suggested that the macroscopic yield point and completion of the so-called "strain hardening" in the beginning of plastic deformation corresponds to the transecting of the shear bands [121]. Also, formation of a dominant shear band must be accompanied by the reduction of the apparent flow stress owing to reduction of the effective cross-section area (X·Y in Fig. 5.11) while many BMGs flow at nearly constant stress after the macroscopic yield point (Fig. 5.4) owing to relatively homogeneous deformation induced by several shear bands.

The plasticity of BMGs was found to depend on the competition between the formation of new shear bands and the reactivation of the existing shear bands during plastic deformation [122]. Clear transition to a dominant shear band was observed in the Zr-Cu-Fe-Al bulk metallic glassy samples [123]. Fig. 5.12 (a) represents the continuous loading stress-plastic strain curve of the cylindrical $Zr_{62.5}Cu_{22.5}Fe_5Al_{10}$ glassy alloy in its plastic deformation range. The stress drop spectrum changes its shape and the flow stress starts to decrease after 2.5 % of deformation. The frequency distribution of the magnitude of stress drops (Fig. 5.12 (b)) calculated from the continuous strain-stress curve in Fig. 5.12 (a) in two ranges of plastic deformation clearly indicates different deformation behaviors in these two ranges. Formation of new shear bands and propagation of the existing shear bands takes place at the first stage while the formation of a major shear band throughout the sample and its propagation takes place at the second stage. Such a behavior is also illustrated by the changes in the stress drop and strain per stress drop plots (Fig. 5.12 (b)). The snapshots from a video file recorded during its deformation shown in the insert in Fig. 5.12 (a) clearly demonstrate the formation and propagation of a major/dominant shear band in the second part of the deformation process. Formation of the dominant shear band is a stochastic process and in various samples it takes place at different strain values. A similar behavior was observed in the case of the $Pd_{40}Ni_{40}Si_4P_{16}$ BMG and $Ti_{43.6}Zr_{7.9}Cu_{40.4}Ni_{7.1}Co_1$ BMG samples containing 1000 ppm B.

Fig. 5.12. (a) Stress and stress drops (diamonds) as a function of plastic strain for the cylindrical $Zr_{62.5}Cu_{22.5}Fe_5Al_{10}$ sample of 2:1 aspect ratio upon the compression test. The inserts in (a) illustrate a sequence of the video file snapshots corresponding to (0 %) the initial sample before deformation, (1.5 %) the sample deformed plastically to 1.5 %, (2.5 %) the sample deformed plastically to 2.5 % and (3.5 %) the sample before fracture (deformed plastically to 3.5 %). (b) The frequency distribution of the magnitude of stress drops calculated from the continuous strain-stress curve in (a) in two ranges, namely, from 0 to 2.5 % and from 2.5 to about 4 % in plastic deformation. Reproduced from [123] with permission of Elsevier.

The two-stage deformation process is schematically illustrated in Fig. 5.13. These BMG samples initially show multiple shear bands and microscopically relatively homogeneous deformation at nearly constant flow stress and the formation of a dominant shear band at the late deformation stage which leads to a significant decrease in the flow stress owing to the decrease in the efficient sample cross-section (Fig 5.11). These two types of the deformation regimes may be responsible for the observed bimodal distribution of the shear offsets measured by nanoindentation [124].

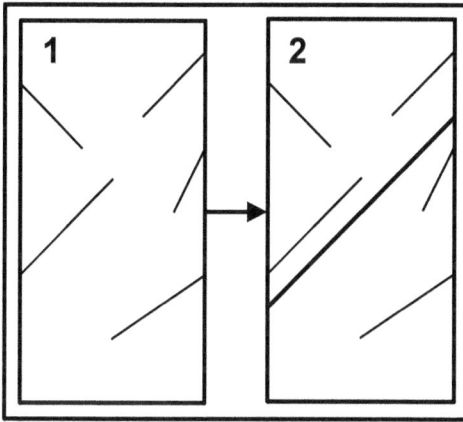

Fig. 5.13. Two-stage deformation mode: 1 – multiple shear bands mostly not crossing the entire sample, 2 – dominant shear band from one side to another leading to stick-slip flow behavior. Reproduced from [123] with permission of Elsevier.

The $Zr_{65}Cu_{15}Ni_{10}Al_{10}$ alloy demonstrated the chaotic nature of the stress serrations by showing the existence of a finite correlation dimension and a positive Lyapunov exponent [125]. In $Cu_{47.5}Zr_{47.5}Al_5$ alloy the distributions of stress drop magnitudes and their time durations appear to follow a power-law scaling form reminiscent of a self-organized critical state. The chaotic dynamics of the less ductile BMG was consistent with a single shear band sliding. In the $Cu_{47.5}Zr_{47.5}Al_5$ alloy, the high ductility appears to be related to concurrent nucleation of a large number of shear bands throughout the sample, which in turn, can give rise to a hierarchy of length scales over which shear bands propagate. In the $Ni_{50}Pd_{30}P_{20}$ bulk metallic glass the serrated flow behavior also appeared to be chaotic and a stochastic model suggested [126] that the underlying serrated slow dynamics initially relies on the appearance of new shear bands on the material, but as the experiment proceeds the nature of these dynamics change, and strain takes place by the extension of the shear bands already present on the BMG sample.

However, the two stage deformation observed in the typical sample having 2:1 aspect ratio (sample height to width or to diameter ratio) may not be the case in the samples of significantly larger or smaller size or different geometry because the engineering plasticity of bulk metallic glasses depends upon their geometry and aspect ratio [127,128]. At low aspect ratio, shear bands are terminated at the interface with the loading platen and become confined. Also, the friction between the sample and the loading platen becomes more efficient. Different specimen geometries influence the shear-band formation and deformation processes, resulting in variations of the mechanical properties.

The samples with 1:2 aspect ratio can be deformed to large strain values without formation of a dominant shear band [81]. Fig. 5.14 and the insert (a) represents the continuous loading stress-plastic strain curve of the cylindrical $Zr_{61}Cu_{27}Fe_2Al_{10}$ glassy alloy. The true stress values were calculated from nominal stress assuming homogeneous deformation and constant volume of the deforming sample throughout the test. As one can see in Fig. 5.14 (b) the samples deformed for 0.24 and 0.68 strain were intact while the sample deformed to 0.80 was destroyed to pieces.

Fig. 5.14. The engineering and calculated compressive true stress-strain curves of the sample having 1:2 aspect ratio tested at room temperature at 5 x 10^{-4} s^{-1} (left axis) and the derivative dσ/de (right axis). The inset (a) a close-up of the curve between 0.10 and 0.12 strain. (b) An optical image of the samples before compression and deformed for 0.24, 0.68 and 0.80 strain, respectively, from left to right. Reproduced from [81] with permission of Elsevier.

The XRD patterns obtained from the deformed samples show no crystalline peaks. As can be seen from shear offsets on the lateral surface in Fig. 5.15 the sample shows formation of multiple shear bands.

Fig. 5.15. Lateral surface of the sample deformed at 5 x 10^{-4} s^{-1} for 0.24 strain at different magnification (a,b). Reproduced from [81] with permission of Elsevier.

The length of the shear steps (L) on the lateral surface per its area (A) at first approximation is supposed to correlate with the fraction of shear bands (shear surfaces) per sample volume. The thermal properties of the samples were tested by DSC and the resulted values (glass-transition temperature (T_g), crystallization temperature (T_x), exothermic peak temperature (T_p), as well as enthalpies of relaxation (H_r), overshoot (H_o) and crystallization (H_c)) are also shown in Table 5.5.

Table 5.5. Strain (e_e and e_t), Vickers Hardness (HV), T_g, T_x, T_p, ΔH_r, ΔH_o and ΔH_c values of the studied alloy. Reproduced from Ref. [81] with permission of Elsevier.

e_e	e_t	L/A nm^{-1}	HV	T_g (K)	T_x (K)	T_p (K)	ΔH_r (J/mol)	ΔH_o (J/mol)	ΔH_c (J/mol)
0	0	0	457±5	670	763	767	204	14	3236
0.24	0.27	90±7	466±4	666	758	765	637	34	3290
0.68	1.14	138±6	497±3	667	759	765	756	36	3313
0.80	1.61	-	-	672	758	765	262	24	3244

The volume fraction of the shear bands is only $9 \cdot 10^{-10}$. This is negligibly small value to cause significant changes in the thermal properties of materials if the rest of the sample is not affected. According to the XRD data obtained with a relatively high signal to noise ratio the samples remain glassy after severe plastic deformation and even failure of the sample. Even after fracture of the sample deformed to 0.8 engineering strain (1.61 true strain) the enthalpy of crystallization is not affected. The reduced values of T_x and T_p after deformation may be connected with ease of heterogeneous nucleation at the shear steps on the surface or inside the shear bands.

The Vickers microhardness measured in the direction of applied stress increases after deformation. This may be connected with large fraction of the elastic stresses in the glassy matrix [38,129] induced by so-called not well developed shear bands, the shear bands which shear front did not cross the entire sample.

5.1.4   Structural rejuvenation by deformation

Structural relaxation (usually observed on heating) decreases the sample volume and causes embrittlement. As has been shown above for the $Zr_{61}Cu_{27}Fe_2Al_{10}$ glassy alloy [81] and also described in other works [130,131] an opposite effect called rejuvenation brings back the glass to a higher energy state. The heat of relaxation increased significantly after deformation (Table 5.5) and the excess heat of relaxation gain as a result of leading to e=0.24 of 433 J/mol is about 9% of the energy used to deform the sample. The stored plastic strain energy at large deformation was also found in a Pd-based alloy [132]. This may indicate that intensive plastic deformation with a large number of the shear bands brought the bulk metallic glass to a higher energy state which is released only after the failure of the sample. On the other hand the heat of the overshoot (ΔH_o) after glass-transition, ordinary indicative of the degree of structural relaxation [133] also increased. Intensive plastic deformation of a bulk metallic glass with a large number of the shear bands before intensive crack initiation has a twofold effect on its properties. In local areas a higher energy state is reached (similar to the glassy phase rejuvenation) which is released after the failure of the sample. The plastically (within the shear bands) and elastically (at the edges of not well developed shear bands) deformed areas release a significant amount of extra heat upon structural relaxation while the decreased T_x may indicate ease of crystallization connected with the preferential states for heterogeneous nucleation in the locally deformed areas. On the other hand another part of the glassy matrix may suffer from the opposite effect of structural relaxation leading to an overall lower T_g and larger heat overshoot of the DSC signal after glass-transition indicative of a more relaxed glass.

Ion irradiation was used to improve the tensile ductility of a metallic glass [134]. Mechanical loading is another method of rejuvenation upon elastostatic loading at stresses well below the yield stress [135] as well as cold rolling [136] and severe plastic deformation [137]. It was also possible to induce residual stresses in a bulk metallic glassy sample by shot peening method which improved the mechanical performance, in particular plasticity [138,139].

Uniaxial compression of a BMG close to the yield strength induced creep of the samples, accompanied by an increase in the stored heat of relaxation [140]. The sample exhibited an increase in compressive plasticity on the subsequent test [141]. This creep is homogeneous and leads to decreases in density and in the elastic moduli. Also, the increase in heat of relaxation and in plasticity in BMGs resulting from static loading are reported to depend on the atomic packing density [142].

As discussed in Chapter 4 cryogenic thermal cycling is another non-destructive method to rejuvenate the glassy structures [143,144]. Contrary to the absence of any effect expected, thermal cycling of metallic glasses induces rejuvenation, reaching less relaxed states of higher energy because of intrinsic non-uniformity of the glassy structure, implying a non-uniform coefficient of thermal expansion. Such a thermal cycling of BMGs also produces desirable improvements in plasticity.

5.1.5   Strain-rate sensitivity

While the strain-rate sensitivity of polycrystalline alloys is generally positive [145] (except for special cases like Hadfield steel described above), for bulk metallic glasses different results were reported.

The value of strain rate sensitivity parameter (m) can be calculated from the true stress-strain curves according to the relationship:

$$m = \left(\frac{\partial \ln \sigma}{\partial \ln \overset{\bullet}{\varepsilon}}\right)_{\varepsilon, T}$$
(5.2)

where σ is the flow stress and e' is the strain rate.

A negative m value was reported for the $Zr_{57}Ti_5Cu_{20}Ni_8Al_{10}$ alloy at strain rates ranging from 10^{-4} to $3 \cdot 10^3$ s^{-1} [146]. m value of about -0.002 at room temperature was reported for $Zr_{52.5}Ti_5Cu_{17.9}Ni_{14.6}Al_{10}$ [147] and $Cu_{50}Zr_{50}$ BMG [148]. The $Zr_{61}Cu_{27}Fe_2Al_{10}$ BMG sample also exhibited negative strain rate sensitivity (Fig. 5.16) [149]. At about 0.06 plastic strain the strain rate was changed from $5 \cdot 10^{-3}$ s^{-1} to $5 \cdot 10^{-2}$ s^{-1} while the flow stress decreased from ~1765 MPa to ~1757 MPa leading to m=-0.002. Opposite changes occur when the strain rate was decreased afterwards. There is a little stress overshoot behavior right after strain rate decrease. Also a negative value was obtained [150] for the $Zr_{52.5}Ti_5Cu_{17.9}Ni_{14.6}Al_{10}$ and for a composite, at strain rates of $3.3 \cdot 10^{-4}$ and $3.7 \cdot 10^{-3}$ s^{-1}. The $Zr_{41.2}Ti_{13.8}Cu_{12.5}Ni_{10}Be_{22.5}$ alloy was reported [151] to be insensitive to the strain rate below 473 K.

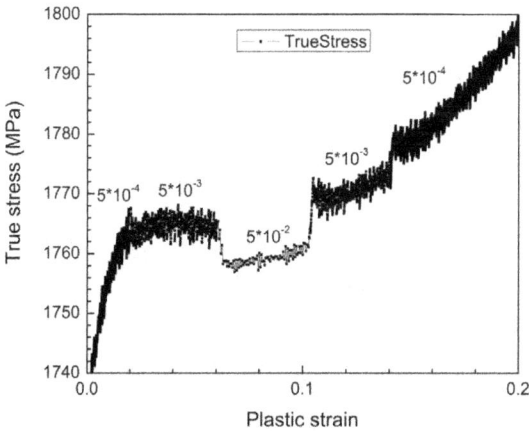

Fig. 5.16. The compressive calculated true stress - plastic strain curves of the $Zr_{61}Cu_{27}Fe_2Al_{10}$ BMG sample of 2 mm in diameter having 1:2 aspect ratio tested at different strain rate as indicated in s^{-1}. Reproduced from [81] with permission of Elsevier.

In Zr-based metallic-glass coatings with micrometer-scale thickness the increase of the penetration rate also leads to the decrease of the hardness [152]. At the same time other nanoindentation studies indicate positive SRS parameter values [153]. Such a difference could be mainly attributed to the stress state which is closer to three-axial compression [154, 155] when tested by indentation compared to uniaxial compression.

The $Zr_{65}Cu_{20}Fe_5Al_{10}$, one of relatively ductile BMG glassy alloys samples, was tested at room temperature and almost no strain rate sensitivity was found at low strain rate while it was negative at higher rates (Fig. 5.17) [156]. The $Zr_{65}Cu_{20}Fe_5Al_{10}$ BMG alloys exhibited negative strain rate sensitivity at high strain rate and the parameter (m=–0.0026) has been calculated from the true stress-strain curves.

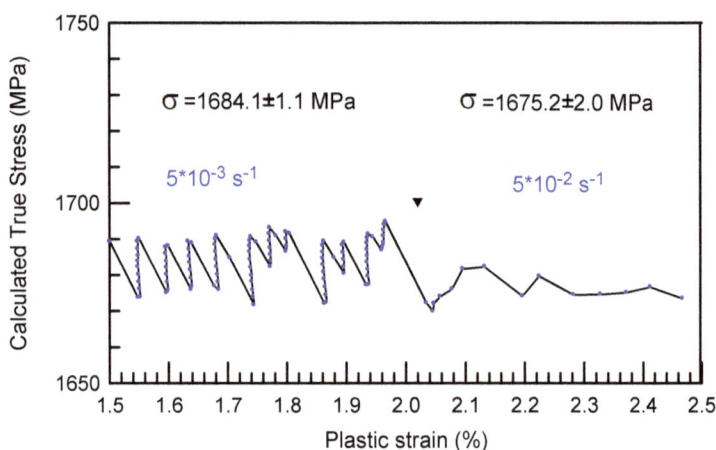

Fig. 5.17. True stress-plastic strain curves (elastic strain is subtracted) performed at strain rates of 5 · 10^{-3} and 5 · 10^{-2} s^{-1}. The data is taken from Ref. [156] with permission of Elsevier.

The mean and maximum flow stresses and the confidence interval (probability P = 0.95) have been calculated in 0.25 % plastic deformation interval using true stress values at both sides of each point of the strain rate change as shown in Table 5.6. Another parameter, which is an average maximum stress before each stress drop (Table 5.6) satisfied normality tests and produced similar results. Not only the flow stress but also the average maximum stress before each stress drop corresponding to a serration for each test is practically the same at each loading rate in the vicinity of the strain rate change point and the differences are practically within the confidence interval which indicates that the $Zr_{65}Cu_{20}Fe_5Al_{10}$ BMG is strain rate insensitive (m ~ 0) in the range of strain rates from $5x10^{-6}$ to $5x10^{-4}$ s^{-1}. However, the SRS becomes meaningfully negative when the strain rate increases from $5 \cdot 10^{-3}$ s^{-1} to $5 \cdot 10^{-2}$ s^{-1} because the strain rate is so fast that the relaxation time is not enough to build up the stress which explains the decrease in the flow stress.

Table 5.6. Mean and mean maximum stress values for test 1 and test 2 calculated considering a 0.25 % plastic deformation interval at both sides of the strain rate change. The mean maximum stress indicates average value of maximum stress before stress drop. The confidence intervals were calculated using 0.95 % probability. Reproduced from [156] with permission of Elsevier.

Test 1				Test 2			
$5 \cdot 10^{-4}$ to $5 \cdot 10^{-5}$ (s^{-1})		$5 \cdot 10^{-5}$ to $5 \cdot 10^{-6}$ (s^{-1})		$5 \cdot 10^{-6}$ to $5 \cdot 10^{-5}$ (s^{-1})		$5 \cdot 10^{-5}$ to $5 \cdot 10^{-4}$ (s^{-1})	
Mean stress (MPa)		Mean stress (MPa)		Mean stress (MPa)		Mean stress (MPa)	
$5 \cdot 10^{-4}$	1609.4 ± 1.2	$5 \cdot 10^{-5}$	1611.9 ± 0.5	$5 \cdot 10^{-6}$	1543.2 ± 0.2	$5 \cdot 10^{-5}$	1543.8 ± 0.3
$5 \cdot 10^{-5}$	1611.9 ± 0.5	$5 \cdot 10^{-6}$	1612.0 ± 0.2	$5 \cdot 10^{-5}$	1543.5 ± 0.4	$5 \cdot 10^{-4}$	1542.4 ± 0.9
Mean Max. stress (MPa)		Mean Max. stress (MPa)		Mean Max. stress (MPa)		Mean Max. stress (MPa)	
$5 \cdot 10^{-4}$	1614.6 ± 1.4	$5 \cdot 10^{-5}$	1618.1 ± 1.2	$5 \cdot 10^{-6}$	1548.2 ± 3.2	$5 \cdot 10^{-5}$	1549.7 ± 1.3
$5 \cdot 10^{-5}$	1619.7 ± 3.0	$5 \cdot 10^{-6}$	1618.1 ± 3.1	$5 \cdot 10^{-5}$	1549.1 ± 2.3	$5 \cdot 10^{-4}$	1547.4 ± 1.1

5.1.6 Size effect and in-situ deformation studies

There is a strong size effect on deformation behavior upon indentation in which the deformed volume is significantly smaller while local plastic deformation is larger than in compression and metallic glasses are proven to deform homogeneously on nanoscale [73,157]. With decreasing specimen size smaller than the equivalent critical shear offset, the shear deformation of metallic glass is changed from unstable to stable, which leads to a transition from global brittleness on the macroscale to large plasticity on the microscale [158].

Plastic deformation in metallic glasses was performed using the in-situ TEM technique [159,160,161]. After initial observation the beam current was reduced and the sample deformed in a straining holder in tension [161]. This was an important precaution to prevent electron-beam heating of the sample upon deformation. This procedure (straining and subsequent observation) has been repeated several times until some cracks appeared. High-resolution in-situ TEM investigation showed that the $Zr_{65}Ni_{10}Cu_5Al_{7.5}Pd_{12.5}$ ribbon sample retains its glassy structure after in-situ deformation and crack propagation (Fig. 5.18). No nanocrystallization was observed even near the crack tip.

Fig. 5.18. High-resolution TEM image obtained upon in-situ deformation in a TEM near the crack in the $Zr_{65}Ni_{10}Cu_5Al_{7.5}Pd_{12.5}$ alloy. The insert shows a nanobeam diffraction (NBD) pattern.

The homogeneous deformation of the thinned $Zr_{65}Ni_{10}Cu_5Al_{7.5}Pd_{12.5}$ alloy is clearly seen in front of the crack tip upon loading. The crack opening speed was quite low – about tens of nanometers per second or even less. A narrow featureless deformation zone formed ahead of the crack tip is much wider than the typical (10-20 nm) width of a shear band. Similar results were obtained for the submicron (about 200 nm) thick $Zr_{52.5}Cu_{17.9}Al_{10}Ni_{14.6}Ti_5$ alloy sample [162].

The in-situ experiments demonstrate a fundamental possibility of a homogeneous-like like plastic flow to occur in a localized area of glassy metal thin foil at room temperature.

According to calculations the thermal energy release and rise in temperature depends on the deformation rate and shear offset, in addition to the thermal conductivity and the strength of the glassy phase [163,164]. However, the deformation areas observed in the sample tested in situ in TEM are much wider compared to the highly localized deformation bands observed in bulk glassy sample. Also no significant heating could have taken place as the heat generated upon the deformation is dissipated over a larger volume. Homogeneous deformation of the $Zr_{62.5}Cu_{22.5}Fe_5Al_{10}$ metallic glass was also observed upon in situ nanoindentation in TEM [165].

5.1.7 Fatigue, cyclic and long-term creep deformation in the elastic region.

BMGs were originally considered to have relatively poor fatigue resistance, possibly also connected to the lack of strain-hardening, but in fact they show a wide range of fatigue behavior influenced by the intrinsic and extrinsic factors [166,167]. The fatigue-endurance limits of some Zr-based BMGs are comparable with those of the high-strength crystalline alloys [168,169]. Under compressive-compressive fatigue of a Cu-Zr-based BMG fracture occurs in shear at roughly the same angle as seen in continuous monotonic compression [170]. It is at 41° to the loading axis (i.e. the normal vector to the fracture surface is at 49° to the loading axis). The fatigue limits in the Co- and Fe-based BMGs exceeded 2 GPa, and those in the Ni- and Ti-based BMG surpassed about 1.5 GPa [171].

In-situ observations of the mechanical-damage behavior of BMGs during both fatigue and tensile testing with a video camera thermography showed the relationship with temperature evolutions. The stress–strain diagrams during high-cycle fatigue indicate three effects: the thermoelastic effect, the inelastic effect, and the heat-transfer effect [105]. The specimen temperature was observed to oscillate regularly corresponding to the stress oscillation at each fatigue cycle. A thermoelastic-degradation behavior was observed at the center of the specimen. This behavior was attributed to the free-volume accumulation inside the specimen during fatigue. However, no shear bands were observed before failure during the fatigue experiments, which indicated that the crack-initiation mechanisms of BMGs during fatigue testing could be surface-voids and defects based, and different from the shear-band mechanisms during tensile testing.

The mechanical properties of a BMG also could be altered by prior cyclic nanoindentation in the nominally elastic range [172] inducing a hardening effect. MD simulations were used to study the mechanisms of the hardening [173], and successfully showed the threshold and saturation effects. It was concluded that during the cycling there are microplastic events. The loading and unloading is apparently elastic, though the atomic rearrangements are not individually reversed; there is an accumulation of permanent structural changes [173].

The beginning of nanocrystallization was observed in the $Zr_{62.5}Fe_5Cu_{22.5}Al_{10}$ bulk metallic glassy cubic samples (edge size equal to 4 mm) under a low stress of 586 ± 242 MPa in the elastic region while no such an effect was seen in the rod shaped samples of 2 mm in diameter [174]. This alloy also showed phase separation on heating prior to crystallization [175]. Later the $Zr_{62.5}Fe_5Cu_{22.5}Al_{10}$ BMG samples of 3 mm diameter were cyclically loaded at higher stresses of 800 ± 200 MPa and 1000 ± 200 MPa which are still significantly lower than the yield strength. It was found that kinetically frozen anelastic deformation accumulates at room temperature and causes crystallization of metallic glassy phase. An increase in the crystallization enthalpy after cyclic loading for 1000 cycles at 1000 ± 200 MPa from 44 J/g to 48 J/g is likely connected with the formation of the nuclei of a non-equilibrium phase which together with the residual glassy phase may represent a state with higher energy (Fig. 5.19) [176]. This also led to a better plasticity of the sample. An anelastic contribution can arise from localized viscous flow of structural defects [177,178]. The volume changes were locally accommodated by cumulative anelastic deformation and led to subsequent partial nanocrystallization of the glass.

Fig. 5.19. DSC traces of the as-cast sample and after cyclic loading at 1000 ± 200 MPa at 1 Hz for 100 cycles. Reproduced from [176] with permission of Elsevier.

The storage modulus (E') and the internal friction modulation upon cyclic loading of the $Zr_{61}Cu_{27}Fe_2Al_{10}$ bulk metallic glassy samples was studied using dynamical mechanical loading (DMA) within quasi-reversible deformation regime [179]. An increase in the storage modulus indicating stiffening of the sample is observed upon cyclic loading. Apparent cyclic softening (decrease in E' in the amplitude dependent bending DMA test) with the number of loading cycles observed at low stress was found to be connected with the instability of the DMA instrument when testing a ceramic sample. Local atomic changes leading to the detectable increase in storage modulus of the BMG sample take place at the critical elastic bending strain value of 0.3 % (Fig. 5.20 (a)). The observed increase in E' with the number of cycles after about 0.3 % deformation most probably indicates structural changes within so-called weakly bonded regions in this BMG which may trigger activation of the local atomic rearrangements leading to the increased stiffness and even local nanocrystallization. After a certain number of cycles the activated zones of larger volume become inactive owing to structural changes and the E' modulus saturates at a certain value.

The structure of as-cast sample and the sample after a series of cyclic loading tests with maximal deflection 500 μm was studied by TEM. A small fraction of nanocrystals of about 5 nm in size is present in the elastically deformed sample (Fig. 5.20 (b)).

The $Zr_{61}Cu_{27}Fe_2Al_{10}$ bulk metallic glass also indicated complex changes in the properties after 10 loading cycles of 50±37.5% of the yield stress in uniaxial compression at room temperature [180]. The surface microhardness measured on the sample faces normal to the loading axis increases by ~8%. This effect progressively decreased on ageing at room temperature and after 45 days the samples have recovered their initial hardness. The resulting stored anelastic strains lead to non-uniformity of the effective hardness measured on the end-faces. The cyclic loading appears to accelerate the development of anelastic strain relative to that on static loading because the overall Young's modulus increases by ~0.8% in 10 cycles. The relaxation and crystallization enthalpies of the sample increased as a result of structural rejuvenation. The excess specific heat capacity, $\Delta C_p = 1.85$ J/mol·K (Fig. 5.21), of an elastically cycled glass over an as-cast glass also suggests that cycling takes the metallic glass into a more excited state.

Fig. 5.20. (a) Evolution of the relative Storage modulus of the $Zr_{61}Cu_{27}Fe_2Al_{10}$ alloy as a function of the deformation cycle at different strain level. High-resolution TEM image of the BMG sample after mechanical cycling (b). The inset shows a Fourier Transform of large central crystalline particle. Reproduced from [179] with permission of Springer.

Fig. 5.21. DSC traces indicating the difference in C_p values between the supercooled liquid and glass in as-cast state, immediately after elastic cycling as well as 3 and 7 days later. Each time the sample was cut from the centre of the elastically cycled cylindrical sample. Reproduced from [180] with permission of Elsevier.

There have been very few studies of low-cycle fatigue (LCF) of metallic glasses and BMGs [181,182]. Samples of the $Zr_{62.5}Fe_5Cu_{22.5}Al_{10}$ bulk metallic glass were tested for LCF. The samples were unloaded, either completely or partially according to the stress ratio $r = \sigma_0/\sigma_{max}$. Mostly $r = 0$ (i.e. complete unloading), but in some cases the applied stress was reduced not to zero but to 200 MPa, giving $r \approx 0.12$. Comparison of the tests in monotonic loading and cyclic loading (repeated loading to the onset of plasticity, then unloading) shows that the compressive plasticity of the glass is drastically reduced under cyclic loading (Fig. 5.22) [183]. This effect arises from stress reversal accelerating concentration of shear in the dominant shear band.

Fig. 5.22. Compression testing of the $Zr_{62.5}Cu_{22.5}Fe_5Al_{10}$ BMG. The samples are cylinders 2 mm in diameter and 4 mm in height (i.e. aspect ratio 2:1). (a) The stress-strain curves (nominal and calculated true stress) for monotonic loading to failure. The arrow indicates the strain at which there appears to be a change in deformation mode (stage 1 to stage 2). (b) The results of cyclic loading (the stress ratio r = 0) are shown by the data points labeled with the cycle number. For each cycle, the maximum stress at the onset of yielding, the minimum stress at the end of the stress drop under load, and the value of the stress drop are given. All the results are shown as a function of plastic strain. Reproduced from [183] with permission of Springer.

5.1.8   Fracture toughness

The relatively large dimensions of bulk metallic glasses with good glass-forming ability permitted performing fracture toughness tests. The fracture toughness, (K_{Ic}) of BMGs varies significantly from about 1 MPa·m$^{1/2}$ for RE-based (rare-earth) and some of Fe-based BMGs, close to that of oxide glasses, up to 100 MPa·m$^{1/2}$ and even more for Zr- and Pd-based BMGs [184]. One of the reasons for low K_{Ic} of RE, Mg and Fe-based BMGs is low solubility of oxygen in the base elements and these glassy alloys which leads to the formation of the oxide particles acting as the stress concentrators [185] whereas Zr- and Pd-based BMGs dissolve a large amount of oxygen interstitially. The $Zr_{61}Ti_2Cu_{25}Al_{12}$ BMG alloy surprisingly showed fatigue pre-cracked fracture toughness in excess of 100 MPa·m$^{1/2}$ [186]. The correlations of the toughness were made with the Poisson's ratio, the product of shear modulus and molar volume (μV_m), and the glass transition temperature. The $Pd_{79}Ag_{3.5}P_6Si_{9.5}Ge_2$ BMG also showed an exceptionally high fracture toughness of 200 MPa·m$^{1/2}$ [187]. A significant reduction in the fracture toughness is found at cryogenic temperature [188]. Toughness of the $Zr_{52.5}Cu_{17.9}Ni_{14.6}Al_{10}Ti_5$ BMG decreases from 101±7 MPa·m$^{1/2}$ at room temperature to 58±10 MPa·m$^{1/2}$ at 123 K.

5.1.9   Fracture mechanisms

The fracture surface of a Zr-based BMG (Fig. 5.23) exhibits alternating areas of vein and flat fracture surfaces formed by penetration of air into a liquidlike layer [189,190] after initial cleavage-like behavior starting from the lateral surface of the rod sample [191]. The cleavage-like and vein patterns also alternate in the body of the sample in the direction normal to shear deformation. The vein fracture patterns, which are specific to metallic glasses, may signify local melting of the material at the surface upon fracture [192,193].

Fig. 5.23. Fracture surface of the $Zr_{61.88}Cu_{18}Ni_{10.12}Al_{10}$ BMG after failure of the sample containing vein pattern and smooth areas.

Brittle BMGs such as Mg-, Fe- or RE-based often destroy to several pieces on fracture while ductile BMGs exhibit smaller number of fragments. Nanoscale wavy steps generated by crack front waves were observed on the fracture surface of the $Pd_{40}Ni_{40}Si_4P_{16}$ bulk metallic glass [194]. The microbranching, also found in other alloys [195,196], results from dissipating energy accumulated at the crack front by creating the additional fracture surface area when the crack speed exceeds the critical value.

5.1.10   Dynamic mechanical properties

As it is schematically shown in Fig. 5.24 [197,198] the fracture strength of the $Zr_{57}Ti_5Cu_{20}Ni_8Al_{10}$ alloy tested at strain rates from 10^{-4} to 10^3 s^{-1} is lowered under dynamic compression while shear-band propagation remains the dominant deformation mechanism. In dynamic loading experiment at high strain rate macroscopic plasticity also dropped to zero due to the lack of time for stress relaxation. High-speed cinematography indicated that at such conditions fracture occurs on a single major plane, accompanied by the emission of visible light which was also observed during an ordinary nearly quasistatic loading testing conditions.

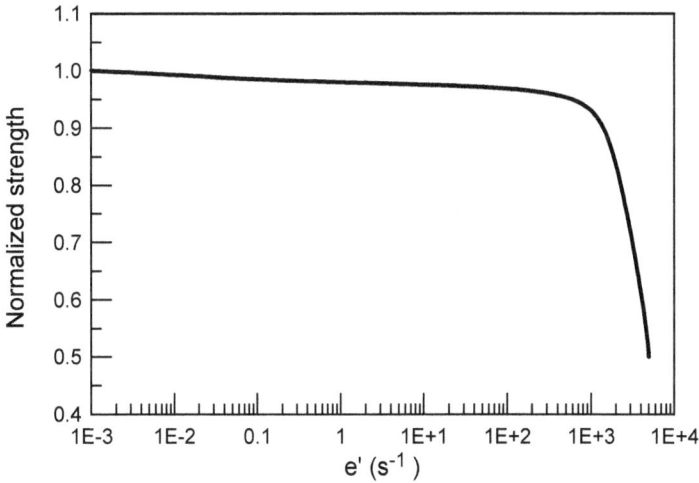

Fig. 5.24. Fracture stress of typical BMGs as a function of the strain rate (e'), schematic representation.

The impact fracture energy of 63 kJ/m^2 and stress of 1615 MPa for the bulk amorphous $Zr_{50}Al_{10}Cu_{30}Ni_5$ alloy were measured by using the Charpy test specimen with U-shape notch [199]. The impact fracture stress was nearly the same as the static tensile fracture strength of 1570 MPa obtained in a wide strain rate range of 10^{-4} to 10^{-1} s^{-1}. The fracture surface appearance changes from the vein pattern in the region near the U-shape notch to an equiaxed dimple-like pattern in the inner region.

5.1.11   Wear resistance

5.1.11.1   Macroscopic wear properties

Macroscopic wear resistance of metallic glasses has been studied extensively [200,201]. Fe-, Ni- and Co-based metallic glasses were found to have high wear resistance [202,203]. La-, Mg-, Pd- and Zr- bulk metallic glasses exhibited classical behavior for abrasive wear and the wear resistance increased linearly with hardness [204]. It was shown that some metallic glasses exhibit better wear resistance compared to the crystalline state [205]. On the other hand, macroscopic wear resistance of other metallic glasses was found to be lower than in the corresponding crystalline states [206,207]. However, usually BMGs embrittle on crystallization owing to the formation of primary intermetallics. The improvement of wear resistance of a Zr-based BMG composite containing β-Zr crystalline phase was attributed to its plastic deformation and work hardening, which decrease strain accumulation and release strain energy in the glassy matrix [208].

5.1.11.2   Nano-scale and atomic-scale wear resistance

Successful applications of metallic glasses in micro-electro-mechanical devices was demonstrated [209,210].   This implies that any mechanical contact between the component surfaces leading to wear is a significant factor limiting their durability. When the size of the mechanical components is reduced to the micrometer and sub-micrometer level, the native surface oxide layer begins to play an important role in the contact mechanical behavior of metallic glasses. The nanoscale tribological properties of the $Ni_{62}Nb_{38}$ and Pt-Cu-Ni-P metallic glasses were studied [211,212,213]. $W_{33}N_{32}B_{35}$ glassy films deposited by sputtering having a high hardness (24 GPa) show wear resistance comparable to industrially used coatings of TiN [214].

The nanoscale and atomic-scale scratch wear resistance of typical engineering material type metallic glasses: $Ti_{43}Zr_{10}Cu_{36}Ni_9Sn_2$, $Zr_{62.5}Cu_{22.5}Fe_5Al_{10}$ and $Mg_{65}Cu_{25}Gd_{10}$ having native surface oxides were studied with the diamond single crystal cantilevers [215]. The $Ti_{43}Zr_{10}Cu_{36}Ni_9Sn_2$ and $Zr_{62.5}Cu_{22.5}Fe_5Al_{10}$ metallic glasses showed nano plowing type wear mechanism with high quality of the wear trace as shown in Fig. 5.25. Isometric projection of a typical scratch is shown in Fig. 5.26.

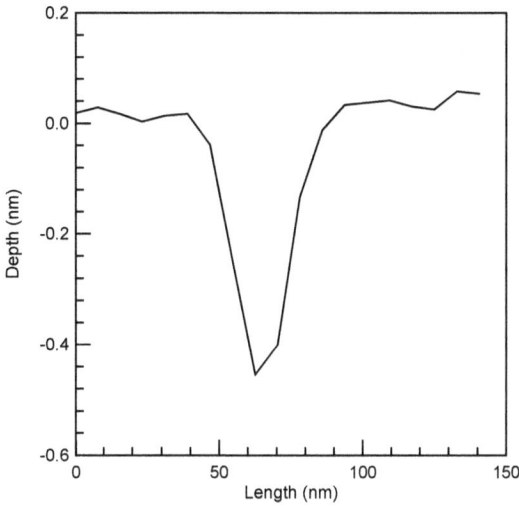

Fig. 5.25. The integrated profile on the surface of the $Ti_{43}Zr_{10}Cu_{36}Ni_9Sn_2$ metallic glass. Reproduced from [215] with permission of Elsevier.

Fig. 5.26. Isometric projection of a scratch on the surface of the $Ti_{43}Zr_{10}Cu_{36}Ni_9Sn_2$ metallic glass.

The $Zr_{62.5}Cu_{22.5}Fe_5Al_{10}$ is the most scratch resistant alloy (average $1/k_w = 0.093 \pm 0.016$ $\mu N/nm^2$); followed by the $Ti_{43}Zr_{10}Cu_{36}Ni_9Sn_2$ (average $1/k_w = 0.051 \pm 0.020$ $\mu N/nm^2$) and $Mg_{65}Cu_{25}Gd_{10}$ (average $1/k_w = 0.033 \pm 0.007$ $\mu N/ nm^2$) metallic glasses. The inverse wear coefficient ($1/k_w$) ($k_w = W_s/P$, where W_s is the wear rate and P is the load) can be used as indicator of wear resistance. At low load the wear resistance mostly correlates with HV_c of the surface oxides (crystalline state values) on the $Ti_{43}Zr_{10}Cu_{36}Ni_9Sn_2$, $Zr_{62.5}Cu_{22.5}Fe_5Al_{10}$ and $Mg_{65}Cu_{25}Gd_{10}$ metallic glasses: TiO_2 ($HV_c = 14$), ZrO_2 ($HV_c = 18$) and MgO ($HV_c = 6$). At high loads, when the oxide film is supposed to be partly destroyed the wear resistance is also connected with a lower hardness of the $Mg_{65}Cu_{25}Gd_{10}$ metallic glass compared to the $Ti_{43}Zr_{10}Cu_{36}Ni_9Sn_2$ and $Zr_{62.5}Cu_{22.5}Fe_5Al_{10}$ metallic glasses. As it was shown, the change in the wear rate as a function of load for the $Ti_{43}Zr_{10}Cu_{36}Ni_9Sn_2$,

$Zr_{62.5}Cu_{22.5}Fe_5Al_{10}$ and $Mg_{65}Cu_{25}Gd_{10}$ metallic glasses takes place at a load of about 1 μN when the oxide film break likely takes place (Fig. 5.27).

Fig. 5.27. The scratch depth as a function of load. Reproduced from [215] with permission of Elsevier.

The wear resistance was found to generally decrease with load in accordance with the idea of surface oxides as a protective medium versus wear (Fig. 5.28) which is destroyed at a higher load.

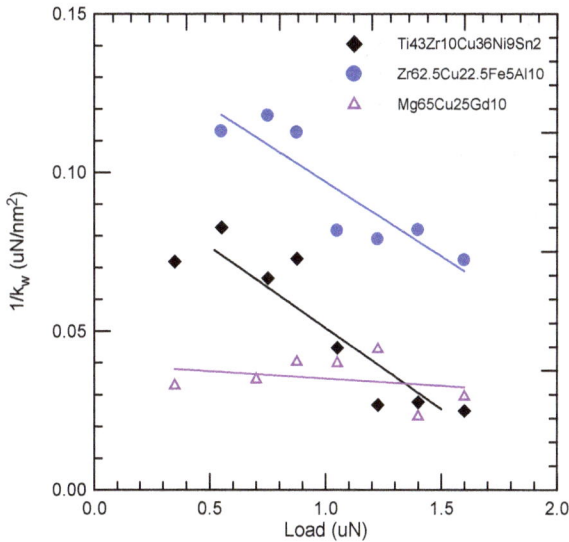

Fig. 5.28. The inverse wear coefficient as wear resistance ($R_w = 1/k_w$) as a function of load.

It is known that the surface oxide can have a beneficial effect on reducing wear during the sliding of metals and alloys, particularly by preventing the metal to metal contact [216]. An oxide thickness of a few nanometers is typical for metallic glassy alloys. It was found that the $Ni_{62}Nb_{38}$ metallic glass forms an amorphous niobium oxide of about 3-4 nm thick at ambient conditions [211,217]. Multicomponent metallic glasses, on the other hand, mostly form complex amorphous oxides on their surface. For example, in the case of the $Cu_{47}Zr_{45}Al_8$ one, it is a mixture of amorphous ZrO_2 and Al_2O_3 with nanoscale inclusions of the Cu_2O crystalline phase [218]. X-ray photoemission spectroscopy the $Ti_{45}Zr_{10}Pd_{10}Cu_{31}Sn_4$ alloy surface showed that the oxides contains all elements except for Pd [219] followed by the Pd and Cu enriched metallic glassy layer underneath. A similar Cu-enriched layer was also observed in the case of the $Cu_{47}Zr_{45}Al_8$ BMG [218]. This is also consistent with a very low Ni content in the surface niobium oxide of the Ni-Nb metallic glass [217].

5.1.12   Bauschinger-type effect

Under constant torque in the $Cu_{46}Zr_{42}Al_7Y_5$ metallic glassy sample [220], the angle of twist shows a time dependence that is interpreted as due to recoverable anelastic deformation. Torque reversal shows a Bauschinger-type effect in the anelastic deformation; in particular, the relative deformation rate increases by a factor of 4.5 on the first reversal. This result considered in terms of the operation of shear transformation zones suggested that stress reversal in the elastic regime is particularly effective in inducing local structural changes.

5.2 Plastic flow of bulk metallic glasses on heating

Structural relaxation on annealing below T_g significantly influences plasticity/toughness of BMGs, their yield strength, elastic moduli and hardness values. Annealing below and above the glass transition temperature leading to structural relaxation showed that the shear modulus was found to increase with the annealing time, while the strain at fracture decreased [221].

When tested in-situ on heating BMGs first show non-Newtonian flow near T_g and slightly above it, and then Newtonian flow in the supercooled liquid region [222,38]. However, the boundaries of all these temperature regions (inhomogeneous deformation, non-Newtonian flow and Newtonian flow) are strain rate dependent, i.e. transition from one region to another can take place with decrease in the strain rate near T_g. Anelastic atomic rearrangements at room temperature induced by applied stress below the yield stress are limited to immediate neighborhood of atoms [223]. At a temperature close to T_g atomic rearrangements are spatially more extended and creep deformation occurs as macroscopically anelastic behavior.

Some of the glassy alloys exhibit "superplasticity" (actually good fluidity) on heating to a supercooled liquid region [224]. Because of this bulk glassy alloys can be thermo-mechanically shaped or welded in the supercooled liquid region before they crystallize. They can be used for creation of micro-gears with diameters ranging from 0.1 to 1 mm by the Newtonian flow. Various micro- and nano-scale patterns [225,226] can be produced on bulk metallic glasses in the supercooled liquid region (Fig. 5.29) [227]. Bulk metallic glasses can be deformed in a manner similar to plastics by blow molding. This allows to net shape complex geometries including shapes, which can not be produced by other ways [228].

Fig. 5.29. (a) and (b) images of Si molds, (c)-(f) the patterns fabricated by using Si molds on the $Pd_{40}Ni_{40}Si_4P_{16}$ metallic glassy surface, (g) is porous alumina and (h) is the corresponding nano rods fabricated on the $Pd_{40}Ni_{40}Si_4P_{16}$ metallic glassy surface with the porous alumina (g). Reproduced from [227] with permission of Springer.

Shaping of BMGs can also be done by the electromechanical shaping technology at low applied stresses due to the high electrical resistivity of glassy alloys [229]. Bonding of glassy alloys can be achieved by Joule heating [230] as well as by laser [231,232], electron-beam [233,234] and friction welding [235]. When the temperature increases to T_g the BMG samples, in general, become more ductile and the yield stress decreases. Above T_g, metallic glasses transform into the supercooled liquid state, and as summarized in Fig. 5.30 (a) [236], only at a high strain rate they exhibit an inhomogeneous deformation.

(a)

(b)

(c)

Fig. 5.30. Transition from inhomogeneous (filled symbols) to homogeneous (open symbols) deformation mode at the elevated temperature for several BMGs in compression as a function of the strain rate and temperature: literature and current research data (a). Reproduced from [236] with permission of the American Institute of Physics. The lateral (b) and fracture surface (c) appearance for the $Au_{49}Cu_{26.9}Ag_{5.5}Pd_{2.3}Si_{16.3}$ BMG rod tested at T_g.

The $Au_{49}Cu_{26.9}Ag_{5.5}Pd_{2.3}Si_{16.3}$ BMG exhibits a mechanical behavior, which is distinct from the other metallic glasses [236]. Even at a low strain rate of $1.25 \cdot 10^{-4}$ s^{-1}, which is generally considered as quasi-static deformation, it shows brittle fracture at temperatures near T_g (Fig. 5.30 (b,c)) and only at the temperature of $1.08\ T_g$ the samples began to show plastic behavior (Fig. 5.30 (a)). A reason for such a behavior may be connected with the structure of the Au-based amorphous phase with large number of the icosahedral clusters [237] and possibly strong covalent bonds with Si. Thus, the internal structure of Au-based BMG can make a strong influence on the temperature of brittle–ductile transition.

5.3 Deformation of bulk metallic glasses at cryogenic temperature

The deformation behavior of the Zr-based BMG alloys was also tested at the boiling liquid nitrogen temperature [238,239,240]. The sample tested in liquid nitrogen exhibits higher compression strength compared to the sample tested at room temperature. Also no clear serrated flow is observed [240]. The typical stress-strain diagram of the $Zr_{64.13}Ni_{10.12}Cu_{15.75}Al_{10}$ glassy alloy tested at boiling liquid nitrogen (LN$_2$) temperature is shown in Fig. 5.31. The mechanical behavior and the kinetics of shear deformation in bulk metallic glasses were also investigated at room and liquid nitrogen temperature using the acoustic emission technique. It was demonstrated that the intensive acoustic emission reflecting the activity of strongly localized shear bands at room temperature vanishes at the transition from serrated to non-serrated plastic flow at low temperature. The disappearance of the acoustic emission signals clearly indicates that the shear band propagation velocity significantly decreases at low temperature, and sliding along the dominant shear band takes place at the machine-driven rate [241].

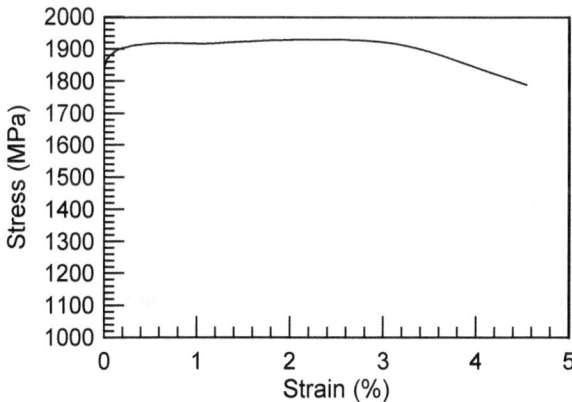

Fig. 5.31. Stress-plastic strain diagram of the $Zr_{64.13}Ni_{10.12}Cu_{15.75}Al_{10}$ glassy alloy tested at boiling LN$_2$ temperature.

The samples tested in LN_2 still exhibited formation of the localized shear deformation bands (Fig. 5.32). The fracture surfaces of the samples tested in LN_2 also show the mirror looking cleavage-like areas mixed up with typical vein patterns in nearly equal proportions like it is shown in Fig. 5.21 for the sample tested at room temperature.

Fig. 5.32. SEM image in secondary electrons of the lateral surface of the $Zr_{64.13}Ni_{10.12}Cu_{15.75}Al_{10}$ alloy sample after mechanical testing at boiling LN_2 temperature.

5.4   Non-uniform materials with glassy phase

5.4.1   Porous glasses

Porous glassy materials are very promising for structural applications owing to their low density and good mechanical properties. Open-pore type Zr-based BMGs [242] were produced by the salt-replication technique while closed-pore type porous Pd-based BMGs [243,244] were produced by holding the melt under pressurized hydrogen which is released when the pressure is reduced. Porous $Zr_{55}Cu_{30}Al_{10}Ni_5$ alloy samples were also produced by spark plasma sintering (SPS) [245]. The porosity was controlled by the sintering temperature and loading pressure in the SPS process. The thermal stability of the sintered porous glassy alloy specimens was similar to that of the original powder. The sintered porous glassy alloy specimens exhibited a larger plastic ductility, a lower Young's modulus and a lower fracture strength than the as-cast alloy specimen.

Closed-pore type glassy alloys were obtained for the $Pd_{42.5}Cu_{30}Ni_{7.5}P_{20}$ glassy alloy which has an extremely high glass-forming ability among metallic alloys. Pd absorbs large quantities of hydrogen. Once the hydrogen pressure is reduced the dissolved hydrogen starts to precipitate from the melt causing its bubbling. The melt was subjected to different hydrogenation treatments following water or oil cooling. In order to produce relatively small finely dispersed pores the foaming pressure before cooling was just 10 % less than the hydrogenation pressure (Fig. 5.33 (a)) while the sample remained glassy (Fig. 5.33 (b)). When the pressure is reduced by more than about 10 % the samples with much larger pores and higher porosity can be produced. As the initial hydrogen pressure at which hydrogen is dissolved increases porosity decreases, pore number density increases and pore size becomes smaller.

Fig. 5.33. SEM (a) image of the porous structure obtained in the $Pd_{42.5}Cu_{30}Ni_{7.5}P_{20}$ BMG with initial pressure of 15 MPa and high-resolution TEM (b) image of the glassy phase with the SAED pattern inserted.

The mechanical properties are summarized in Table 5.7. As one can see the yield strength at 3.7 % porosity is 0.93 of the monolithic glass while its plasticity reached 18%.

Table 5.7. Porosity, 0.2 % proof stress ($\sigma_{0.2}$), the Young's modulus (E) and plastic strain (e) of the porous $Pd_{42.5}Cu_{30}Ni_{7.5}P_{20}$ glassy rods tested in compression. The data from Ref. [246].

Porosity (%)	$\sigma_{0.2}$ (MPa)	E (GPa)	e (%)
0	1630	102	0
1.7	1580	94	9.5
2.2	1550	98	4.5
3.0	1520	98	7.5
3.7	1520	88	18

Pore size and distribution can be controlled by the preparation condition depending on temperature and pressure of the melt. Another method to control the pore size and distribution is subsequent annealing of the porous alloy within the supercooled liquid region which causes further expansion of the pores leading to the materials with porosity over 10 %. The expansion of pores takes place owing to significantly reduced viscosity of a liquid compared to a glass.

5.4.2   Glassy-crystalline dual phase alloys

The formation of heterogeneous microstructures combining a glassy matrix with crystalline particles enhances plasticity of BMGs [247,248]. The mechanisms of ductilization of bulk metallic glassy alloys by crystalline and quasicrystalline inclusions are associated with blocking and branching of the propagating localized shear bands by the structural inhomogeneities for example, crystalline particles [248]. This enables multiplication, branching and termination of the shear bands similar to the composites where cracks are either blocked or blunted by the secondary phases.

Various bulk metallic glass composites with inclusions of a crystalline phase(s) were produced starting from the glassy sample [249,250,251,252]. Ductile BMGs samples containing dendritic crystals were produced in the La-Al-(Cu,Ni) [253] (hcp α-La phase as the dendritic phase) and Zr-Ti-Nb-Cu-Be (in which the dendritic phase is a body-centered cubic (Zr,Ti,Nb) solid solution) with large Be content [254,255] as well as $Ti_{46}Zr_{20}V_{12}Cu_5Be_{17}$ [256] alloys. These materials were created using the strategy of microstructural toughening and ductility enhancement in metallic glasses. Although the Zr-Ti-Nb-Cu-Be β-Zr/BMG dual phase alloys show necking owing to stress softening of the glassy phase, it was reported that the Cu-Zr-Al-Co [257,258] and Cu-Zr-Al system composites [259] show strain hardening owing to transformation-induced plasticity upon martensitic transformation. 1 mm diameter rods of Ti-Cu-Ni glassy/crystal alloys also exhibited tensile plasticity [260]. Thus, BMG alloys and the composites containing glassy phase are promising structural and functional materials of the present century. Some BMGs undergo crystallization during deformation, which also leads to their ductilization [261]. Deformation-induced nanocrystallization [262,263] is also helpful in ductilization of BMGs.

The $Ni_{40}Cu_{10}Ti_{33}Zr_{17}$ alloy [264] produced by copper mold casting shows a superior combination of strength and ductility not only due to a composite effect resulting from a multiphase structure, but also due to a strain-induced martensitic transformation. The microstructure consisted of the austenitic cP2 cubic (Ni,Cu)(Ti,Zr) phase embedded in the glassy matrix in the as-cast alloy, while mP4 martensitic (Ni,Cu)(Ti,Zr) phase is formed during plastic deformation. The crystals further act as strong barriers for shear band propagation, promoting increase in stress and subsequent shear band formation and branching, thus extending the ductility and preventing a premature brittle fracture. The sample showed the ultimate strength of 2000 MPa, the yield stress of 650 MPa and the compressive ductility of 15 % [264]. Unlike common observations of serrated plastic flow associated with the large scale shear bands in monolithic metallic glasses, propagation of the shear bands in the studied alloy does not exhibit the stress drops, although the shear bands were readily seen on the lateral surface of the samples tested to fracture in addition to multiple slip bands in cP2 crystalline grains.

Amplitude dependent internal friction measurement was preformed for the $Ni_{40}Cu_{10}Ti_{33}Zr_{17}$ alloy samples. The internal friction values at initial loading were always higher than the subsequent ones. These facts are in line with the irreversible martensitic cP2→mP4 transformation mechanism, which is also responsible for the formation of a hysteresis loop. Owing to relatively high sensitivity of the DMA machine to tiny structural changes the occurrence of martensitic transformation is detected even in the linear quasi-elastic deformation mode [265].

Various Ni-Cu-Ti-Zr alloys were further studied [266] and more complex systems alloys were produced [267] (Table 5.8). The formation of the mixed glassy-crystal structure is illustrated in Fig. 5.34. The samples demonstrate effects of Transformation-Induced Plasticity (TRIP) and Superelasticity. It was found that the addition of Y does not lead to increase in the glass forming ability of Ni-Ti-Cu-Zr system alloys while a large amount of Y (more than 0.5 at.%) induces $NiTi_2$ phase precipitates. The additions of Co and Nb increase yield strength of investigated alloys by 350 and 600 MPa respectively. At the same time the alloys exhibit large plasticity. The additions of Al and B lead to double increase of the yield strength, but fully suppress the martensitic transformation. These alloys have lower plasticity compared to the alloys with Nb.

Table 5.8. The values of maximum stress σ_{max}, stress of the beginning of plastic deformation σ_p, the stress initiating the martensitic transformation σ_m and the relative strain ε. Phase composition cP2 + glassy phase, $Ni_{50}Zr_{17}Ti_{33}$ also contains the $Ni_{10}Zr_7$ one.

Alloy	σ_{max} (MPa)	σ_p (MPa)	σ_m (MPa)	ε (%)	Ref
$Ni_{25}Cu_{25}Ti_{33}Zr_{17}$	1605	1455	-	1.1	266
$Ni_{35}Cu_{15}Ti_{33}Zr_{17}$	1790	1475	635	9.4	266
$Ni_{40}Cu_{10}Ti_{35}Zr_{15}$	1870	1350	270	18.2	266
$Ni_{40}Cu_{10}Ti_{40}Zr_{10}$	1810	1325	420	16.4	266
$Ni_{45}Cu_5Ti_{42}Zr_8$	2700	1600	165	22	266
$Ni_{50}Zr_{17}Ti_{33}$	2450	2200	900	2	266
$Ti_{42.0}Ni_{39.0}Cu_9Zr_{10}$	2150	1970	320	18.3	267
$Ti_{40.0}Ni_{39.5}Cu_8Zr_{10}Co_2Y_{0.5}$	2620	1955	780	25	267
$Ti_{39.5}Ni_{39.0}Cu_8Zr_8Co_2Y_{0.5}Nb_3$	2520	1730	1210	25	267

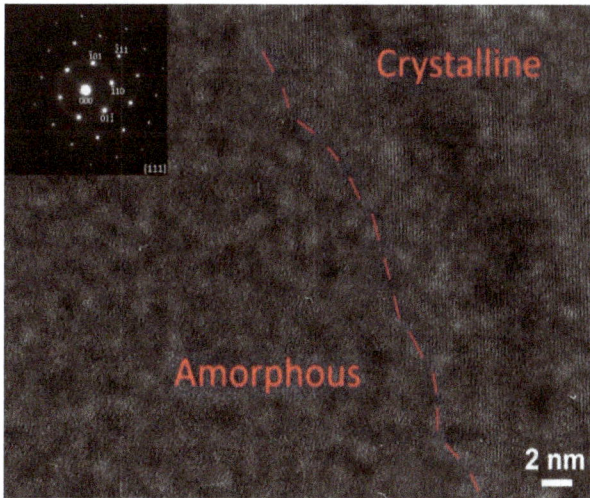

Fig. 5.34. HRTEM of the as-cast $Ti_{40}Ni_{39.5}Cu_8Zr_{10}Co_2Y_{0.5}$ with a typical electron diffraction pattern containing both sharp rings from the crystalline particle and a halo pattern from the surrounding glassy phase (inset). Reproduced from [267] with permission of Elsevier.

The $Ti_{42.3}Zr_{7.7}Cu_{41.7}Ni_{7.3}Co_1$ sample with 1000 ppm of B containing a crystalline phase embedded in the glassy matrix (Fig. 5.35) [268] also exhibited formation of multiple shear offsets on the lateral surface of the sample (Fig. 5.35, insert). Strain hardening and no detectable serrated flow were observed at the initial stage of deformation while weak stress drops of 1-3.5 MPa are seen after 4.5 % of plastic deformation. Fracture occurs when strain hardening ($d\sigma/d\varepsilon$) (stress σ and strain ε) mechanisms in the crystalline phase are exhausted. Deformation of such a dual-phase material likely takes place by a series of smaller sample displacements in the shear bands though the shear banding mechanism is still active in this case (Fig. 5.35 insert).

Fig. 5.35. Stress and stress drop values as a function of plastic strain for the cylindrical $Ti_{42.3}Zr_{7.7}Cu_{41.7}Ni_{7.3}Co_1$ crystal-glassy sample with 1000 ppm of B of 2:1 aspect ratio upon the compression test. The insert represents a lateral surface of the sample tested to fracture. The diamond symbols denote the stress drop values exceeding 1 MPa. Reproduced from [123] with permission of Elsevier.

A strong influence of particles distributed in the amorphous matrix on shear band propagation was confirmed using the numerical methods [269,270,271,272] and experimental studies [273,274]. For example, Ta-rich particles with bimodal distribution act as discrete obstacles, separating and restricting the highly localized shear-banding, avoiding catastrophic shear-through of the whole sample and enhancing plasticity of the monolithic Zr-Cu-Ni-Al BMG [273].

The Zr-based bulk metallic glassy samples containing the rare earth elements with a maximum diameter of 5 mm were successfully prepared by copper mold casting method at low vacuum conditions [275]. The structure of the alloys consisted of the amorphous matrix with a small fraction of RE oxides. The maximum compression strength was achieved in the $Zr_{62}Cu_{22.5}Fe_5Al_{10}Dy_{0.5}$ alloy. At the same time, the optimal combination of strength and plasticity (fracture strength of 1890 MPa and plastic deformation of 3.3 %) was observed in the $Zr_{62}Cu_{22}Fe_5Al_{10}Gd_1$ alloy.

(a)

(d)

(b)

(e)

(c)

(f)

(c)

(f)

Fig. 5.36. Finite element modeling of plastic deformation of the non-deformable sample matrix with a shear band (a,b,c) and of the non-deformable sample matrix with both a shear band and a non-deformable particle (d,e,f): strain distribution at 1.5 % strain (a,d), stress distribution at 1.5 % strain (b,e) and stress-strain curves (c,f). Reproduced from [275] with permission of Elsevier.

The finite element model FEM analysis [276,277] indicated that the oxide particles with equiaxed shape can block propagation of a shear band and promote generation of new shear bands facilitating plasticity. The results of finite element deformation modeling of two microstructure types: (1) with a shear band and (2) with both a shear band and a non-deformable oxide particle indicated that, strain is localized in shear band in both cases, while the local stress maxima are, as expected, on the surface of the sample in the region close to shear band in the first model and on the surface of the particle in the second model (Fig. 5.36). The deformation behavior of the first model sample was similar to experimental behavior of the BMG: after elastic deformation the plastic flow takes place without significant strain hardening, which usually leads to catastrophic shear band propagation and failure. The presence of a non-deformable particle of a favorable (round) shape hampers the shear band propagation. The fracture strength of RE oxides under soft deformation condition is more than 2500 MPa [278,279] which is enough for this purpose. At the same time in the other regions of the sample the level of stress will be increased to generate new shear bands which will improve overall plasticity.

The sample with 1 at. % of Gd containing in the RE metal oxide particles showed the formation of multiple shear bands in contradistinction to the alloy without oxide particles. Bulk glassy-crystal composites with enhanced ductility have been produced by proper alloying in Zr- [280,281] and Cu-based alloys [282].One should also mention that glassy-polymer composite materials developed recently combine attractive properties of both [283].

5.4.3   Mechanical properties of nanocomposite alloys

Mechanical strength and ductility of glassy alloys can be improved by the precipitates of nanocrystalline or nanoquasicrystalline phase. The blockage of the crack-tip area by nanoparticles stabilizing plastic deformation was observed in the $Cu_{50}Zr_{50}$ and the $Zr_{65}Al_{7.5}Ni_{10}Pd_{17.5}$ glassy alloys [284]. The $Zr_{65}Al_{7.5}Cu_{7.5}Ni_{10}Pd_{10}$ alloy having nanoscale icosahedral phase particles embedded in the glassy matrix showed a better combination of the mechanical properties compared to the as-cast glassy sample [285] without precipitates. Because the single-phase icosahedral phase alloys are extremely brittle it can be considered that the inclusions of a precipitating phase in this alloy cause deviation and blockage of the operating shear bands which improve the plasticity of the alloy. Thus, good mechanical properties are attributed to the existence of the residual intergranular glassy phase while the icosahedral particles can act as a resisting medium against the shear deformation. Nanocrystalline precipitates increase the room-temperature mechanical strength of the Zr-Al-Cu-Pd [286], Zr-Al-Cu-Pd-Fe and (Zr/Ti)-Cu-Al-Ni [287] bulk glassy alloys. The compressive room-temperature stress-strain curves of the

$Zr_{65}Al_{7.5}Ni_{10}Cu_{17.5}$ and $Zr_{65}Al_{7.5}Ni_{10}Pd_{17.5}$ bulk metallic glassy alloys revealed significantly different plasticity of these alloys, though both of the alloys possess a glassy structure [288].

An alloy, which crystallizes by the nucleation and growth mechanism, is supposed to be less prone to nanocrystallization during deformation while the alloy having pre-existing nuclei is predisposed to show such a behavior [289]. Precipitation of the nanoscale α-Al particles at room temperature was observed in the Al-TM-RE glass-formers (like $Al_{90}Fe_5Gd_5$ or $Al_{90}Fe_5Ce_5$) within the shear bands on deformation [290,291], while a very limited fraction or complete absence of nanocrystals (depending on RE metal) was observed after significant plastic deformation in the $Al_{85}RE_8Ni_5Co_2$ amorphous and glassy samples [292] (~30 % reduction in thickness) which undergo a clear nucleation and growth behavior.

The nanocrystallization within the shear bands was also observed in the $Cu_{47.5}Zr_{47.5}Al_5$ glassy alloy which was reported to be strain-hardenable [262] and a Ti-based alloy [293]. The effect of strain rate on the flow stress [294] and ductility was studied [295,296] and the results showed that the higher strain rate used facilitates plasticity. The $Cu_{60}Zr_{20}Ti_{20}$ bulk metallic glass was rolled at room temperature and at cryogenic temperature [297]. Nanocrystallization and the free volume decrease were found at room temperature. Cu-enriched glassy regions as precursors to nanocrystallization were observed at the cryogenic temperature. A strong composition dependence of plasticity of $(ZrCu)_{100-x}Al_x$ BMGs was observed. The compositions with primary nanoscale phase exhibit a large plastic strain facilitated by deformation-induced nanocrystallization in the shear bands [298]. Yavari et al. [299] using a synchrotron X-ray beam of 2 μm in diameter showed formation of deformation-induced nanocrystals on the compressed side of the $Pd_{40}Cu_{30}Ni_{10}P_{20}$ alloy tested in bending. As crystallization is accompanied by a decrease in the molar volume the applied pressure increases the driving force for crystallization.

Al-RE-TM amorphous alloys posses a high tensile strength exceeding 1200 MPa and good bend ductility [300,301]. The homogeneous dispersion of the nanoscale fcc α-Al particles in the amorphous matrix causes a drastic increase in the tensile fracture strength to 1560 MPa [302]. It is considered that the nanoscale α-Al particles can act as an effective barrier against shear deformation of the amorphous matrix [303]. At the same time, it was also suggested [304] that the hardening could be attributed mainly to solute enrichment of the residual glassy matrix due to lowering of the Al content. It is suggested that not only α-Al particles but also the amorphous matrix has some role in the hardening and embrittlement of the alloy. The $(Al_{0.84}Y_{0.09}Ni_{0.05}Co_{0.02})_{95}Sc_5$ amorphous alloy has an ultra-high tensile fracture strength slightly exceeding 1500 MPa which surpasses those for all other Al-based fully crystalline and fully amorphous alloys reported up to date [305] and can be likely further strengthened by nanoscale crystallization.

However, in some cases, formation of the primary α-Al particles was found to deteriorate the mechanical properties: decrease the tensile strength and hardness. Partial substitution of Ni by Cu in the $Al_{85}Y_8Ni_5Co_2$ metallic glass caused formation of the nanoscale α-Al particles and drastically decreased the tensile strength and hardness values of the alloy [306]. Cu has much lower absolute value of heat of mixing with Al, Y and Co than Ni may be responsible for such a decrease. The α-Al inter-particular distances in the $Al_{85}Y_8Ni_3Co_2Cu_2$ metallic glass significantly exceed the particle size. Thus, here, α-Al particles cannot act as an effective barriers against the shear deformation in the amorphous matrix.

References

[1] W.H. Wang, The elastic properties, elastic models and elastic perspectives of metallic glasses, Progress in Materials Science, 57 (2012) 487. https://doi.org/10.1016/j.pmatsci.2011.07.001

[2] D.V. Louzguine-Luzgin, M. Fukuhara and A. Inoue, Specific volume and elastic properties of glassy, icosahedral quasicrystalline and crystalline phases in Zr-Ni-Cu-Al-Pd alloy, Acta Materialia, 55 (2007) 1009. https://doi.org/10.1016/j.actamat.2006.09.013

[3] H. Kato, J. Saida and A. Inoue, Influence of hydrostatic pressure during casting on as cast structure and mechanical properties in $Zr_{65}Al_{7.5}Ni_{10}Cu_{17.5-x}Pd_x$ (x = 0, 17.5) alloys, Scripta Mater. 51 (2004) 1063. https://doi.org/10.1016/j.scriptamat.2004.08.004

[4] A.S. Argon, Mechanisms of inelastic deformation in metallic lasses, Journal of Physics and Chemistry of Solids. 43 (1982) 945. https://doi.org/10.1016/0022-3697(82)90111-1

[5] N. Togashi, M. Ishida, N. Nishiyama and A. Inoue, Wear resistance of metallic glass bearings, Rev. Adv. Mater. Sci. 18 (2008) 93.

[6] A.R. Yavari, J.J. Lewandowski and J. Eckert, Mechanical properties of bulk metallic, MRS Bull. 32 (2007) 635. https://doi.org/10.1557/mrs2007.125

[7] A. Inoue, S. Sobu, D.V. Louzguine, H. Kimura and K. Sasamori, Ultrahigh strength Al-based amorphous alloys containing, J. Mater. Res. 19 (2004) 1539. https://doi.org/10.1557/JMR.2004.0206

[8]   M. Wakeda, Y. Shibutani, S. Ogata and J.Y. Park, Relationship between local geometrical factors and mechanical properties for Cu–Zr amorphous alloys, Intermetallics 15 (2007) 139. https://doi.org/10.1016/j.intermet.2006.04.002

[9]   A. Inoue, Stabilization of metallic supercooled liquid and bulk amorphous alloys, Acta Mater. 48 (2000) 279. https://doi.org/10.1016/S1359-6454(99)00300-6

[10]   D.V. Louzguine and A. Inoue, Proceedings of the ninth international symposium on metastable, mechanically alloyed and nanocrystalline materials, Mater. Sci. Forum/J. Metastable Nanocryst. Mater. 15–16 (2003) 31. https://doi.org/10.1016/j.stam.2003.09.003

[11]   D.V. Louzguine, H. Kato and A. Inoue, Investigation of mechanical properties and devitrification of Cu-based bulk glass formers alloyed with noble metals, Science and Technology of Advanced Materials. 4 (2003) 327. https://doi.org/10.1080/09500830310000066126

[12]   D.V Louzguine and A. Inoue, Structural and thermal investigations of a high-strength Cu-Zr-Ti-Co bulk metallic glass, Philosophical Magazine Letters. 83 (2003) 191. https://doi.org/10.1063/1.119580

[13]   A. Inoue, T. Zhang and A. Takeuchi, Bulk amorphous alloys with high mechanical strength and good soft magnetic properties in Fe–TM–B (TM= IV–VIII group transition metal) system, Appl. Phys. Lett. 71 (1997) 464. https://doi.org/10.1557/jmr.2011.187

[14]   K. Amiya and A. Inoue, Fe-(Cr, Mo)-(C, B)-Tm bulk metallic glasses with high, strength and high glass-forming ability, Reviews on Advanced Materials Science 18 (2008) 27.

[15]   J.F. Wang, R. Li, H. Nengbin and T. Zhang, Co-based ternary bulk metallic glasses with ultrahigh strength and plasticity, Journal of Materials Research 26 (2011) 2072. https://doi.org/10.1016/0036-9748(75)90177-5

[16]   L.A. Davis, Hardness/strength ratio of metallic glasses, Scripta Metallurgica 9 (1975) 431. https://doi.org/10.1016/0036-9748(75)90177-5

[17]   Y. Yokoyama and A. Inoue, Compositional dependence of thermal and mechanical properties of quaternary Zr-Cu-Ni-Al bulk glassy alloys, Materials Transactions 48 (2007) 1282. https://doi.org/10.2320/matertrans.MF200622

[18]   E. Ma and J. Ding, Tailoring structural inhomogeneities in metallic glasses to enable tensile ductility at room temperature, Mater. Today 19 (2016) 568. https://doi.org/10.1016/j.mattod.2016.04.001

[19]   A. Inoue, W. Zhang, T. Zhang and K. Kurosaka, Formation and mechanical properties of Cu–Hf–Ti bulk glassy alloys, J. Mater. Res. 16 (2001) 2836. https://doi.org/10.1557/JMR.2001.0391

[20]   A. Inoue, W. Zhang, T. Zhang and K. Kurosaka, High-strength Cu-based bulk glassy alloys in Cu–Zr–Ti and Cu–Hf–Ti ternary systems, Acta Mater. 49 (2001) 2645. https://doi.org/10.1016/S1359-6454(01)00181-1

[21]   K.B. Kim, J. Das, M.H. Lee, S. Yi, E. Fleury, Z.F. Zhang, W.H. Wang and J. Eckert, Propagation of shear bands in a $Cu_{47.5}Zr_{47.5}Al_5$ bulk metallic glass, J. Mater. Res. 23 (2008) 6. https://doi.org/10.1557/JMR.2008.0025

[22]   Q. Zhang, W. Zhang, G. Xie and A. Inoue, Glass-forming ability and mechanical properties of the ternary Cu-Zr-Al and quaternary Cu-Zr-Al-Ag bulk metallic glasses, Materials Transactions 48 (2007) 1626. https://doi.org/10.2320/matertrans.MJ200704

[23]   J. Chen, Y. Zhang, J.P. He, K.F. Yao, B.C. Wei and G.L. Chen, Metallographic analysis of Cu–Zr–Al bulk amorphous alloys with yttrium addition, Scripta Materialia 54 (2006) 1351. https://doi.org/10.1016/j.scriptamat.2005.12.002

[24]   J. Bhatt, W. Jiang, X. Junhai, W. Qing, C. Dong and B.S. Murty, Optimization of bulk metallic glass forming compositions in Zr-Cu-Al system by thermodynamic modeling, Intermetallics 15 (2007) 716. https://doi.org/10.1016/j.intermet.2006.10.018

[25]   H. Men, J. Fu, C. Ma, S. Pang and T. Zhang, Bulk glass formation in ternary Cu-Zr-Ti system, Journal of University of Science and Technology Beijing, 14 (2007) 19. https://doi.org/10.1016/S1005-8850(07)60100-5

[26]   A. Inoue, T. Zhang, T. Masumoto, Al-La-Ni amorphous alloys with a wide supercooled liquid region, Materials Transactions JIM, 30 (1989) 965.

[27]   A. Inoue, A. Kato, T. Zhang, S. Kim and T. Masumoto, Mg-Cu-Y amorphous alloys with high mechanical strengths produced by metallic mold casting method, Materials Transactions JIM, 32 (1991) 609.

[28]   A. Inoue, T. Zhang and T. Masumoto, Zr-Al-Ni amorphous alloys with high class transition temperature and significant supercooled liquid region, Materials Transactions JIM, 31 (1990) 177.

[29]   Q. Jing, Y. Zhang, D. Wang and Y. Li, A study of the glass forming ability in Zr-Ni-Al alloys, Materials Science and Engineering A, 441 (2006) 106. https://doi.org/10.1016/j.msea.2006.08.109

[30] A. Inoue, T. Zhang and T. Masumoto, Zr-Al-Ni amorphous alloys with high glass transition temperature and significant supercooled liquid region, Materials Transactions JIM. 31 (1990) 177.

[31] C. Qin, W. Zhang, H. Kimura and A. Inoue, Excellent mechanical properties of Cu-Hf-Ti-Ta bulk glassy alloys containing in-situ dendrite Ta-based BCC phase, Materials Transactions, 45 (2004) 2936. https://doi.org/10.2320/matertrans.45.2936

[32] T. Zhang, K. Kurosaka and A. Inoue, Thermal and mechanical properties of Cu-based Cu-Zr- Ti-Y bulk glassy alloys, Materials Transactions, 42 (2001) 2042. https://doi.org/10.2320/matertrans.42.2042

[33] L. Liu, A Inoue and T. Zhang, Formation of bulk Pd-Cu-Si-P glass with good mechanical properties, Materials Transactions, 46 (2005) 376. https://doi.org/10.2320/matertrans.46.376

[34] T. Wada, T. Zhang and A. Inoue, Formation and high mechanical strength of bulk glassy alloys in Zr–Al–Co–Cu system, Materials Transactions, 44 (2003) 1839. https://doi.org/10.2320/matertrans.44.1839

[35] Q.S. Zhang, W. Zhang, D.V. Louzguine-Luzgin and A. Inoue, High glass-forming ability and unusual deformation behavior of new Zr-Cu-Fe-Al bulk metallic glasses, Materials Science Forum, 654-656 (2010) 1042. https://doi.org/10.4028/www.scientific.net/MSF.654-656.1042

[36] Q.S. Zhang, W. Zhang, G.Q. Xie, D.V. Louzguine-Luzgin and A. Inoue, Stable flowing of localized shear bands in soft bulk metallic glasses, Acta Materialia 58 (2010) 904. https://doi.org/10.1016/j.actamat.2009.10.005

[37] A. L. Greer and E. Ma, Bulk metallic glasses: at the cutting edge of metals research, MRS Bull. 32 (2007) 611. https://doi.org/10.1557/mrs2007.121

[38] C. A. Schuh, T. C. Hufnagel and U. Ramamurty, Mechanical behavior of amorphous alloys, Acta Mater. 55 (2007) 4067. https://doi.org/10.1016/j.actamat.2007.01.052

[39] A. R. Yavari, J. J. Lewandowski and J. Eckert, Mechanical properties of bulk metallic glasses, MRS Bul. 32 (2007) 635. https://doi.org/10.1557/mrs2007.125

[40] J. Pan, K.C. Chan, Q. Chen and L. Liu, Enhanced plasticity by introducing icosahedral medium-range order in ZrCuNiAl metallic glass, Intermetallics 24 (2012) 79. https://doi.org/10.1016/j.intermet.2012.01.006

[41] Q.S. Zhang, W Zhang, G.Q Xie, D.V Louzguine-Luzgin and A. Inoue, Stable flowing of localized shear bands in soft bulk metallic glasses, Acta Materialia, 58 (2010) 904. https://doi.org/10.1016/j.actamat.2009.10.005

[42] H. Kato, J. Saida and A. Inoue, Influence of hydrostatic pressure during casting on as cast structure and mechanical properties in $Zr_{65}Al_{7.5}Ni_{10}Cu_{17.5-x}Pd_x$ (x= 0, 17.5) alloys, Scripta Mater. 51 (2004) 1063. https://doi.org/10.1016/j.scriptamat.2004.08.004

[43] J. Schroers and W.L. Johnson, Ductile bulk metallic glass, Phys. Rev. Lett. 93 (2004) 255506. https://doi.org/10.1103/PhysRevLett.93.255506

[44] Y. Yokoyama, K. Fujita, A.R. Yavari and A. Inoue, Malleable hypoeutectic Zr–Ni–Cu–Al bulk glassy alloys with tensile plastic elongation at room temperature, Philos. Mag. Lett. 89 (2009) 322. https://doi.org/10.1080/09500830902873575

[45] K.K. Song, S. Y. Pauly, S. Zhang, P. Scudino, K.B. Gargarella, U. Surreddi, K. Kühn and J. Eckert, Significant tensile ductility induced by cold rolling in $Cu_{47.5}Zr_{47.5}Al_5$ bulk metallic glass, Intermetallics 19 (2011) 1394. https://doi.org/10.1016/j.intermet.2011.05.001

[46] N. Chen, D.V. Louzguine-Luzgin, G.Q. Xie, T. Wada and A. Inoue, Influence of minor Si addition on glass forming ability and mechanical properties of $Pd_{40}Ni_{40}P_{20}$ alloy, Acta Materialia 57 (2009) 2775. https://doi.org/10.1016/j.actamat.2009.02.028

[47] N. Chen, D. Pan, D.V. Louzguine-Luzgin, G.Q. Xie, M.W. Chen and A. Inoue, Improved thermal stability and ductility of flux-treated $Pd_{40}Ni_{40}Si_4P_{16}$ BMG, Scripta Materialia 62 (2010) 17. https://doi.org/10.1016/j.scriptamat.2009.09.013

[48] D.V. Louzguine-Luzgin, G. Xie, Q. Zhang and A. Inoue, Effect of Fe on the glass-forming ability, structure and devitrification behavior of Zr-Cu-Al bulk glass-forming alloys, Philosophical Magazine 90 (2010) 1955. https://doi.org/10.1080/14786430903571495

[49] J.J. Lewandowski, W.H. Wang and A.L. Greer, Intrinsic plasticity or brittleness of metallic glasses, Philos. Mag. Lett. 85(2005) 77. https://doi.org/10.1080/09500830500080474

[50] K. Hajlaoui, A.R. Yavari, B. Doisneau, A. LeMoulec, W.J. Botta, G. Vaughan, A. Greer L., A. Inoue, W. Zhang and A. Kvick, Shear delocalization and crack blunting of a metallic glass containing nanoparticles: In situ deformation in TEM analysis, Scripta Mater. 54 (2006) 1829. https://doi.org/10.1016/j.scriptamat.2006.02.030

[51] K.B. Kim, J. Das, F. Baier, M.B. Tang, W.H. Wang and J. Eckert, Heterogeneity of a $Cu_{47.5}Zr_{47.5}Al_5$ bulk metallic glass, Appl. Phys. Lett. 88 (2006) 051911. https://doi.org/10.1063/1.2171472

[52] G.N. Greaves, A.L. Greer, R.S. Lakes and T. Rouxel, Poisson's ratio and modern materials Nature mater., 10 (2011) 823. https://doi.org/10.1038/nmat3134

[53] S.V. Ketov, N. Chen, A. Caron, A. Inoue and D.V. Louzguine-Luzgin, Structural features and high quasi-static strain rate sensitivity of $Au_{49}Cu_{26.9}Ag_{5.5}Pd_{2.3}Si_{16.3}$ bulk metallic glass, Applied Physics Letters, 101 (2012) 241905. https://doi.org/10.1063/1.4770072

[54] S.V. Madge, D.V. Louzguine-Luzgin, J.J. Lewandowski and A.L. Greer, Toughness, extrinsic effects and Poisson's ratio of bulk metallic glasses, Acta Materialia 60 (2012) 4800. https://doi.org/10.1016/j.actamat.2012.05.025

[55] W.F. Gale and T.C. Totemeier, editors. Smithells Metals Reference Book 8-th Edition, Elsevier Butterworth-Heinemann Ltd., Oxford UK (2004) 11.

[56] Y.N. Dastur and W.C. Leslie, Mechanism of work hardening in Hadfield manganese steel, Metallurgical Transactions A, 12 (1981)749. https://doi.org/10.1007/BF02648339

[57] W.J. Wright, R.B Schwarz and W.D. Nix, Localized heating during serrated plastic flow in bulk metallic glasses, Materials Science and Engineering A, 219 (2001) 319. https://doi.org/10.1016/S0921-5093(01)01066-8

[58] F.H. Dalla Torre, A. Dubach, M.E. Siegrist and J.F. Löffler, Negative strain rate sensitivity in bulk metallic glass and its similarities with dynamic strain aging effect during deformation, Appl. Phy. Lett. 89 (2006) 091918. https://doi.org/10.1063/1.2234309

[59] A. Caron, R. Wunderlich, D.V. Louzguine-Luzgin, G. Xie, A. Inoue and H.-J. Fecht, Influence of minor aluminum concentration changes in zirconium-based bulk metallic glasses on the elastic, anelastic, and plastic properties, Acta Materialia 58 (2010) 2004. https://doi.org/10.1016/j.actamat.2009.11.043

[60] H.B. Yu, W.H. Wang, J.L. Zhang, C.H. Shek and H.Y. Bai, Statistic analysis of the mechanical behavior of bulk metallic glasses, Adv. Eng. Mater. 11 (2009) 370. https://doi.org/10.1002/adem.200800380

[61] Ch.E. Lekka, Cu-Zr and Cu-Zr-Al clusters: Bonding characteristics and mechanical properties, Journal of Alloys and Compounds, 504 (2010) 190. https://doi.org/10.1016/j.jallcom.2010.02.067

[62]   R.D. Conner, Y. Li, W.D. Nix and W.L. Johnson, Shear band spacing under bending of Zr-based metallic glass plates, Acta Materialia, 52 (2004) 2429. https://doi.org/10.1016/j.actamat.2004.01.034

[63]   T. Masumoto, R. Maddin, The mechanical properties of palladium 20 a/o silicon alloy quenched from the liquid state, Acta Metall. 19 (1971) 725. https://doi.org/10.1016/0001-6160(71)90028-9

[64]   P.E. Donovan and W.M. Stobbs, The structure of shear bands in metallic glasses, Acta Metall. 29 (1981) 1419. https://doi.org/10.1016/0001-6160(81)90177-2

[65]   D.P.Wang, B. A.Sun, X.R. Niu, Y. Yang, W.H. Wang and C.T. Liu, Mutual interaction of shear bands in metallic glasses, Intermetallics 85 (2017) 48. https://doi.org/10.1016/j.intermet.2017.01.015

[66]   N. Jia, P. Eisenlohr, F. Roters, D. Raabe and X. Zhao, Orientation dependence of shear banding in face-centered-cubic single crystals, Acta Materialia, 60 (2012) 3415. https://doi.org/10.1016/j.actamat.2012.03.005

[67]   H.S. Chen, Plastic flow in metallic glasses under compression, Scripta Metallurgica, 7 (1973) 931. https://doi.org/10.1016/0036-9748(73)90143-9

[68]   F.O. Mear, T Wada, D.V Louzguine-Luzgin and A. Inoue, Highly inhomogeneous compressive plasticity in nanocrystal-toughened Zr–Cu–Ni–Al bulk metallic glass, Philosophical Magazine Letters, 89 (2009) 276. https://doi.org/10.1080/09500830902817861

[69]   D. Klaumünzer, R. Maaß and J.F. Löffler, Stick-slip dynamics and recent insights into shear banding in metallic glasses, Journal of Materials Research, 26 (2011) 1453. https://doi.org/10.1557/jmr.2011.178

[70]   S. Gonzalez, N. Chen, Q.S. Zhang, D.V. Louzguine-Luzgin, J.H. Perepezko and A. Inoue, Effect of shear bands initiated in the pre-yield region on the deformation behaviour of Zr-based metallic glasses, Scripta Materialia, 64 (2011) 713. https://doi.org/10.1016/j.scriptamat.2010.12.025

[71]   J.H. Na, M.D. Demetriou, M. Floyd, A. Hoff, G.R. Garrett and W.L. Johnson, Compositional landscape for glass formation in metal alloys, 111, (2014) 9031. https://doi.org/10.1073/pnas.1407780111

[72]   Y. Yokoyama, H. Tokunaga, A.R. Yavari, T. Kawamata, T. Yamasaki, K. Fujita, K. Sugiyama, P.K. Liaw and A. Inoue, Tough hypoeutectic Zr-based bulk metallic glasses, Metallurgical and Materials Transactions A, 42 (2011) 1468. https://doi.org/10.1007/s11661-011-0631-1

[73] H. Guo, P.F. Yan, Y.B. Wang, J. Tan, Z.F. Zhang, M.L. Sui and E. Ma, Tensile ductility and necking of metallic glass, Nature Materials, 6 (2007) 735. https://doi.org/10.1038/nmat1984

[74] D.V. Louzguine-Luzgin, A.R. Yavari, G.Q. Xie, S. Madge, S. Li, J. Saida, A.L. Greer and A. Inoue, Tensile deformation behaviour of Zr-based glassy alloys, Phil. Mag. Letters, 90 (2010) 139. https://doi.org/10.1080/09500830903485544

[75] M. Aljerf, K. Georgarakis and A.R. Yavari, Shaping of metallic glasses by stress-annealing without thermal embrittlement, Acta Materialia, 59(2011) 3817. https://doi.org/10.1016/j.actamat.2011.02.039

[76] Y.H. Liu, G. Wang, R.J. Wang, D.Q. Zhao, M.X. Pan and W.H. Wang, Super plastic bulk metallic glasses at room temperature, Science 315 (2007) 1385. https://doi.org/10.1126/science.1136726

[77] P.E. Donovan, A yield criterion for $Pd_{40}Ni_{40}P_{20}$ metallic glass, Acta Metall. 37 (1989) 445. https://doi.org/10.1016/0001-6160(89)90228-9

[78] F. Spaepen, A microscopic mechanism for steady state inhomogeneous flow in metallic glasses, Acta Mater. 25 (1977) 407. https://doi.org/10.1016/0001-6160(77)90232-2

[79] A. Argon, Plastic deformation in metallic glasses, Acta Metall. 27 (1979) 47. https://doi.org/10.1016/0001-6160(79)90055-5

[80] Y. Zhang and A.L. Greer, Thickness of shear bands in metallic glasses, Appl. Phys. Lett. 89 (2006) 071907. https://doi.org/10.1063/1.2336598

[81] D.V. Louzguine-Luzgin, S.V. Ketov, Z. Wang, M.J. Miyama, A.A. Tsarkov and A.Yu. Churyumov, Plastic deformation studies of Zr-based bulk metallic glassy samples with a low aspect ratio, Materials Science and Engineering A, 616 (2014) 288. https://doi.org/10.1016/j.msea.2014.08.006

[82] T. Egami, T. Iwashita and W. Dmowski, Mechanical properties of metallic glasses, Metals 3 (2013) 77. https://doi.org/10.3390/met3010077

[83] A. Vinogradov, M. Seleznev and I. S. Yasnikov, Dislocation characteristics of shear bands in metallic glasses, Scripta Materialia, 130 (2017) 138. https://doi.org/10.1016/j.scriptamat.2016.11.017

[84] F. Delogu, Identification and characterization of potential shear transformation zones in metallic glasses, Physical Review Letters, 100 (2008) 255901. https://doi.org/10.1103/PhysRevLett.100.255901

[85]   D. Srolovitz, V. Vitek and T. Egami, An atomistic study of deformation of amorphous metals, Acta Metallurgica, 31 (1983) 335. https://doi.org/10.1016/0001-6160(83)90110-4

[86]   J. Ding, S. Patinet, M.L. Falk, Y. Cheng and E. Ma, Soft spots and their structural signature in a metallic glass, Proceedings of the National Academy of Sciences of the United States of America 111 (2014) 14052. https://doi.org/10.1073/pnas.1412095111

[87]   Z.Y. Liu, Y. Yang and C.T. Liu, Yielding and shear banding of metallic glasses, Acta Materialia, 61 (2013) 5928. https://doi.org/10.1016/j.actamat.2013.06.025

[88]   A.L. Greer, Y.Q. Cheng and E. Ma, Shear bands in metallic glasses, Materials Science and Engineering R, 74 (2013) 71. https://doi.org/10.1016/j.mser.2013.04.001

[89]   F. Shimizu, S. Ogata and J. Li, Yield point of metallic glass, Acta Materialia, 54 (2006) 4293. https://doi.org/10.1016/j.actamat.2006.05.024

[90]   D. Klaumünzer, R. Maaß and J.F. Löffler, Journal of Materials Research, 26 (2011) 1453.

[91]   J.Q.Wang and J.H.Perepezko, Nucleation kinetics of shear bands in metallic glass, Journal of Chemical Physics 145 (2016) 211803. https://doi.org/10.1063/1.4966662

[92]   J.J. Lewandowski, Effects of annealing and changes in stress state on fracture toughness of bulk metallic glass, Mater. Trans. JIM, 42 (2001) 633. https://doi.org/10.2320/matertrans.42.633

[93]   U. Ramamurty, M.L. Lee, J. Basu and Y. Li, Embrittlement of a bulk metallic glass due to low-temperature annealing, Scripta Mat. 47 (2002) 107. https://doi.org/10.1016/S1359-6462(02)00102-1

[94]   Y.Q. Cheng, A.J. Cao and E. Ma, Correlation between the elastic modulus and the intrinsic plastic behavior of metallic glasses: The roles of atomic configuration and alloy composition, Acta Mat. 57 (2009) 3253. https://doi.org/10.1016/j.actamat.2009.03.027

[95]   A.J. Cao, Y.Q. Cheng and E. Ma, Structural processes that initiate shear localization in metallic glass, Acta Mater. 57 (2009) 5146. https://doi.org/10.1016/j.actamat.2009.07.016

[96]   Y. Ritter and K. Albe, Chemical and topological order in shear bands of $Cu_{64}Zr_{36}$ and $Cu_{36}Zr_{64}$ glasses, J. Appl. Phys. 111 (2012) 103527. https://doi.org/10.1063/1.4717748

[97]   X.D. Wang, S. Aryal, C. Zhong, W.Y. Ching, H.W. Sheng, H. Zhang, D.X. Zhang, Q.P. Cao and J.Z. Jiang, Atomic picture of elastic deformation in a metallic glass, Scientific Reports. 5 (2015) 9184. https://doi.org/10.1038/srep09184

[98]   W.H. Jiang and M. Atzmon, The effect of compression and tension on shear-band structure and nanocrystallization in amorphous $Al_{90}Fe_5Gd_5$: a high-resolution transmission electron microscopy study, Acta Materialia. 51 (2003) 4095. https://doi.org/10.1016/S1359-6454(03)00229-5

[99]   C. Zhong, H. Zhang, Q.P. Cao, X.D. Wang, D.X. Zhang, U. Ramamurty and J.Z. Jiang, Deformation behavior of metallic glasses with shear band like atomic structure: a molecular dynamics study, Scientific Reports 6 (2016) 30935. https://doi.org/10.1038/srep30935

[100] Y. Shao, G.N. Yang, K.F. Yao and X. Liu, Direct experimental evidence of nano-voids formation and coalescence within shear bands, Appl. Phys. Lett. 105, (2014) 181909. https://doi.org/10.1063/1.4901281

[101] Q. An, G. Garrett, K. Samwer, Y. Liu, S.V. Zybin, S.-N. Luo, M.D. Demetriou, W.L. Johnson and W.A. Goddard III, Atomistic characterization of stochastic cavitation of a binary metallic liquid under negative pressure. J. Phys. Chem. Lett. 2, (2011) 1320. https://doi.org/10.1021/jz200351m

[102] Q. An, K. Samwer, M.D. Demetriou, M.C. Floyd, D.O. Duggins, W.L. Johnson and W.A. Goddard III, How the toughness in metallic glasses depends on topological and chemical heterogeneity, PNAS 113, (2016) 7053. https://doi.org/10.1073/pnas.1607506113

[103] J.E. Velasco, A. Concustell, E. Pineda and D. Crespo, Plastic deformation induced anisotropy in metallic glasses: A molecular dynamics study, Journal of Alloys and Compounds, 707 (2017) 102. https://doi.org/10.1016/j.jallcom.2016.12.233

[104] S.X. Song, X.L. Wang and T.G. Nieh, Capturing shear band propagation in a Zr-based metallic glass using a high-speed camera, Scripta Mater., 62 (2010) 847. https://doi.org/10.1016/j.scriptamat.2010.02.017

[105] B. Yang, P.K. Liaw, G. Wang, M. Morrison, C.T. Liu, R.A. Buchanan and Y. Yokoyama, In-situ thermographic observation of mechanical damage in bulk-metallic glasses during fatigue and tensile experiments, Intermetallics, 12 (2004) 1265. https://doi.org/10.1016/j.intermet.2004.04.006

[106] G. Wang, Q. Feng, B. Yang, W. Jiang, P.K. Liaw and C.T. Liu, Thermographic studies of temperature evolutions in bulk metallic glasses, Intermetallics, 30 (2012) 1. https://doi.org/10.1016/j.intermet.2012.03.022

[107] J.J. Lewandowski and A.L. Greer, Temperature rise at shear bands in metallic glasses, Nature Materials, 5 (2006) 15. https://doi.org/10.1038/nmat1536

[108] P. Guan, M.W. Chen and T. Egami, Stress-temperature scaling for steady-state flow in metallic glasses, Phys. Rev. Lett. 104 (2010) 205701. https://doi.org/10.1103/PhysRevLett.104.205701

[109] Y. Zhang, N.A. Stelmashenko, Z.H. Barber, W.H. Wang, J.J. Lewandowski and A.L. Greer, Local temperature rises during mechanical testing of metallic glasses, J. Mater. Res, 22 (2007) 419. https://doi.org/10.1557/jmr.2007.0068

[110] K.M. Flores and R.H. Dauskard, Local heating associated with crack tip plasticity in Zr–Ti–Ni–Cu–Be bulk amorphous metals, J. Mater. Res. 14 (1999) 638. https://doi.org/10.1557/JMR.1999.0642

[111] M. Zhao and M. Li, Local heating in shear banding of bulk metallic glasses, Scr. Mater. 65 (2011) 493. https://doi.org/10.1016/j.scriptamat.2011.06.007

[112] S.V. Ketov and D.V. Louzguine-Luzgin, Localized shear deformation and softening of bulk metallic glass: stress or temperature driven? Scientific Reports 3 (2013) 2798. https://doi.org/10.1038/srep02798

[113] S.X. Song and T.G. Nieh, Flow serration and shear-band viscosity during inhomogeneous deformation of a Zr-based bulk metallic glassIntermetallics, 17 (2009) 762.

[114] Q.K. Li and M. Li, Free volume evolution in metallic glasses subjected to mechanical deformation, Materials Transactions, 48 (2007) 1816. https://doi.org/10.2320/matertrans.MJ200785

[115] J. Bokeloh, S.V. Divinski, G. Reglitz and G. Wilde, Tracer measurement of atomic diffusion inside shear bands of a bulk metallic glass, Phys. Rev. Lett. 107 (2011) 235503. https://doi.org/10.1103/PhysRevLett.107.235503

[116] J. Pan, Q. Chen, L. Liu and Y. Li, Softening and dilatation in a single shear band, Acta Materialia, 59 (2011) 5146. https://doi.org/10.1016/j.actamat.2011.04.047

[117] S.V. Ketov, H.K. Nguyen, A.S. Trifonov, K. Nakajima and D.V. Louzguine-Luzgin, Huge reduction of Young's modulus near a shear band in metallic glass, Journal of Alloys and Compounds, 687 (2016) 221. https://doi.org/10.1016/j.jallcom.2016.06.116

[118] W.J. Wright, M.W. Samale, T.C. Hufnagel, M.M. LeBlanc and J.N. Florando, Studies of shear band velocity using spatially and temporally resolved measurements of strain during quasistatic compression of a bulk metallic glass, Acta Mater., 57 (2009) 4639. https://doi.org/10.1016/j.actamat.2009.06.013

[119] B. Hou, M. Zhao, P. Yang and Y.L. Li, Capture of shear crack propagation in metallic glass by high-speed camera and in situ SEM, Key Engineering Materials, 626 (2015) 162. https://doi.org/10.4028/www.scientific.net/KEM.626.162

[120] W.J. Wright, R.R. Byer and X. Gu, High-speed imaging of a bulk metallic glass during uniaxial compression, Appl. Phys. Lett. 102 (2013) 241920. https://doi.org/10.1063/1.4811744

[121] R.T. Qu, Z.Q. Liu, G. Wang and Z.F. Zhang, Progressive shear band propagation in metallic glasses under compression, Acta Materialia. 91 (2015) 19. https://doi.org/10.1016/j.actamat.2015.03.026

[122] Y.H. Liu, C.T. Liu, A. Gali, A. Inoue and M.W. Chen, Evolution of shear bands and its correlation with mechanical response of a ductile $Zr_{55}Pd_{10}Cu_{20}Ni_5Al_{10}$ bulk metallic glass, Intermetallics. 18 (2010) 1455. https://doi.org/10.1016/j.intermet.2010.03.037

[123] D.V. Louzguine-Luzgin, V.Yu. Zadorozhnyy, N. Chen and S.V. Ketov, Evidence of the existence of two deformation stages in bulk metallic glasses, Journal of Non-Crystalline Solids. 396–397 (2014) 20. https://doi.org/10.1016/j.jnoncrysol.2014.04.014

[124] J.H. Perepezko, S.D. Imhoff, M.W. Chen, S. Gonzalez and A. Inoue, Nucleation reactions during deformation and crystallization of metallic glass, Journal of Alloys and Compounds, 536 (2012) 55. https://doi.org/10.1016/j.jallcom.2011.12.064

[125] R. Sarmah, G. Ananthakrishna, B.A. Sun and W.H. Wang, Hidden order in serrated flow of metallic glasses, Acta Materialia, 59 (2011) 4482. https://doi.org/10.1016/j.actamat.2011.03.071

[126] D.V. Louzguine-Luzgin, D.M. Packwood, G. Xie and A.Yu. Churyumov, On deformation behavior of a Ni-based bulk metallic glass produced by flux treatment, Journal of Alloys and Compounds, 561 (2013) 241. https://doi.org/10.1016/j.jallcom.2013.01.193

[127] F.X. Liu, P.K. Liaw, G.Y. Wang, C.L. Chiang, D.A Smith, P.D Rack, P.P.Chu and R.A. Buchanan, Specimen-geometry effects on mechanical behavior of metallic glasses, Intermetallics 14 (2006) 1014. https://doi.org/10.1016/j.intermet.2006.01.043

[128] M.D. Uchic, D.M. Dimiduk, N. Florando and W.D. Nix, Sample dimensions influence strength and crystal plasticity, Science 305 (2004) 986. https://doi.org/10.1126/science.1098993

[129] C.E. Packard and C.A. Schuh, Initiation of shear bands near a stress concentration in metallic glass, Acta Mater., 55 (2007), 5348. https://doi.org/10.1016/j.actamat.2007.05.054

[130] F. Meng, K. Tsuchiya, S. Il and Y. Yokoyama, Reversible transition of deformation mode by structural rejuvenation and relaxation in bulk metallic glass, Appl. Phys. Lett. 101 (2012) 121914. https://doi.org/10.1063/1.4753998

[131] Y. Sun, A. Concustell and A.L. Greer, Thermomechanical processing of metallic glasses: extending the range of the glassy state, Nature Reviews Materials. 1 (2016) 16039. https://doi.org/10.1038/natrevmats.2016.39

[132] F.O. Mear, B. Lenk, Y. Zhang and A.L. Greer, Structural relaxation in a heavily cold-worked metallic glass, Scripta Materialia. 59 (2008) 1243. https://doi.org/10.1016/j.scriptamat.2008.08.023

[133] D.V. Louzguine-Luzgin, I. Seki, T. Yamamoto, H. Kawaji, C. Suryanarayana and A. Inoue, Structural relaxation and crystallization processes in $Cu_{55}Hf_{25}Ti_{15}Pd_5$ metallic glassy alloy, Intermetallics. 23 (2012) 177. https://doi.org/10.1016/j.intermet.2011.11.019

[134] R. Liontas, X.W. Gu, E. Fu, Y. Wang, N. Li, N. Mara and J.R. Greer, Effects of helium implantation on the tensile properties and microstructure of $Ni_{73}P_{27}$ metallic glass nanostructures, Nano Letters. 14 (2014) 5176. https://doi.org/10.1021/nl502074d

[135] K.W. Park, C.M. Lee, M. Wakeda, Y. Shibutani, M.L. Falk and J.C. Lee, Elastostatically induced structural disordering in amorphous alloys, Acta Materialia. 56 (2008) 5440. https://doi.org/10.1016/j.actamat.2008.07.033

[136] H.S. Chen, Stored energy in a cold-rolled metallic glass, Appl. Phys. Lett. 29 (1976) 328. https://doi.org/10.1063/1.89084

[137] D. Dmowski, Y.Yokoyama, A. Chuang, Y. Ren, M. Umemoto, K. Tuchiya, A. Inoue and T. Egami, Structural rejuvenation in a bulk metallic glass induced by severe plastic deformation. Acta Mater 58 (2010) 429. https://doi.org/10.1016/j.actamat.2009.09.021

[138] A. Concustell, F.O. Méar, S. Suriñach, M.D. Baró and A.L. Greer, Structural relaxation and rejuvenation in a metallic glass induced by shot-peening, Philos. Mag. Lett. 89 (2009) 831. https://doi.org/10.1080/09500830903337919

[139] Y. Zhang, W.H. Wang and A.L. Greer, Making metallic glasses plastic by control of residual stress, Nature Mater. 5 (2006) 857. https://doi.org/10.1038/nmat1758

[140] H.B Ke, P. Wen, H.L. Peng, W.H. Wang and A.L. Greer, Homogeneous deformation of metallic glass at room temperature reveals large dilatation, Scripta Materialia, 64 (2011) 966. https://doi.org/10.1016/j.scriptamat.2011.01.047

[141] S.C Lee, C.M. Lee, J.W. Yang and J.C. Lee, Microstructural evolution of an elastostatically compressed amorphous alloy and its influence on the mechanical properties, Scripta Mater. 58 (2008) 591.
https://doi.org/10.1016/j.scriptamat.2007.11.036

[142] K.W. Park, C.M. Lee, M. Wakeda, Y. Shibutani, M.L. Falk and J.C. Lee, Elastostatically induced structural disordering in amorphous alloys, Acta Mater. 56 (2008) 5440. https://doi.org/10.1016/j.actamat.2008.07.033

[143] S.V. Ketov, Y.H. Sun, S. Nachum, Z. Lu, A. Checchi, A.R. Beraldin, H.Y. Bai, W.H. Wang, D.V. Louzguine-Luzgin, M.A. Carpenter and A.L. Greer, Rejuvenation of metallic glasses by non-affine thermal strain, Nature, 524 (2015) 200. https://doi.org/10.1038/nature14674

[144] F. Shimizu, S. Ogata and J. Li, Theory of shear banding in metallic glasses and molecular dynamics calculations, Materials Transactions 48 (2007) 2923.
https://doi.org/10.2320/matertrans.MJ200769

[145] D.A. Woodford, Strain-rate sensitivity as a measure ductility, Transactions of the American Society for Metals, 62(1969) 291.

[146] T. Hufnagel, C.Jiao, T.Li, Y. Xing and L.Q. Ramesh, Deformation and failure of $Zr_{57}Ti_5Cu_{20}Ni_8Al_{10}$ bulk metallic glass under quasi-static and dynamic compression, J. Mater. Res. 17 (2002) 1441.
https://doi.org/10.2320/matertrans.MJ200769

[147] S. Song, H. Bei, J, Wadsworth and T.G. Nieh, Flow serration in a Zr-based bulk metallic glass in compression at low strain rates, Intermetallics, 16 (2008) 813.
https://doi.org/10.1016/j.intermet.2008.03.007

[148] D. Torre, F.H. Dubach, A. Nelson and J.F. Löffler, Temperature, strain and strain rate dependence of serrated flow in bulk metallic glasses, Mater. Trans. 48 (2007) 1774. https://doi.org/10.2320/matertrans.MJ200782

[149] D. V. Louzguine-Luzgin, S.V. Ketov, Z. Wang, M. J. Miyama, A. A. Tsarkov and A.Yu. Churyumov, Plastic deformation studies of Zr-based bulk metallic glassy samples with a low aspect ratio, Materials Science and Engineering A. 616 (2014) 288. https://doi.org/10.1016/j.msea.2014.08.006

[150] D.Torre, F.H. Dubach, A. Siegrist and J.F. Löffler, Shear striations and deformation kinetics in highly deformed Zr-based bulk metallic glasses, Appl. Phys. Lett. 89 (2006) 091918.

[151] J. Lu, G. Ravichandran and W.L. Johnson, Deformation behavior of the $Zr_{41.2}Ti_{13.8}Cu_{12.5}Ni_{10}Be_{22.5}$ bulk metallic glass over a wide range of strain-rates and temperatures, Acta Mater. 51 (2003) 3429. https://doi.org/10.1016/S1359-6454(03)00164-2

[152] F.X. Liu, Y.F. Gao and P.K. Liaw, Rate-dependent deformation behavior of zr-based metallic-glass coatings examined by nanoindentation, Metallurgical and Materials Transactions A, 39 (2008) 1862. https://doi.org/10.1007/s11661-007-9399-8

[153] D. Pan and M.W. Chen, Rate-change instrumented indentation for measuring strain rate sensitivity, Journal of Materials Research, 24 (2009) 1466. https://doi.org/10.1557/jmr.2009.0168

[154] D.S. Sanditov, V.V. Mantatov and S.Sh. Sangadiev, Microhardness and plastic deformation of glasses upon microindentation, Glass Physics and Chemistry, 30 (2004) 415. https://doi.org/10.1023/B:GPAC.0000045921.17106.64

[155] G. Srikant, N. Chollacoop and U. Ramamurty, Plastic strain distribution underneath a Vickers Indenter: Role of yield strength and work hardening exponent, Acta Mater.54 (2006) 5171. https://doi.org/10.1016/j.actamat.2006.06.032

[156] S. González, G.Q. Xie, D.V. Louzguine-Luzgin, J.H. Perepezko and A. Inoue, Deformation and strain rate sensitivity of a Zr–Cu–Fe–Al metallic glass, Materials Science and Engineering A, 528 (2011) 3506. https://doi.org/10.1016/j.msea.2011.01.049

[157] H. Guo, P.F. Yan, Y.B. Wang, J. Tan, Z.F. Zhang, M.L. Sui and E. Ma, Tensile ductility and necking of metallic glass, Nature Materials, 6 (2007) 735. https://doi.org/10.1038/nmat1984

[158] F.F. Wu, Z.F. Zhang and S.X. Mao, Size-dependent shear fracture and global tensile plasticity of metallic glasses, Acta Mater. 57 (2009) 257. https://doi.org/10.1016/j.actamat.2008.09.012

[159] K. Hajlaoui, A.R Yavari, B. Doisneau, A. LeMoulec, W.J.F. Botta, G. Vaughan, A.L. Greer, A. Inoue, W. Zhang and A. Kvick, Shear delocalization and crack blunting of a metallic glass containing nanoparticles: In situ deformation in TEM

analysis, Scripta Materialia. 54 (2006) 1829.
https://doi.org/10.1016/j.scriptamat.2006.02.030

[160] H. Guo P. Yan, Y.B Wang, J. Tan, Z. Zhang, M. Sui and E. Ma, Tensile ductility and necking of metallic glass, Nature Materials 6 (2007) 735. https://doi.org/10.1038/nmat1984

[161] D.V. Louzguine-Luzgin, A.R.Yavari, G.Q. Xie, S. Li, S. Madge, J. Saida, A. Greer and A. Inoue, Tensile deformation behaviour of Zr-based glassy alloys, Phil. Mag. Letters. 90 (2010) 139. https://doi.org/10.1080/09500830903485544

[162] H. Guo, P.P. Yan, Y.B. Wang, J. Tan, Z.F. Zhang, M.L. Sui and E. Ma, Tensile ductility and necking of metallic glass, Nature Materials 6 (2007) 735. https://doi.org/10.1038/nmat1984

[163] K. Georgarakis, M. Aljerf, Y. Li, A. Lemoulec, F. Charlot, A.R. Yavari, K. Chornokhvostenko, E. Tabachnikova, G.A. Evangelakis, D.B. Miracle, A.L. Greer and T. Zhang, Shear band melting and serrated flow in metallic glasses, Applied Physics Letters, 93 (2008) 031907. https://doi.org/10.1063/1.2956666

[164] J.J. Lewandowski, M. Shazly and A.S. Nouri, Intrinsic and extrinsic toughening of metallic glasses, Scripta Materialia, 54 (2008) 337. https://doi.org/10.1016/j.scriptamat.2005.10.010

[165] L. Gu, L. Xu, Q.S. Zhang, D. Pan, N. Chen, D.V. Louzguine-Luzgin, K.-F. Yao, W.H. Wang and Y. Ikuhara, Direct in situ observation of metallic glass deformation by real-time nano-scale indentation, Scientific Reports, 5 (2015) 9122. https://doi.org/10.1038/srep09122

[166] G.Y. Wang, P.K. Liaw and M.L. Morrison, Progress in studying the fatigue behavior of Zr-based bulk-metallic glasses and their composites, Intermetallics 17 (2009) 579. https://doi.org/10.1016/j.intermet.2009.01.017

[167] J.J. Kruzic, Understanding the problem of fatigue in bulk metallic glasses, Metall. Mater. Trans. A. 42 (2011) 1516. https://doi.org/10.1007/s11661-010-0413-1

[168] Z.F. Zhang, J. Eckert and L. Schultz, Fatigue and fracture behavior of bulk metallic glass, Metallurgical and Materials Transactions A, 35 (2004) 3489. https://doi.org/10.1007/s11661-004-0186-5

[169] G.Y. Wang, P.K. Liaw, W.H. Peter, B. Yang, Y. Yokoyama, M.L. Benson, B.A. Green, M.J. Kirkham, S.A. White, T.A. Saleh, R.L. McDaniels, R.V. Steward, R.A. Buchanan, C.T. Liu and C.R. Brook, Fatigue behavior of bulk metallic glasses, Intermetallics 12 (2004) 885. https://doi.org/10.1016/j.intermet.2004.02.043

[170] M. Freels, G.Y. Wang, W. Zhang, P.K. Liaw amd A. Inoue, Cyclic compression behavior of a Cu-Zr-Al-Ag bulk metallic glass, Intermetallics, 19 (2011) 1174. https://doi.org/10.1016/j.intermet.2011.03.023

[171] K. Fujita, W. Zhang, B.L. Shen, K. Amiya, C.L. Ma and N. Nishiyama, Fatigue properties in high strength bulk metallic glasses, Intermetallics 30 (2012) 12. https://doi.org/10.1016/j.intermet.2012.03.021

[172] C.E. Packard, L.M. Witmer and C.A. Schuh, Hardening of a metallic glass during cyclic loading in the elastic range, Appl. Phys. Lett. 92 (2008) 171911. https://doi.org/10.1063/1.2919722

[173] C. Deng and C.A. Schuh, Atomistic mechanisms of cyclic hardening in metallic glass, Appl. Phys. Lett. 100 (2012) 251909. https://doi.org/10.1063/1.4729941

[174] A. Caron, A. Kawashima, H.J. Fecht, D.V. Louzguine-Luzguin and A. Inoue, On the anelasticity and strain induced structural changes in a Zr-based bulk metallic glass, Appl. Phys. Lett. 99 (2011) 171907. https://doi.org/10.1063/1.3655999

[175] D.V. Louzguine-Luzgin, G. Xie, Q. Zhang and A.Inoue, Effect of Fe on the glass-forming ability, structure and devitrification behavior of Zr-Cu-Al bulk glass-forming alloys. Philosophical Magazine, 90 (2010) 1955. https://doi.org/10.1080/14786430903571495

[176] A.Yu. Churyumov, A.I. Bazlov, V.Yu. Zadorozhnyy, A.N. Solonin, A. Caron and D.V. Louzguine-Luzgin, Phase transformations in Zr-based bulk metallic glass cyclically loaded before plastic yielding, Materials Science and Engineering A, 550 (2012) 358. https://doi.org/10.1016/j.msea.2012.04.087

[177] J.Y. Cavaille, L David and J.Perez, Relaxation phenomena in non crystalline solids: case of polymeric materials, Material Science Forum, 366–368 (2001) 499.

[178] J.M. Pelletier, D.V. Louzguine-Luzgin, S. Li and A Inoue, Elastic and viscoelastic properties of glassy, quasicrystalline and crystalline phases in $Zr_{65}Cu_5Ni_{10}Al_{7.5}Pd_{12.5}$ alloys, Acta Materialia. 59 (2011) 2797. https://doi.org/10.1016/j.actamat.2011.01.018

[179] V. Yu. Zadorozhnyy, M. Yu. Zadorozhnyy, A.Yu. Churyumov, S.V. Ketov, I.S. Golovin and D.V. Louzguine-Luzgin, Room-temperature dynamic quasi-elastic mechanical behavior of a Zr–Cu–Fe–Al bulk metallic glass, Phys. Status Solidi A, 213 (2016) 450. https://doi.org/10.1002/pssa.201532638

[180] D.V. Louzguine-Luzgin, V.Yu. Zadorozhnyy, S.V. Ketov, Z. Wang, A.A. Tsarkov and A.L. Greer, On room-temperature quasi-elastic mechanical behaviour of bulk

metallic glasses, Acta Materialia, 129 (2017) 343.
https://doi.org/10.1016/j.actamat.2017.02.049

[181] H. Zhang, K.Q. Qiu, Z.G. Wang, Q.S. Zang and H.F. Zhang, Low-cycle fatigue of a bulk amorphous alloy, Acta Metall. Sinica, 39 (2003) 405.

[182] C.K. Huang and J.J. Lewandowski, Effects of changes in chemistry on the flex bending fatigue behavior of Al-based amorphous alloy ribbons, Metall. Mater. Trans. A, 43A. (2012) 2687. https://doi.org/10.1007/s11661-011-0853-2

[183] D.V. Louzguine-Luzgin, L.V. Louzguina-Luzgina, S.V. Ketov, V.Yu. Zadorozhnyy and A.L. Greer, Influence of cyclic loading on the onset of failure in a Zr-based bulk metallic glass, Journal of Materials Science, 49 (2014) 6716. https://doi.org/10.1007/s10853-014-8276-2

[184] S.V. Madge, Toughness of bulk metallic glasses, Metals, 5 (2015) 1279. https://doi.org/10.3390/met5031279

[185] S.V. Madge, P. Sharma, D.V Louzguine-Luzgin, A.L Greer and A Inoue, New La-based glass-crystal ex situ composites with enhanced toughness, Scr. Mater. 62 (2010) 210. https://doi.org/10.1016/j.scriptamat.2009.10.029

[186] Q. He, Y.Q. Cheng, E. Ma and J. Xu, Locating bulk metallic glasses with high fracture toughness: Chemical effects and composition optimization, Acta Materialia. 59 (2011) 202. https://doi.org/10.1016/j.actamat.2010.09.025

[187] M.D. Demetriou, M.E. Launey, G. Garrett, P.J. Schramm, D.C. Hofmann, W.L. Johnson and R.O. Ritchie, A damage-tolerant glass, Nature Materials, 10 (2011) 123. https://doi.org/10.1038/nmat2930

[188] Y. Zhou, J. Liu, D. Han, X. Chen, G. Wang and Q. Zhai, Reduced fracture toughness of metallic glass at cryogenic temperature, Metals. 7 (2017) 151. https://doi.org/10.3390/met7040151

[189] A.S. Argon and M. Salama, The mechanism of fracture in glassy materials capable of some inelastic deformation. Mater. Sci. Eng. 23 (1976) 219. https://doi.org/10.1016/0025-5416(76)90198-1

[190] D.V Louzguine-Luzgin, A. Vinogradov, A.R. Yavari, S. Li, G. Xie and A. Inoue, On the deformation and fracture behaviour of a Zr-based glassy alloy, Philosophical Magazine, 88 (2008) 2979. https://doi.org/10.1080/14786430802446674

[191] C.A. Pampillo, Flow and fracture in amorphous alloys, J. Mater. Sci. 10 (1975) 1194. https://doi.org/10.1007/BF00541403

[192] F. Spaepen, Metallic glasses: Must shear bands be hot? Nature Materials 5, (2006) 7. https://doi.org/10.1038/nmat1552

[193] B. Yang, M.L. Morrison, P.P. Liaw, K.R.A. Buchanan, G.Wang, C.T. Liu and M. Denda, Dynamic evolution of nanoscale shear bands in a bulk-metallic glass, Appl. Phys. Lett. 86 (2005) 141904. https://doi.org/10.1063/1.1891302

[194] N. Chen, D.V. Louzguine-Luzgin, G.Q. Xie and A. Inoue, Nanoscale wavy fracture surface of a Pd-based bulk metallic glass, Applied Physics Letters, 94 (2009) 131906. https://doi.org/10.1063/1.3109797

[195] D.V. Louzguine-Luzgin, A. Vinogradov, A.R. Yavari, S. Li, G. Xie and A. Inoue, On the deformation and fracture behaviour of a Zr-based glassy alloy, Philosophical Magazine, 23 (2008) 2979. https://doi.org/10.1080/14786430802446674

[196] J.X. Meng, Z. Ling, M.Q. Jiang, H.S. Zhang and L.H. Dai, Dynamic fracture instability of tough bulk metallic glass, Appl. Phys. Lett., 92 (2008) 171909. https://doi.org/10.1063/1.2913206

[197] H.A. Bruck, A.J. Rosakis and W.L. Johnson, The dynamic compressive behavior of beryllium bearing bulk metallic glasses, J. Mater. Res. 11 (1996)503. https://doi.org/10.1557/JMR.1996.0060

[198] T. Hufnagel, T. Jiao, Y. Li, L. Xing and K. Ramesh, Deformation and failure of $Zr_{57}Ti_5Cu_{20}Ni_8Al_{10}$ bulk metallic glass under quasistatic and dynamic compression, Journal of Materials Research, 17 (2002) 1441. https://doi.org/10.1557/JMR.2002.0214

[199] A. Inoue and T. Zhang, Impact fracture energy of bulk amorphous $Zr_{55}Al_{10}Cu_{30}Ni_5$ alloy, Materials Transactions, JIM, 37 (1996) 1726.

[200] E. Fleury, S.M. Lee, H.S. Ahn, W.T. Kim and D.H. Kim, Tribological properties of bulk metallic glasses, Materials Science and Engineering A, 375–377 (2004) 276. https://doi.org/10.1016/j.msea.2003.10.065

[201] A.L. Greer and W.N. Myung, Abrasive wear resistance of bulk metallic glasses, MRS Proceedings, 644 (2000) L10.4.

[202] R. Moreton and J.K. Lancaster, The friction and wear behaviour of various metallic glasses, Journal of Materials Science Letters, 4 (1985) 133. https://doi.org/10.1007/BF00728057

[203] B. Prakash, Abrasive wear behaviour of Fe, Co and Ni based metallic glasses, Wear, 258 (2005) 217. https://doi.org/10.1016/j.wear.2004.09.010

[204] A.L. Greer and W.N. Myung, Abrasive wear resistance of bulk metallic glasses, Mat. Res. Soc. Symp. Proc. 644 (2001) L10.4.1

[205] J. Bhatt, S. Kumar, C. Dong and B.S. Murty, Tribological behaviour of $Cu_{60}Zr_{30}Ti_{10}$ bulk metallic glass, Materials Science and Engineering A, 458 (2007) 290. https://doi.org/10.1016/j.msea.2006.12.060

[206] M. Anis, W.M. Rainforth and H.A. Davies, Wear behavior of rapidly solidified $Fe_{68}Cr_{18}Mo_2B_{12}$, Wear 172 (1994) 135. https://doi.org/10.1016/0043-1648(94)90281-X

[207] H.W. Jin, R. Ayer, J.Y. Koo, R. Raghavan and U. Ramamurty, Reciprocating wear mechanisms in a Zr-based bulk metallic glass, J. Mater. Res. 22 (2007) 264. https://doi.org/10.1557/jmr.2007.0048

[208] X.F. Wu, G.A. Zhang and F.F Wu, Wear behaviour of Zr-based in situ bulk metallic glass matrix composites, Bull Mater Sci. 39 (2016) 703. https://doi.org/10.1007/s12034-016-1187-x

[209] Y. Saotome, S. Miwa, T. Zhang and A. Inoue, The micro-formability of Zr-based amorphous alloys in the supercooled liquid state and their application to micro-dies, Journal of Materials Processing Technology, 113 (2001) 64. https://doi.org/10.1016/S0924-0136(01)00605-7

[210] J.W. Lee, Y.C. Lin, N. Chen, D. V. Louzguine, M. Esashi and T. Gessner, Development of the large scanning mirror using Fe-based metallic glass ribbon, Japanese Journal of Applied Physics. 50 (2011) 087301. https://doi.org/10.7567/JJAP.50.087301

[211] A. Caron, C.L. Qin, L. Gu, S. Gonzalez, A. Shluger, H.J. Fecht, D.V. Louzguine-Luzgin and A. Inoue, Structure and nano-mechanical characteristics of surface oxide layers on a metallic glass, Nanotechnology 22 (2011) 095704. https://doi.org/10.1088/0957-4484/22/9/095704

[212] A. Caron, P. Sharma, A. Shluger, H.J. Fecht, D.V. Louzguine-Luzguin and A. Inoue, Effect of surface oxidation on the nm-scale wear behavior of a metallic glass, Journal of Applied Physics. 109 (2011) 083515. https://doi.org/10.1063/1.3573778

[213] A. Caron, D.V. Louzguine-Luzgin and R. Bennewitz, Structure vs chemistry: friction and wear of Pt-based metallic surfaces, ACS Appl. Mater. Interfaces. 5 (2013) 11341. https://doi.org/10.1021/am403564a

[214] S.V. Madge, A. Caron, R. Gralla, G. Wilde and S.K. Mishra, Novel W-based metallic glass with high hardness and wear resistance, Intermetallics 47 (2014) 6. https://doi.org/10.1016/j.intermet.2013.12.003

[215] D.V. Louzguine-Luzgin, H.K. Nguyen, K. Nakajima, S.V. Ketov and A.S. Trifonov, A study of the nanoscale and atomic-scale wear resistance of metallic glasses, Materials Letters. 185 (2016) 54. https://doi.org/10.1016/j.matlet.2016.08.035

[216] F.H. Stott, The role of oxidation in the wear of alloys, Tribology International. 31 (1998) 61. https://doi.org/10.1016/S0301-679X(98)00008-5

[217] A.S. Trifonov, A.V. Lubenchenko, V.I. Polkin, A.B. Pavolotsky, S.V. Ketov and D.V. Louzguine-Luzgin, Difference in charge transport properties of Ni-Nb thin films with native and artificial oxide, Journal of Applied Physics. 117 (2015) 125704. https://doi.org/10.1063/1.4915935

[218] D.V. Louzguine-Luzgin, C.L. Chen, L.Y. Lin, Z.C. Wang, S.V. Ketov, M.J. Miyama, A.S. Trifonov, A.V. Lubenchenko and Y. Ikuhara, Bulk metallic glassy surface native oxide: Its atomic structure, growth rate and electrical properties, Acta Materialia, 97 (2015) 282. https://doi.org/10.1016/j.actamat.2015.06.039

[219] C.L. Qin, J.J. Oak, N. Ohtsu, K. Asami and A. Inoue, XPS study on the surface films of a newly designed Ni-free Ti-based bulk metallic glass, Acta Materialia, 55 (2007) 2057. https://doi.org/10.1016/j.actamat.2006.10.054

[220] Y.H. Sun, D.V. Louzguine-Luzgin, S. Ketov and A.L. Greer, Pure shear stress reversal on a Cu-based bulk metallic glass reveals a Bauschinger-type effect, Journal of Alloys and Compounds. 615 (2014) 75. https://doi.org/10.1016/j.jallcom.2013.11.104

[221] G. Kumar, D. Rector, R.D. Conner and J. Schroers, Embrittlement of Zr-based bulk metallic glasses, Acta Mater. 57 (2009) 3572. https://doi.org/10.1016/j.actamat.2009.04.016

[222] Y. Kawamura, T. Nakamura, A. Inoue and T. Masumoto, High-strain-rate superplasticity due to Newtonian viscous flow in $La_{55}Al_{25}Ni_{20}$ metallic glass, Mater Trans JIM, 40 (1999) 794. https://doi.org/10.2320/matertrans1989.40.794

[223] T. Egami, Y. Tong and W. Dmowski, Deformation in metallic glasses studied by synchrotron X-ray diffraction, Metals, 6 (2016) 22. https://doi.org/10.3390/met6010022

[224] Y. Kawamura, T. Shibata, A. Inoue and T. Masumoto, Superplastic deformation of $Zr_{65}Al_{10}Ni_{10}Cu_{15}$ metallic glass, Scripta Mater. 37 (1997) 431.

[225] Y. Saotome, K. Ito, T. Zhang and A. Inoue, Superplastic nanoforming of Pd-based amorphous alloy, Scripta Mater. 44 (2001) 1541. https://doi.org/10.1016/S1359-6462(01)00837-5

[226] G. Kumar, H.X. Tang and J. Schroers, Nanomoulding with amorphous metals, Nature. 457 (2009) 868. https://doi.org/10.1038/nature07718

[227] N. Chen, H.A. Yang, A. Caron, P.C. Chen, Y.C. Lin, D.V. Louzguine-Luzgin, K.F. Yao, M. Esashi and A. Inoue, Glass-forming ability and thermoplastic formability of a $Pd_{40}Ni_{40}Si_4P_{16}$ glassy alloy, Journal of Materials Science 46 (2011) 2091. https://doi.org/10.1007/s10853-010-5043-x

[228] J. Schroers, T.M. Hodges, G. Kumar, H. Raman, A.J. Barnes, Q. Pham and T.A. Waniuk, Thermoplastic blow molding of metals, Materials Today 14 (2011) 14. https://doi.org/10.1016/S1369-7021(11)70018-9

[229] A.R. Yavari, M.F. Oliveira, C.S. Kiminami, A. Inoue and W.J. Botta, Electromechanical shaping assembly and engraving of bulk metallic glasses, Mater. Sci. Eng. A 375 (2004) 227. https://doi.org/10.1016/j.msea.2003.10.267

[230] M. Oliveria, W.J. Botta and A.R. Yavari, Connecting, assemblage and electromechanical shaping of bulk metallic glasses, Mater. Trans. JIM, 41 (2000) 1501. https://doi.org/10.2320/matertrans1989.41.1501

[231] D.V. Louzguine-Luzgin, G.Q. Xie, T. Tsumura, H. Fukuda, K. Nakata, H.M. Kimura and A Inoue, Structural investigation of Ni–Nb–Ti–Zr–Co–Cu glassy samples prepared by different welding techniques, Materials Science and Engineering B, 148 (2008) 88. https://doi.org/10.1016/j.mseb.2007.09.034

[232] T. Tsumura and K. Nakata, Laser welding of Ni-based metallic glass foil, Welding International. 25 (2011) 491. https://doi.org/10.1080/09507111003655366

[233] Y. Yokoyama, N. Abe, K. Fukaura and A. Inoue, Electron beam welding of $Zr_{50}Cu_{30}Ni_{10}Al_{10}$ bulk glassy alloys, Mater. Trans. JIM. 43 (2002) 2509. https://doi.org/10.2320/matertrans.43.2509

[234] D.V. Louzguine-Luzgin, Y. Yokoyama, G. Xie, N. Abe and A. Inoue, Transmission electron microscopy investigation of the structure of a welded $Zr_{50}Cu_{30}Ni_{10}Al_{10}$ glassy alloy sample Philosophical Magazine Letters. 87 (2007) 549. https://doi.org/10.1080/09500830701320299

[235] T. Shoji, Y. Ohno and Y. Kawamura, Joining of $Zr_{41}Be_{23}Ti_{14}Cu_{12}Ni_{10}$ bulk metallic glasses by a friction welding method, Materials Transactions. 44 (2003) 1809. https://doi.org/10.2320/matertrans.44.1809

[236] S.V. Ketov, N. Chen, A. Caron, A. Inoue and D.V. Louzguine-Luzgin, Structural features and high quasi-static strain rate sensitivity of $Au_{49}Cu_{26.9}Ag_{5.5}Pd_{2.3}Si_{16.3}$ bulk metallic glass, Applied Physics Letters, 101 (2012) 241905. https://doi.org/10.1063/1.4770072

[237] S. Mechler, E. Yahel, P.S. Pershan, M. Meron and B. Lin, Crystalline monolayer surface of liquid Au–Cu–Si–Ag–Pd: Metallic glass former, App. Phys. Let. 98 (2011) 251915. https://doi.org/10.1063/1.3599515

[238] A. Kawashima, Y. Zeng, M. Fukuhara, H. Kurishita, N. Nishiyama, H. Miki, A. Inoue, Mechanical properties of a $Ni_{60}Pd_{20}P_{17}B_3$ bulk glassy alloy at cryogenic temperatures, Materials Science and Engineering A, 498 (2008) 475. https://doi.org/10.1016/j.msea.2008.08.033

[239] E.D. Tabachnikova, A.V Podol'ski, V.Z Bengus, S.N Smirnov, D.V. Luzgin and A. Inoue, Low-temperature plasticity anomaly in the bulk metallic glass $Zr_{64.13}Cu_{15.75}Ni_{10.12}Al_{10}$, Low Temp. Phys. 34 (2008) 675. https://doi.org/10.1063/1.2967517

[240] D.V. Louzguine-Luzgin, A. Vinogradov, S. Li, A. Kawashima, G. Xie, A.R. Yavari and A. Inoue, Deformation and fracture behavior of metallic glassy alloys and glassy-crystal composites, Metallurgical and Materials Transactions A, 42 (2011) 1504. https://doi.org/10.1007/s11661-010-0391-3

[241] A. Vinogradov, A. Lazarev, D.V. Louzguine-Luzgin, Y. Yokoyama, S. Li, A.R. Yavari and A. Inoue, Propagation of shear bands in metallic glasses and transition from serrated to non-serrated plastic flow at low temperatures, Acta Materialia, 58 (2010) 6736. https://doi.org/10.1016/j.actamat.2010.08.039

[242] A.H. Brothers and D.C. Dunand, Ductile bulk metallic glass foams, Adv. Mater., 17 (2005). 484. https://doi.org/10.1002/adma.200400897

[243] T. Wada and A. Inoue, Formation of porous pd-based bulk glassy alloys by a high hydrogen pressure melting-water quenching method and their mechanical properties, Mater. Trans., 45 (2004) 2761. https://doi.org/10.2320/matertrans.45.2761

[244] A. Inoue, T. Wada and D.V. Louzguine-Luzgin, Improved mechanical properties of bulk glassy alloys containing spherical pores, Materials Science and Engineering: A, 471 (2007)144. https://doi.org/10.1016/j.msea.2006.10.172

[245] G. Xie, W. Zhang, D.V. Louzguine-Luzgin, H. Kimura and A. Inoue, Fabrication of porous Zr–Cu–Al–Ni bulk metallic glass by spark plasma sintering process, Scripta Materialia, 55 (2006) 687. https://doi.org/10.1016/j.scriptamat.2006.06.034

[246] T. Wada, A. Inoue and A.L. Greer, Enhancement of room-temperature plasticity in a bulk metallic glass by finely dispersed porosity, Appl. Phys. Lett. 86 (2005) 251907. https://doi.org/10.1063/1.1953884

[247] A. Inoue, W. Zhang, T. Tsurui, A.R Yavari and A.L. Greer, Unusual room-temperature compressive plasticity in nanocrystal-toughened bulk copper-zirconium glass, Philos. Mag. Lett. 85 (2005) 221. https://doi.org/10.1080/09500830500197724

[248] D.V Louzguine, H. Kato and A. Inoue, High-strength Cu- based crystal-glassy composite with enhanced ductility. Applied Physics Letters, 84 (2004) 1088. https://doi.org/10.1063/1.1647278

[249] C.C. Hays, C.P. Kim and W.L. Johnson, Microstructure controlled shear band pattern formation and enhanced plasticity of bulk metallic glasses containing in situ formed ductile phase dendrite dispersions, Phys. Rev. Lett. 84 (2000) 2901. https://doi.org/10.1103/PhysRevLett.84.2901

[250] U. Kuhn, J. Eckert, N. Mattern and L. Schultz, ZrNbCuNiAl bulk metallic glass matrix composites containing dendritic bcc phase precipitates, Appl. Phys. Lett. 80 (2002) 2478. https://doi.org/10.1063/1.1467707

[251] S. Pauly S. Pauly, G. Liu, G. Wang and J. Eckert, Microstructural heterogeneities governing the deformation of $Cu_{47.5}Zr_{47.5}Al_5$ bulk metallic glass composites, Acta Mater. 57 (2009) 5445. https://doi.org/10.1016/j.actamat.2009.07.042

[252] P. Gargarella, S. Pauly, K.K. Song, J. Hu, N.S. Barekar, M. Khoshkhoo, A. Teresiak, H. Wendrock, U. Kühn, C. Ruffing, E. Kerscher and J. Eckert, Ti–Cu–Ni shape memory bulk metallic glass composites, Acta Materialia 61 (2013) 151. https://doi.org/10.1016/j.actamat.2012.09.042

[253] M.L. Lee, Y. Li and C.A. Schuh, Effect of a controlled volume fraction of dendritic phases on tensile and compressive ductility in La-based metallic glass matrix composites, Acta Mater. 52 (2004) 4121. https://doi.org/10.1016/j.actamat.2004.05.025

[254] D.C. Hofmann, J.Y. Suh, A. Wiest, M.L. Lind, M.D. Demetriou and W.L. Johnson, Development of tough, low-density titanium-based bulk metallic glass matrix composites with tensile ductility, Proc. Natl. Acad. Sci. USA. 105 (2008) 20136. https://doi.org/10.1073/pnas.0809000106

[255] D.C. Hofmann, J.Y. Suh, A. Wiest, G. Duan, M.L. Lind, M.D. Demetriou and W.L. Johnson, Designing metallic glass matrix composites with high toughness and tensile ductility, Nature 451 (2008) 1085. https://doi.org/10.1038/nature06598

[256] J.W. Qiao, T. Zhang, F.Q. Yang, P.K. Liaw, S. Pauly and B.S. Xu, A tensile deformation model for in-situ dendrite/metallic glass matrix composites, Scientific Reports. 3 (2011) 2816. https://doi.org/10.1038/srep02816

[257] Y. Wu, Y. Xiao, G. Chen, C.T. Liu and Z. Lu, Bulk metallic glass composites with transformation-mediated work-hardening and ductility, Adv. Mater. 22 (2010) 2270. https://doi.org/10.1002/adma.201000482

[258] D.C. Hofmann, Shape memory bulk metallic glass composites, Science, 329 (2010) 1294. https://doi.org/10.1126/science.1193522

[259] S. Pauly, S. Gorantla, G. Wang, U. Kühn and J. Eckert, Transformation-mediated ductility in CuZr-based bulk metallic glasses, Nat. Mater. 9 (2010) 473. https://doi.org/10.1038/nmat2767

[260] P. Gargarella, S. Pauly, K.K. Songa, J. Hu, N.S. Barekar, M.S. Khoshkhoo, A. Teresiak, H. Wendrock, U. Kühn, C. Ruffing, E. Kerscher and J. Eckert, Ti-Cu-Ni shape memory bulk metallic glass composites, Acta Materialia, 61 (2013) 151. https://doi.org/10.1016/j.actamat.2012.09.042

[261] S. Pauly, S. Gorantla and G. Wang, Transformation-mediated ductility in CuZr-based bulk metallic glasses, Nature Mater. 9 (2010) 473. https://doi.org/10.1038/nmat2767

[262] J. Das, M.B. Tang, K.B. Kim, R. Theissmann, F. Baier, W.H. Wang and J. Eckert, Work-hardenable ductile bulk metallic glass, Phys. Rev. Letters, 94 (2005) 205501. https://doi.org/10.1103/PhysRevLett.94.205501

[263] F.O. Mear, T. Wada, D.V. Louzguine-Luzgin and A. Inoue, Highly inhomogeneous compressive plasticity in nanocrystals toughened Zr–Cu–Ni–Al bulk metallic glass, Philosophical Magazine Letters, 4 89 (2009) 276. https://doi.org/10.1080/09500830902817861

[264] D.V. Louzguine-Luzgin, A. Vinogradov, G. Xie, S. Li, A. Lazarev, S. Hashimoto and A. Inoue, High-strength and ductile glassy-crystal Ni-Cu-Zr-Ti composite exhibiting stress-induced martensitic transformation, Philosophical Magazine, 89 (2009) 2887. https://doi.org/10.1080/14786430903128577

[265] I.S. Golovin, V.Yu. Zadorozhnyy, A.Yu. Churyumov and D.V. Louzguine-Luzgin, Internal friction in a Ni-Ti-based glassy-crystal alloy, Journal of Alloys and Compounds, 579 (2013) 633. https://doi.org/10.1016/j.jallcom.2013.07.102

[266] A.Yu. Churyumov, A.I. Bazlov, A.N. Solonin, V.Yu. Zadorozhnyi, G.Q. Xie, S. Li and D.V. Louzguine-Luzgin, Structure and mechanical properties of Ni-Cu-Ti-

Zr composite materials with amorphous phase, The Physics of Metals and Metallography 114 (2013) 773. https://doi.org/10.1134/S0031918X13090044

[267] A.A. Tsarkov, A.Y. Churyumov, V.Y. Zadorozhnyy and D.V. Louzguine-Luzgin, High-strength and ductile (Ti-Ni)-(Cu-Zr) crystalline/amorphous composite materials with superelasticity and TRIP effect, Journal of Alloys and Compounds. 658 (2016) 402. https://doi.org/10.1016/j.jallcom.2015.10.175

[268] V.Yu. Zadorozhnyy, A. Inoue and D.V. Louzguine-Luzgin, Formation and investigation of the structure and mechanical properties of bulk metallic glassy composite (Ti-Zr)–(Cu-Ni-Co) alloys, Intermetallics, 31 (2012) 173. https://doi.org/10.1016/j.intermet.2012.07.008

[269] Y. Jiang, X. Shi and K. Qiu, Numerical study of shear banding evolution in bulk metallic glass composites, Materials & Design, 77 (2015) 15. https://doi.org/10.1016/j.matdes.2015.04.010

[270] G.P. Zheng and Y. Shen, Simulation of shear banding and crack propagation in bulk metallic glass matrix composites, Journal of Alloys and Compounds, 509 (2011) 136. https://doi.org/10.1016/j.jallcom.2010.08.131

[271] Y. Shen and G.P. Zheng, Modeling of shear band multiplication and interaction in metallic glass matrix composites, Scripta Materialia, 63 (2010) 181. https://doi.org/10.1016/j.scriptamat.2010.03.046

[272] H.Y. Zhang and G.P. Zheng, Simulation of shear banding in bulk metallic glass composites containing dendrite phases, Journal of Alloys and Compounds, 586 (2014) 262. https://doi.org/10.1016/j.jallcom.2013.03.159

[273] J.S.C. Jang, T.H. Li, P.H. Tsai, J.C. Huang and T.G. Nieh, Critical obstacle size to deflect shear banding in Zr-based bulk metallic glass composites, Intermetallics, 64 (2015) 102. https://doi.org/10.1016/j.intermet.2015.05.001

[274] Y. Xue, X. Zhong, L. Wang, Q. Fan, L. Zhu, B. Fan, H. Zhang and H. Fu, Effect of W volume fraction on dynamic mechanical behaviors of W fiber/Zr-based bulk metallic glass composites, Materials Science and Engineering: A, 639 (2015) 417. https://doi.org/10.1016/j.msea.2015.05.086

[275] A.Yu. Churyumov, A.I. Bazlov, A.A. Tsarkov, A.N. Solonin and D.V. Louzguine-Luzgin, Microstructure, mechanical properties, and crystallization behavior of Zr-based bulk metallic glasses prepared under a low vacuum, Journal of Alloys and Compounds, 654 (2016) 87. https://doi.org/10.1016/j.jallcom.2015.09.003

[276] D.V. Louzguine-Luzgin, S.V. Ketov, Z. Wang, M.J. Miyama, A.A. Tsarkov and A.Yu. Churyumov, Plastic deformation studies of Zr-based bulk metallic glassy

samples with a low aspect ratio, Mat. Sci. Eng. A, 616 (2014) 288.
https://doi.org/10.1016/j.msea.2014.08.006

[277] A.I. Bazlov, A.Yu. Churyumov, A.A. Tsar'kov and D.M. Khazhina, Studies of the structure and mechanical properties of $Ti_{43.2}Zr_{7.8}Cu_{40.8}Ni_{7.2}Co_1$ alloy containing amorphous and crystalline phases, Phys. Met. Metallogr. 116 (2015) 684.
https://doi.org/10.1134/S0031918X15050038

[278] L. Guo, H. Guo, S. Gong and H. Xu, Improvement on the phase stability, mechanical properties and thermal insulation of Y_2O_3-stabilized ZrO_2 by Gd_2O_3 and Yb_2O_3 co-doping, Ceram. Int. 39 (2013) 9009.
https://doi.org/10.1016/j.ceramint.2013.04.103

[279] S. Jiang, J. Liu, C. Lin, L. Bai, Y. Zhang, X. Li, Y. Li, L. Tang and H. Wang, Structural transformations in cubic Dy_2O_3 at high pressures, Solid. State. Commun. 169 (2013) 37. https://doi.org/10.1016/j.ssc.2013.06.027

[280] A. Inoue, C. Fan and A Takeuchi, Synthesis and properties of Ti-based bulk amorphous alloys with a large supercooled liquid region, Mater Sci Forum 312-314 (1999) 307. https://doi.org/10.4028/www.scientific.net/MSF.312-314.307

[281] T.C. Hufnagel, C. Fan, R.T. Ott, J. Li and S. Brennan, Controlling shear band behavior in metallic glasses through microstructural design, Intermetallics 10 (2002) 1163. https://doi.org/10.1016/S0966-9795(02)00157-7

[282] D. Wu, K.K.Song, C.D. Cao, R. Li, G. Wang, Y. Wu, F. Wan, F. Ding, Y. Shi, X.J. Bai, I. Kaban and J. Eckert, Deformation-induced martensitic transformation in Cu-Zr-Zn bulk metallic glass composites metals, 5, (2015), 2134.
https:// doi.org/10.3390/met5042134

[283] V.Yu. Zadorozhnyy, M.V. Gorshenkov, M.N. Churyukanova, M.Yu. Zadorozhnyy, A.A. Stepashkin, D.O. Moskovskikh, S.V. Ketov, L.Kh. Zinnurova, A. Sharma, D.V. Louzguine-Luzgin and S.D. Kaloshkin, Investigation of structure and thermal properties in composite materials based on metallic glasses with small addition of polytetrafluoroethylene, Journal of Alloys and Compounds 707 (2017) 264. https://doi.org/10.1016/j.jallcom.2016.11.359

[284] K. Hajlaoui, A.R. Yavari, A. LeMoulec, W.J. Botta, F.G. Vaughan, J. Das, A.L Greer and A. Kvick, Plasticity induced by nanoparticle dispersions in bulk metallic glasses, Journal of Non-Crystalline Solids, 353 (2007) 327.
https://doi.org/10.1016/j.jnoncrysol.2006.10.011

[285] A. Inoue, T. Zhang, M. W. Chen, T. Sakurai, J. Saida and M. Matsushita, Ductile quasicrystalline alloys, Appl. Phys. Lett. 76 (2006) 967. https://doi.org/10.1063/1.125907

[286] A. Inoue, C. Fan and A. Takeuchi, High-strength bulk nanocrystalline alloys in a Zr-based system containing compound and glassy phases, J. Non-Cryst. Sol. 250-252 (1999) 724. https://doi.org/10.1016/S0022-3093(99)00168-4

[287] J. Eckert, U. Kühn, N. Mattern, A. Reger-Leonhard and M. Heilmaier, Bulk nanostructured Zr-based multiphase alloys with high strength and good ductility, Scripta Mater, 44 (2001) 1587. https://doi.org/10.1016/S1359-6462(01)00779-5

[288] J. Saida, H. Kato, A.D. Setyawan and A. Inoue, Characterization and properties of nanocrystal-forming Zr-based bulk metallic glasses, Rev. Adv. Mater. Sci. 10 (2005) 34. https://doi.org/10.1557/jmr.2007.0126

[289] D.V. Louzguine-Luzgin, Y. Zeng, A.D.H. Setyawan, N. Nishiyama, H. Kato, J. Saida and A. Inoue, Deformation behavior of Zr- and Ni-based bulk glassy alloys, J. Mater. Res. 22 (2007) 1087. https://doi.org/10.2320/matertrans.42.1741

[290] M.C. Gao, R.E. Hackenberg and G.J. Shiflet, Deformation-induced nanocrystal precipitation in Al-base metallic glasses, Materials Transactions, 42 (2001) 1741. https://doi.org/10.1016/j.intermet.2006.01.013

[291] W.H. Jiang and M. Atzmon, Plastic flow of a nanocrystalline/amorphous $Al_{90}Fe_5Gd_5$ composite formed by rolling, Intermetallics, 14 (2006) 962. https://doi.org/10.1016/j.jnoncrysol.2006.06.022

[292] D.V. Louzguine-Luzgin and A. Inoue, Comparative study of the effect of cold rolling on the structure of Al–RE–Ni–Co (RE=rare-earth metals) amorphous and glassy alloys, Journal of Non-Crystalline Solids, 352 (2006) 3903. https://doi.org/10.1016/j.scriptamat.2004.12.014

[293] A. M. Glezer, M. R. Plotnikova, R. V. Sundeev, and N. A. Shurygina, Self blocking of shear bands and the delocalization of plastic flows in amorphous alloys upon megaplastic deformation, Bulletin of the Russian Academy of Sciences, Physics, 77 (2013) 1391–1396. https://doi.org/10.3103/S1062873813110129

[294] D. Torre, F.H. Dubach, A. Schallibaum and J. Loffler, Shear striations and deformation kinetics in highly deformed Zr-based bulk metallic glasses, Acta Materialia, 56 (2008) 4635. https://doi.org/10.1016/j.scriptamat.2004.02.019

[295] A.V. Sergueeva, N.A. Mara, D.J. Branagan and A.K. Mukherjee, Strain rate effect on metallic glass Ductility, Scr. Mater. 50 (2004) 1303. https://doi.org/10.1016/j.actamat.2006.05.030

[296] K. Hajlaoui, M. Stoica, A.LeMoulec, F. Charlot and A.R. Yavari, Fe-Nb-B bulk metallic glass with high boron content, Rev. Adv. Mater. Sci., 18 (2008) 23.

[297] Q.P. Cao, J.F. Li, Y.H. Zhou, A. Horsewell and J.Z. Jiang, Effect of rolling deformation on the microstructure of bulk $Cu_{60}Zr_{20}Ti_{20}$ metallic glass and its crystallization, Acta Materialia. 54 (2006) 4373. https://doi.org/10.1016/j.actamat.2006.05.030

[298] G. Kumar, T. Ohkubo, T. Mukai and K. Hono, Plasticity and microstructure of Zr–Cu–Al bulk metallic glasses, Scr. Mater. 57 (2007) 173. https://doi.org/10.1016/j.scriptamat.2007.02.013

[299] A.R. Yavari, K. Georgarakis, J. Antonowicz, M. Stoica, N. Nishiyama, G. Vaughan, M. Chen and M. Pons, Crystallization during bending of a Pd-based metallic glass detected by X-ray microscopy, Physical Review Letters, 109 (2012) 085501. https://doi.org/10.1103/PhysRevLett.109.085501

[300] A. Inoue, K Ohtera, A.P. Tsai and T. Masumoto, Glass transition behavior of Al-Y-Ni and Al-Ce-Ni amorphous alloys, Jpn. J. Appl. Phys. 27 (1988) 479. https://doi.org/10.1143/JJAP.27.L479

[301] G.J. Shiflet, Y. He and S.J. Poon, Mechanical properties of a new class of metallic glasses based on aluminum, J. Appl. Phys. 64 (1988) 6863. https://doi.org/10.1063/1.341978

[302] Y.H. Kim, A. Inoue and T. Masumoto, Ultrahigh tensile strengths of $Al_{88}Y_2Ni_9M_1$ (M=Mn or Fe) amorphous alloys containing finely dispersed fcc-Al particles, Mater. Trans. JIM, 31 (1990) 747. https://doi.org/10.2320/matertrans1989.31.747

[303] A. Inoue, H. Kimura and K. Amiya, Developments of aluminum-and magnesium-based nanophase high-strength alloys by use of melt quenching-induced metastable phase, Mater. Trans. 43 (2002) 2006. https://doi.org/10.2320/matertrans.43.2006

[304] Z.C. Zhong, X.Y. Jiang and A.L Greer, Microstructure and hardening of Al-based nanophase composites, Mater. Sci. Eng. A, 226–228 (1997) 531. https://doi.org/10.1016/S0921-5093(97)80062-7

[305] A. Inoue, S. Sobu, D.V. Louzguine, H. Kimura and K. Sasamori, Ultrahigh strength Al-based amorphous alloys containing Sc, J. Mater. Res. 19 (2004) 1539. https://doi.org/10.1557/JMR.2004.0206

[306] D.V. Louzguine and A. Inoue, Influence of a supercooled liquid on crystallization behaviour of Al–Y–Ni–Co metallic glass, J. Mater. Res. 17 (2002) 1014. https://doi.org/10.1557/JMR.2002.0149

Chapter 6

Magnetic properties

Magnetic materials compose another very important and developed application field of metallic glassy and nanostructured alloys. The connection between nanostructure and magnetic properties is a topic of intensive investigations. Ferromagnetic alloys can exhibit hard or soft magnetism depending on their coercive force (H_c) also called coercivity. Magnetic materials having a coercivity above about 10^4 A/m are considered to be hard while soft magnetic materials have a coercivity below 10^3 A/m. The current chapter is devoted to this subject.

Contents

6.1 Soft magnetic alloys

Fe-, Ni- and Co-based bulk metallic glassy alloys exhibit good soft magnetic properties [1,2,3] with high saturation magnetization (M_s), high permeability (μ) and a coercivity of about 10 A/m and even lower while Nd-based alloys show hard magnetic properties with a coercivity above 10^4 A/m. Compared to other bulk metallic glasses, (Fe-Co)-based alloys are also particularly attractive for engineering applications due to their combination of ultrahigh strength, good wear resistance, good glass forming ability and rather low cost. A typical magnetization curve of the soft magnetic alloy (also called magnetically soft alloy) is given in Fig. 6.1.

Fe-Zr-B soft magnetic materials also exhibit high magnetization saturation (M_s) of 1.60-1.70 T as well as high effective permeability of 15000 at 1 kHz [4]. In the case of the (Fe, Co, Ni)$_{70}$Zr$_{10}$B$_{20}$ glassy alloys H_c is as low as 3 A/m [5]. The saturation magnetization increases to 0.9 T with increasing Fe content, while the saturated magnetostriction (λ_s) equals zero in the Co-rich composition range and increases monotonously to 15×10^{-6}

with increasing Fe content. μ reaches the maximum of about 20000 in the Fe- and Co-rich composition ranges. These glassy alloys exhibit good soft magnetic properties including the saturation magnetization of 0.9 T, H_c of 3–6 A/m, λ_s of $12–15 \times 10^{-6}$ and μ of 20000 in the Fe-rich range and the saturation magnetization of 0.5 T, H_c of 6 A/m, nearly zero λ_s and μ of 20000 in the Co-rich range.

Large-scale cores are made from metallic glassy powders by sintering thus overcoming limitation connected with low glass-forming ability of some metallic glassy alloys. The addition of the SiC was effective in improving the high frequency magnetic properties [6]. This approach is promising for creation net shape products.

Fig. 6.1. Magnetization curve of a Co-Fe-Ta-B metallic glassy sample.

The coercivity of soft magnetic materials changes with grain size (d) and both coarse grain and nanostructured alloys are known to have low H_c. Fig. 6.2 shows the dependence of coercivity on the grain size. Atomic clusters of sub-nanometer size as the structural units of glassy alloys make them ideal soft magnetic materials. Nanocrystallization is typical for various Fe-based glassy alloys [7,8] and can be used for further improving their magnetic properties.

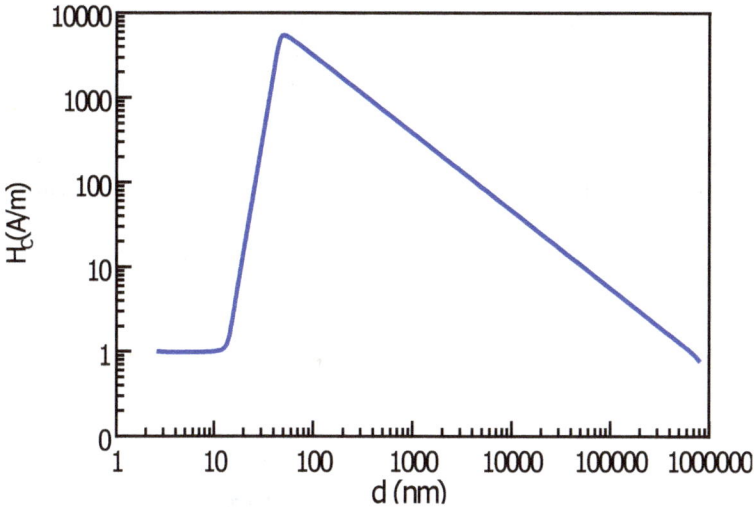

Fig. 6.2. Schematic representation of the changes in H_c of soft magnetic materials as a function of grain size (d).

Well known Fe-based soft magnetic materials with mixed nanocrystalline and amorphous structure are: $Fe_{73.5}Cu_1Nb_3Si_{13.5}B_9$ [9] and $Fe_{84}Zr_{3.5}Nb_{3.5}B_8Cu_1$ [10] alloys. The structure consists of fine BCC Fe nanocrystals dispersed in the amorphous matrix. The $Fe_{84}Zr_{3.5}Nb_{3.5}B_8Cu_1$ alloy shows high permeability (μ) of 100000, high magnetization saturation up to 1.5 T and low hysteresis losses [11]. Cu, Nb and Zr are responsible for BCC Fe grain refinement and formation of a nanostructure in these alloys. 3D atom probe field ion microscopy and high resolution transmission electron microscopy studies showed [12] that Cu formed nano clusters in the $Fe_{73.5}Si_{13.5}B_9Nb_3Cu_1$ amorphous matrix which work as heterogeneous nucleation sites for bcc Fe particles on devitrification. The studies for X-ray absorption fine structure also showed that Cu clusters with nearly FCC structure were present from very early stages of the devitrification process [13]. The density of the clusters is of the order of 10^{-24} m^{-3} while the average cluster size is about 2 nm [14]. This analysis also showed a low Fe content in the grain boundaries between the bcc Fe solid solution nano grains of the nanocrystalline $Fe_{85}Zr_7B_6Cu_2$ alloy. Yavari and Negri [15] discussed nanocrystallization process of soft magnetic Fe-based amorphous alloys using the concentration gradients of the elements that are insoluble in the primary crystalline phase. The origin of the good soft magnetic properties is connected with the formation of the nanoscale BCC Fe structure and a strong magnetic coupling between the BCC grains through the intergranular ferromagnetic amorphous phase.

Nanocrystalline $(Fe_{1-x}Co_x)_{90}Zr_7B_3$ (x=0, 0.1, 0.2, and 0.3) showed a low coercivity of 9.1A/m and high permeability values of 4500 at 1 kHz along with a high saturation magnetization of 1.74 T [16]. The Fe–M–B (M=Zr, Hf, or Nb) alloys also show low core losses [17]. They also form a nanostructure upon initial crystallization on heating. Nanocrystalline $Fe_{89}(Hf,Nb)_7B_4$ alloys subjected to the optimum annealing exhibit high magnetization saturation above 1.5 T as well as high effective permeability at 1 kHz above 20000 [18]. The $Fe_{82}(Zr,Hf,Nb)_7B_{10}Cu_1$ alloys exhibited good soft magnetic properties especially in high-frequency range [19].

Nanocrystalline $Fe_{42.5}Co_{42.5}Nb_7B_8$ alloy with a structure consisting of nearly spherical bcc grains with a size ranged from 5 to 10 nm dispersed in the residual amorphous matrix exhibits a high saturation magnetization of 1.9 T and a low coercivity (H_c) [20]. It also exhibits a very high Curie temperature (T_c) exceeding 1173 K.

For long time soft magnetic glassy alloys were limited to marginal glass-formers. However, as soon as a ferromagnetic Fe-(Al,Ga)-metalloid bulk glassy alloy was produced in 1995 [21], Fe-based bulk metallic glasses have attracted significant interest in the scientific community owing to their high strength and good soft magnetic properties [22]. Bulk glassy alloys exhibiting a wide supercooled liquid region before crystallization were also obtained in the Fe–(Co,Ni)–(Zr,Nb,Ta)–(Mo,W)–B system [23]. These alloys have a high T_g of about 870 K and the supercooled liquid region close to 90 K. The critical diameter is 6 mm. The alloys also exhibit high compressive strength of 3800 MPa, high Vickers hardness of HV1360, and high corrosion resistance. These glassy alloys exhibit a large magnetization saturation of 0.74–0.96 T, low coercivity of 1.1–3.2 A/m, high permeability exceeding 12000 at 1 kHz and low magnetostriction. Magnetization curves for as-cast the $Fe_{56}Co_7Ni_7Nb_{10}B_{20}$ rod samples of different diameter were obtained in [24].

The $(Fe,Ni,Co)_{70}Mo_5P_{10}C_{10}B_5$ bulk metallic glasses with a critical sample diameter exceeding 4 mm showed good thermoplastic formability and good soft magnetic and mechanical properties. These glasses exhibit a wide supercooled liquid region and low viscosity of about 10^7 Pa·s in the supercooled liquid state [25]. The Fe–(Al,Ga)-(P,C,B) and Fe–(Al–Ga)-(P,C,B,Si) bulk glassy alloys also exhibited good soft magnetic properties [26,27]. For example, a ring-shaped sample of the $Fe_{70}Al_5Ga_2P_{9.65}C_{5.75}B_{4.6}Si_3$ glassy alloy with a thickness of 1 mm, an outer diameter of 10 mm and an inner diameter of 6 mm formed by the copper mold casting method exhibits the saturation magnetization of 1.2 T, very low coercivity of 2.2 A/m and rather low λ_s of 21×10^{-6}. The maximum permeability is 110 000. The remarkable improvement of the soft magnetic properties has been demonstrated to result from the significant difference in magnetic domain structure. The domain walls are arranged along the circumference direction for the cast-ring alloy and radial direction for the ribbon ring sheet. The difference in the domain wall structure was reported to originate from the difference in residual stress during the preparations of the ring sample and the melt-spun ribbon. The $Co_{46}Fe_{20}B_{23}Si_5Nb_6$ BMG with a critical diameter of 5 mm exhibited B_s, H_c, μ and fracture strength of ~0.7 T, ~2 A/m, ~20000 and 4400 MPa, respectively [28].

The $Co_{43}Fe_{20}Ta_{5.5}B_{31.5}$ glassy alloy exhibited an extremely low value of the coercive force as low as 0.25 A/m together and a high permeability of 550,000. The Fe–(Co,Ni)–(Zr,Nb,Ta)–(Mo,W)–B system glassy alloys exhibit a large saturation magnetization of 0.74–0.96 T, low coercivity of 1.1–3.2 A/m, high permeability exceeding $1.2 \cdot 10^4$ at 1 kHz, and low magnetostriction of about 12×10^{-6} [29].

The $Fe_{76}Si_9B_{10}P_5$ alloy with a critical diameter of D_{cr} 2.5 mm exhibited saturation magnetization of 1.51 T and H_c below 1 A/m [30]. Minor addition of Cu improved plasticity of this alloy [31]. Fe, C, Si, and P, are the constituent elements in cast-iron produced in a blast furnace. As in expensive Fe–Si and Fe–P ferroalloys are in mass production there is no restriction in availability of such materials which offers advantages of lower material cost for industry.

A recently developed $Fe_{80}P_{11}C_9$ BMG alloy exhibiting high strength of 3.2 GPa [32] in the partially devitrified state also exhibited good soft-magnetic properties including the saturation magnetization of 1.5 T and coercivity of 4 A/m. The $(Fe_{1-x}Ni_x)_{72}B_{20}Si_4Nb_4$ (x=0.0–0.5) bulk metallic glasses showed the saturation magnetization up to 1.15 T. The maximum Curie temperature was 598 K [33]. The $(Fe_{1-x}Co_x)_{76}Si_9B_{10}P_5$ (x = 0~0.4) ferromagnetic bulk glassy alloys demonstrated excellent combination of high GFA, good soft-magnetic properties as well as high strength [34]. These BMGs exhibited a high saturation magnetization of 1.5 T and low coercive force of 1.2 A/m.

The M_s and μ of the $Fe_{76.5}C_6Si_{3.3}B_{5.5}P_{8.7}$ BMG alloy were enhanced from 1.35 T and 3500 in the as-cast state to 1.57 T and 9890, respectively, upon annealing at 873 K for 30 s [35]. The MRO zones of 1–3 nm size were observed in the glassy matrix and the coercivity was below 20 A/m. The saturation magnetization of the $(Fe,Co,Ni)_{75.5}C_{7.0}Si_{3.3}B_{5.5}P_{8.7}$ bulk metallic glass continuously decreases with increasing Co or Ni content, while the Curie temperature and the permeability increase. The Co-bearing alloys show smaller coercivity and larger permeability than the Ni-bearing alloys [36].

Y addition was confirmed to improve GFA of the $Fe_{72}B_{24}Nb_4$ alloy. Saturation magnetization and coercivity of the as-cast $(Fe_{0.72}B_{0.24}Nb_{0.04})_{95.5}Y_{4.5}$ bulk metallic glassy ring were found to be 0.8 T and 0.8 A/m, respectively, in the relaxed state after annealing at 821 K [37]. The $Fe_{76+x}Si_{9-x}B_{10}P_5$ bulk metallic glasses showed the saturation magnetization above 1.6 T, low coercive force of ~2 A/m and high effective permeability of 17000 [38]. Soft magnetic powder core with a higher saturation magnetization of 1.3 T were developed in the Fe–Nb–B–Si and Fe–Nb–Cr–P–B–Si systems [39]. The Fe–Ni–Cr–Mo–B–Si glassy alloy powders produced by water atomization have also been commercialized with the commercial name AMO-beads owing to the high glass-forming ability for the Fe-based alloy.

Table 6.1. Composition, critical diameter (D_{cr}) and magnetic properties (B_s, H_c) of typical ferromagnetic BMGs.

Nr	Alloy	D_{cr} (mm)	B_s (T)	H_c (A/m)	*Ref.*
1	$Fe_{76}Si_9B_{10}P_5$	2.5	1.51	0.8	*30*
2	$(Fe_{0.75}Si_{0.1}B_{0.15})_{96}Nb_4$	1.5	1.47	2.9	*30*
3	$Fe_{73}Al_5Ga_2P_{11}C_5B_4$	1	1.29	6.3	*30*
4	$Fe_{30}Co_{30}Ni_{15}Si_8B_{17}$	1.2	0.92	3.4	*30*
5	$Fe_{70}Mo_5P_{10}C_{10}B_5$	3	0.93	2.36	*40*
6	$Fe_{74}Nb_6Y_3B_{17}$	2	0.81	15	*41*
7	$Fe_{65}Ni_5Mo_5P_{10}C_{10}B_5$	3	0.88	3.2	*25*
8	$Fe_{55}Co_{10}Ni_5Mo_5P_{10}C_{10}B_5$	4	0.85	2.56	*25*
10	$(Fe_{0.75}B_{0.15}Si_{0.1})_{96}Nb_4$	1.5	1.2	3.7	*42*
11	$[(Fe_{0.8}Co_{0.1}Ni_{0.1})_{0.75}B_{0.2}Si_{0.05}]_{96}Nb_4$	2.5	1.1	3	*42*
12	$Co_{43}Fe_{20}Ta_{5.5}B_{31.5}$	3	0.49	0.25	*43*

These values are plotted in the diagram in Fig. 6.3.

Ferromagnetic Co-based BMGs [44] possess low H_c, high μ, nearly zero λ_s, low high-frequency loss, high fracture strength 4 GPa, high glass transition and crystallization temperatures, large supercooled liquid region, high corrosion resistance, and high wear resistance. The $(Co_{0.942}Fe_{0.058})_{67}Nb_5B_{22.4}Si_{5.6}$ alloy exhibits high GFA (critical diameter 4 mm) and good combination properties, i.e., 3600 MPa for fracture strength, 0.43 T for B_s, 0.2 A/m for H_c and 47165 for μ_e at 1 kHz and 1 A/m [45]. Furthermore, the addition of lanthanide element (Er, Tb, Y or Dy) enables a significant increase of the supercooled liquid region and the largest interval reaches 130 K for $(Co_{0.5}Fe_{0.5})_{62}Nb_6Dy_2B_{30}$. In the supercooled liquid region, the alloy exhibits good superplastic formability as well as good nanoscale imprintability [46].

The properties of some alloys are summarized in Table 6.1.

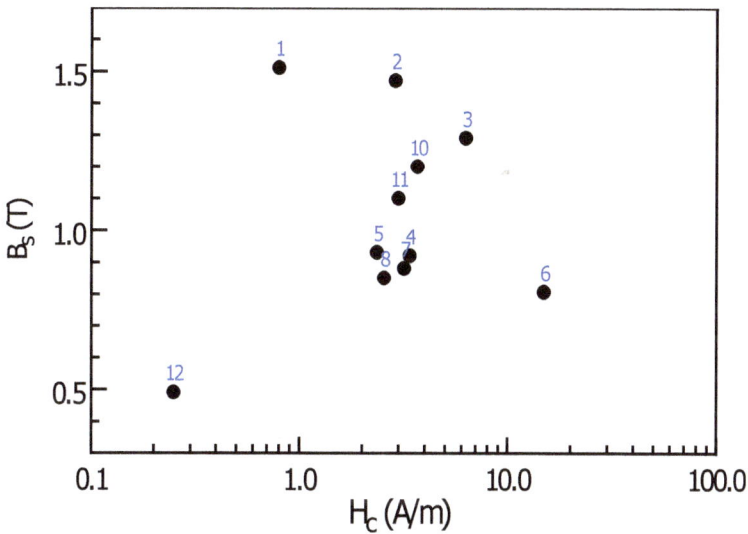

Fig. 6.3. Magnetic properties of the BMGs listed in Table 1.

In addition to the massive glassy samples metallic glassy micro- [47,48] and nano-wires [49,50] and nanoglasses attract increasing attention. Here there is an area for further application of magnetic materials. Arrays of magnetic nanowires attract interest for perpendicular magnetic recording [51]. The nanowires are potentially capable of producing recording densities in excess of several tens of Gbit per square inch. It has been also reported [52] that the crystalline magnetic nanowires of Fe, Co and Ni show significantly enhanced magnetic coercivity than that of their bulk counterparts. Such an approach can be applied to metallic glassy nanowires as well.

The irreversible changes of the magnetic properties and the Curie temperature in a Co-Cr-Zr metallic glass were found after thermal cycling between room and liquid nitrogen temperature [53]. After thermal cycling the coercive force, increased by 35%, magnetization increased by 6%, and remanent magnetization increased by 3%. Aging of thermally cycled samples at 300 K for 10 h led to the further increasing of these values by 60%, 16%, and 10%, respectively, compared to the as-cast state. Subsequent thermal cycling did not change the properties. The observed dependencies were explained as a sequence of phase transitions in the amorphous state.

6.2   Hard magnetic alloys

Hard magnetic alloys (also called magnetically hard alloys) have sufficiently high coercive force as a resistance to demagnetizing fields exceeding 10 kA/m. High magnetic induction is retained because of a strong resistance to demagnetization, for example, as a result of high anisotropy. Permanent magnets must have a large coercive force, high saturation magnetization and high energy product BH_{max}. These alloys can be used as permanent magnet materials with high magnetic induction.

Ferromagnetic $Nd_{90-x}Fe_xAl_{10}$ bulk amorphous alloys with high coercive force at room temperature were obtained by a copper mold casting method. The maximum diameter of the cylindrical amorphous samples with a length of 50 mm was about 7 mm. Neither glass transition nor supercooled liquid region was observed in these alloys in the temperature range before crystallization, which makes them different from classical bulk glassy alloys exhibiting a wide supercooled liquid region before crystallization [54]. The bulk amorphous $Nd_{70}Fe_{20}Al_{10}$ alloy exhibits ferromagnetism with the Curie temperature (T_c) of about 600 K which is much higher than the highest T_c (about 480 K) for the Nd-Fe binary amorphous alloy ribbons. The remanence (B_r) and H_c for the bulk $Nd_{60}Fe_{30}Al_{10}$ alloy are 0.122 T and 277 kA/m, respectively, in the as-cast state and 0.128 T and 277 kA/m, respectively, after annealing for 600 s at 600 K [55]. The B_r and H_c decrease to 0.045 T and 265 kA/m, respectively, for the crystallized sample. The hard magnetic properties for the bulk amorphous alloys are presumably due to the homogeneous development of ferromagnetic clusters with large random magnetic anisotropy.

Hard magnetic alloys can be produced by crystallization of the glassy phase. For example, a hard magnetic material consisting of the Fe_3B and $Nd_2Fe_{14}B$ phases was obtained by annealing the $Nd_{4.5}Fe_{77}B_{18.5}$ rapidly solidified alloy [56]. The microstructure is composed of the magnetically soft Fe_3B phase nanoscale grains and the magnetically hard $Nd_2Fe_{14}B$ phase. High remanence (B_r) of 0.8 T is obtained due to the permanence enhancement effect of exchange-coupled magnetic grains, the maximum energy product $(BH)_{max}$ is 97 kJ/m^3 while its coercivity H_c=240 kJ/m. The influence of heating rate on the microstructure of Fe$_3$B/Nd$_2$Fe$_{14}$B nanocomposite magnets was also studied [57]. High coercivity values exceeding 300 kA/m were obtained in the amorphous Nd$_5$Fe$_{72}$Cr$_5$B$_{18}$ alloy crystallized into the Fe$_3$B/Nd$_2$Fe$_{14}$B state [58]. Fe-Nd-B amorphous alloys containing 88 - 90 at% Fe at 923-1023 K form a nanostructure consisting of BCC-Fe, Fe$_{14}$Nd$_2$B and the residual amorphous phase. The Fe$_{89}$Nd$_7$B$_4$ glassy alloy exhibits good hard magnetic properties, i.e., B$_r$ of 1.28 T, H$_c$ of 252 kA/m and $(BH)_{max}$ of 146 kJ/m^3 for [59].

The α-Fe/Nd$_2$Fe$_{14}$B nanomaterials have higher coercivity than those for Fe$_3$B/Nd$_2$Fe$_{14}$B magnets. H$_c$ of 480 kA/m and $(BH)_{max}$ of 160 kJ/m^3 are reported. The structure consists of two phases: magnetically hard Nd$_2$Fe$_{14}$B with nanoparticles of α-Fe on grain boundaries. Grain size of the Nd$_2$Fe$_{14}$B phase is below 30 nm and particle size of the α-Fe phase is below 10 nm [60]. α-Fe/Nd$_2$Fe$_{14}$B nanostructured magnet of Fe$_{89}$Nd$_7$B$_4$ composition contains residual amorphous phase and shows B$_r$=1.22 T, H$_c$=240 kA/m and $(BH)_{max}$=130 kJ/m^3. This alloy has high iron and low boron concentration [61].

The rapidly solidified $(Fe_{0.65}Pt_{0.35})_{83}B_{17}$ alloy possesses higher coercivity in the annealed state compared to the binary Fe-Pt alloys [62]. B$_r$, M$_r$/M$_s$, H$_c$, and $(BH)_{max}$ of the rapidly-solidified Fe$_{80-x}$Pt$_x$B$_{20}$ (x = 20,22,24) ribbons in the annealed state are in the range of 0.93–1.05 T, 0.79–0.82, 375–487 kA/m, and 118–127 kJ/m^3, respectively (Fig. 6.4). The good hard magnetic properties result from the exchange magnetic coupling between the nanoscale magnetically hard γ_1 tP4 FePt and magnetically soft γ cF4 Fe(Pt) solid solution as well as Fe$_2$B phases [63]. Fe-Pt-P rapidly solidified alloys were also found to possess good magnetic properties [64]. Although these are rapidly solidified samples, they can be compacted by SPS or hot pressing. The $(Fe_{0.75}Pt_{0.25})_{75-70}B_{25-30}$ alloys were also found to possess good hard magnetic properties including high intrinsic coercivity values up to 400 kA/m in the nano-crystallized state [65]. They are promising candidates for nanocomposite permanent magnets. The structure of the rapidly solidified $(Fe_{0.75}Pt_{0.25})_{75}B_{25}$ alloy contains a limited volume fraction of the nanoscale cubic cF4 Fe(Pt) solid solution particles of about 4 nm in size embedded in the amorphous matrix. The nanoparticles of cF4 Fe(Pt) phase start growing at the elevated temperature and then undergo a transformation forming of the tP4 FePt compound of about 15 nm in size which is followed by the formation of the tI12 Fe$_2$B phase from the residual amorphous matrix [66].

Fig. 6.4. Hysteresis loops of the $Fe_{80-x}Pt_xB_{20}$ (x=20, 22, and 24) metallic glassy samples annealed at 798 K for 900 s. Reproduced from [63] with permission of the American Institute of Physics.

6.3 The magnetocaloric effect

There is also an interest in the magnetic cooling [67]. The magnetocaloric effect is related to the change in temperature of a material as a result of the alignment of its magnetic spins on exposure to an external magnetic field. Magnetic refrigeration could be more efficient and environmentally friendly than conventional refrigeration systems. In addition to the $Gd_5Ge_2Si_2$ compound and its derivatives [68,69] RE-based BMGs could be promising candidates [70]. Magnetization of the Gd-based BMGs exhibits a sharp magnetization change at the ordering temperature, though the transition temperatures for the RE-based BMGs up to 100 K are quite low [71].

References

[1]   A. Makino, A. Inoue and T. Masumoto, Nanocrystalline soft magnetic Fe–M–B (M=Zr, Hf, Nb) alloys produced by crystallization of amorphous phase, Mater. Trans. JIM, 36 (1995) 924. https://doi.org/10.2320/matertrans1989.36.924

[2]   K. Amiya, A. Urata, N. Nishiyama and A. Inoue, Magnetic properties of Co–Fe–B–Si–Nb bulk glassy alloy with zero magnetostriction, J. Appl. Phys, 101 (2007) 112. https://doi.org/10.1063/1.2718353

[3]   A. Inoue and B.L. Shen, New Fe-based bulk glassy alloys with high saturated magnetic flux density of 1.4–1.5 T, Mater. Sci. Eng., A 375–377 (2004) 302.

[4]   K. Suzuki and A. Makino, The role of boron in nanocrystalline Fe-Zr-B soft magnetic alloys, Materials Science and Engineering: A, 179–180 (1994) 501. https://doi.org/10.1016/0921-5093(94)90255-0

[5]   A. Inoue, T. Zhang and T. Itoi, New Fe-Co-Ni-Zr-B amorphous alloys with wide supercooled liquid regions and good soft magnetic properties, Mater. Trans. JIM, 38 (1997) 359. https://doi.org/10.2320/matertrans1989.38.359

[6]   G. Xie, H.M. Kimura, D.V. Louzguine-Luzgin, H. Men and A. Inoue, SiC dispersed Fe-based glassy composite cores produced by spark plasma sintering and their high frequency magnetic properties, Intermetallics 20 (2012) 76. https://doi.org/10.1016/j.intermet.2011.08.023

[7]   A. Aronin, G. Abrosimova, D. Matveev and O. Rybchenko, Structure and properties of nanocrystalline alloys prepared by high pressure torsion, Rev. Adv. Mater. Sci. 25 (2010) 52.

[8]   G. Abrosimova, A. Aronin, D. Matveev and E. Pershina, Nanocrystal formation, structure and magnetic properties of Fe–Si–B amorphous alloy after deformation, Materials Letters 97 (2013) 15. https://doi.org/10.1016/j.matlet.2013.01.092

[9]   H. Yoshizawa, K. Yamauchi, T. Yamane and H. Sugihara, Common mode choke cores using the new Fe - based alloys composed of ultrafine grain structure, J. Appl. Phys. 64 (1988) 6047. https://doi.org/10.1063/1.342150

[10]   K. Suzuki, N. Kataoka, A. Inoue, A. Makino and T. Masumoto, High saturation magnetization and soft magnetic properties of bcc Fe-Zr-B alloys with ultrafine grain structure, Mater. Trans., JIM 31 (1990) 743.

[11]   A. Makino, K. Suzuki, A. Inoue and T. Masumoto, Common mode choke cores using the new Fe - based alloys composed of ultrafine grain structure, Mater. Trans. JIM 32 (1991) 551. https://doi.org/10.2320/matertrans1989.32.551

[12]   K. Hono, K. Hiraga, Q. Wang, A. Inoue and T. Sakurai, The microstructure evolution of a $Fe_{73.5}Si_{13.5}B_9Nb_3Cu_1$ nanocrystalline soft magnetic material, Acta Metall. et Mater. 40 (1992) 2137. https://doi.org/10.1016/0956-7151(92)90131-W

[13]   J.D. Ayers, V.G. Harris, J.A. Sprague, W.T. Elam and H.N. Jones, On the formation of nanocrystals in the soft magnetic alloy $Fe_{73.5}Nb_3Cu_1Si_{13.5}B_9$, Acta Materialia 46 (1998) 1861. https://doi.org/10.1016/S1359-6454(97)00436-9

[14]   M. Ohnuma, K. Hono, H. Onodera, J.S Pedersen and S. Linderoth, Cu clustering stage before the crystallization in Fe–Si–B–Nb–Cu amorphous alloys, Nanostr. Mater. 12 (1999) 693. https://doi.org/10.1016/S0965-9773(99)00219-6

[15]   A.R. Yavari and D. Negri, Effect of concentration gradients on nanostructure development during primary crystallization of soft-magnetic iron-based amorphous alloys and its modelling, Nanostruct. Mater. 8 (1997) 969. https://doi.org/10.1016/S0965-9773(98)00047-6

[16]   K. Suzuki, N. Ito, J.S. Garitaonandia and J.D. Cashion, High saturation magnetization and soft magnetic properties of nanocrystalline $(Fe,Co)_{90}Zr_7B_3$ alloys annealed under a rotating magnetic field, Journal of Applied Physics 99 (2006) 08F114. https://doi.org/10.1063/1.2199753

[17]   K. Suzuki, A. Makino, A. Inoue and T. Masumoto, Low core losses of nanocrystalline Fe–M–B (M=Zr, Hf, or Nb) alloys, Journal of Applied Physics, 74 (1993) 3316. https://doi.org/10.1063/1.354555

[18]   A. Makino, A. Inoue and T. Masumoto, Nanocrystalline soft magnetic Fe–M–B (M=Zr, Hf, Nb) alloys produced by crystallization of amorphous phase, Materials Transactions, JIM 36 (1995) 924.

[19]   Y.M. Moon, K.S. Kim, S.C. Yu and K.V. Rao, Magnetization study of glassy and nanocrystalline $Fe_{82}(Zr, Hf, Nb)_7B_{10}Cu_1$ alloys, Journal of Magnetism and Magnetic Materials, 177 (1998) 968.

[20]   B. Shen, H. Kimura and A. Inoue, Structure and magnetic properties of $Fe_{42.5}Co_{42.5}Nb_7B_8$ nanocrystalline alloy, Materials Transactions 43 (2002) 589. https://doi.org/10.2320/matertrans.43.589

[21]   A. Inoue, Y. Shinohara and J.S. Gook, Thermal and magnetic properties of bulk fe-based glassy alloys prepared by copper mold casting, Mater. Trans. JIM 36 (1995) 1427. https://doi.org/10.2320/matertrans1989.36.1427

[22]   A. Inoue, A. Makino and T. Mizushima, Ferromagnetic bulk glassy alloys, Journal of Magnetism and Magnetic Materials, 215-216 (2000) 246. https://doi.org/10.1016/S0304-8853(00)00127-X

[23]   A. Inoue, T. Zhang and A. Takeuchi, Bulk amorphous alloys with high mechanical strength and good soft magnetic properties in Fe–TM–B (TM=IV–VIII group transition metal) system, Appl. Phys. Lett. 71 (1997) 464. https://doi.org/10.1063/1.119580

[24]   H. Chiriac and N. Lupu, New FeNbB-based bulk amorphous and nanocomposite soft magnetic alloys, IEEE Transactions on Magnetics, 41 (2005) 3289. https://doi.org/10.1109/TMAG.2005.854721

[25]   W. Zhang, C.F. Fang and Y.H. Li, Ferromagnetic Fe-based bulk metallic glasses with high thermoplastic formability, Scripta Materialia 69 (2013) 77. https://doi.org/10.1016/j.scriptamat.2013.03.003

[26]   T. Mizushima, K. Ikarashi, M. Yoshida, A. Makino and A. Inoue, Soft magnetic properties of ring shape bulk glassy Fe-Al-Ga-P-C-B-Si alloy prepared by copper mold casting, Mater. Trans. JIM, 40 (1999) 1019. https://doi.org/10.2320/matertrans1989.40.1019

[27]   T. Bitoh, A. Makino and A. Inoue, Magnetization process and coercivity of Fe-(Al, Ga)-(P, C, B, Si) soft magnetic glassy alloys, Materials Transactions, 45 (2004) 1219. https://doi.org/10.2320/matertrans.45.1219

[28]   Q.K. Man, H.J. Sun, Y.Q. Dong, B.L. Shen, H. Kimura, A. Makino and A. Inoue, Enhancement of glass-forming ability of CoFeBSiNb bulk glassy alloys with excellent soft-magnetic properties and superhigh strength, Intermetallics, 18 (2010) 1876. https://doi.org/10.1016/j.intermet.2010.02.047

[29]   A. Inoue, T. Zhang and A. Takeuchi, Bulk amorphous alloys with high mechanical strength and good soft magnetic properties in Fe–TM–B (TM=IV–VIII group transition metal) system, Appl. Phys. Lett., 71 (1997) 464. https://doi.org/10.1063/1.119580

[30]   A. Makino, T. Kubota, C. Chang, M. Makabe and and A. Inoue, Fe-metalloids bulk glassy alloys with high Fe content and high glass-forming ability, J. Mater. Res., 23 (2008) 1339. https://doi.org/10.1557/JMR.2008.0180

[31]   A. Makino, X. Li, K. Yubuta, C. Chang, T. Kubota and A. Inoue, The effect of Cu on the plasticity of Fe–Si–B–P-based bulk metallic glass, Scripta Materialia, 60 (2009) 277–280. https://doi.org/10.1016/j.scriptamat.2008.09.008

[32]   J.F. Wang, R. Li, N.B. Hua, L. Huang and T. Zhang, Ternary Fe–P–C bulk metallic glass with good soft-magnetic and mechanical properties, Scripta Materialia 65 (2011) 536. https://doi.org/10.1016/j.scriptamat.2011.06.020

[33]   W. Yang, H. Liu, L. Xue, J. Li, C. Dun, J. Zhang, Y. Zhao and B. Shen, Magnetic properties of $(Fe_{1-x}Ni_x)_{72}B_{20}Si_4Nb_4$ (x=0.0–0.5) bulk metallic glasses, Journal of Magnetism and Magnetic Materials 35 (2013) 172. https://doi.org/10.1016/j.jmmm.2013.02.004

[34]   X. Li, Y. Zhang, H. Kato, A. Makino and A. Inoue, The effect of Co addition on glassy forming ability and soft magnetic properties of Fe-Si-B-P bulk metallic glass, Key Engineering Materials 508 (2012) 112. https://doi.org/10.4028/www.scientific.net/KEM.508.112

[35]   H.Y. Jung and S. Yi, Nanocrystallization and soft magnetic properties of $Fe_{23}M_6$ (M: C or B) phase in Fe-based bulk metallic glass, Intermetallics 49 (2014) 18. https://doi.org/10.1016/j.intermet.2014.01.005

[36]   H.Y. Jung, M. Stoica, S. Yi, D.H. Kim and J. Eckert, Electrical and magnetic properties of Fe-based bulk metallic glass with minor Co and Ni addition, Journal of Magnetism and Magnetic Materials, 364, (2014) 80. https://doi.org/10.1016/j.jmmm.2014.04.028

[37]   S. Lee, H. Kato, T. Kubota, K. Yubuta, A. Makino and A. Inoue, Excellent thermal stability and bulk glass forming ability of Fe-B-Nb-Y soft magnetic metallic glass, Materials Transactions, 49 (2008) 506. https://doi.org/10.2320/matertrans.MBW200732

[38]   J.H. Zhang, C.T. Chang, A. Wang and B.L. Shen, Development of quaternary Fe-based bulk metallic glasses with high saturation magnetization above 1.6 T, Journal of Non-Crystalline Solids, 358 (2012) 1443. https://doi.org/10.1016/j.jnoncrysol.2012.03.023

[39]   H. Matsumoto, A. Urata, Y. Yamada and A. Inoue, Novel $Fe_{(97-x-y)}P_xB_yNb_2Cr_1$ glassy alloys with high magnetization and low loss characteristics for inductor core materials, IEEE Trans. Magn., 46 (2010) 373. https://doi.org/10.1109/TMAG.2009.2033708

[40]   W. Zhang, C.F. Fang and Y.H. Li, Ferromagnetic Fe-based bulk metallic glasses with high thermoplastic formability, Scripta Materialia 69 (2013) 77. https://doi.org/10.1016/j.scriptamat.2013.03.003

[41]   D.S. Song, J.H. Kim, E. Fleury, W.T. Kim and D.H. Kim, Synthesis of ferromagnetic Fe-based bulk glassy alloys in the Fe–Nb–B–Y system, J. Alloys Compd., 389 (2005) 159. https://doi.org/10.1016/j.jallcom.2004.08.014

[42]   A. Inoue and B.L. Shen, Soft magnetic bulk glassy Fe–B–Si–Nb alloys with high saturation magnetization above 1.5 T, Materials Transactions, 43, (2002) 766. https://doi.org/10.2320/matertrans.43.766

[43]   A. Inoue, B. Shen, H. Koshiba, H. Kato and A.R. Yavari, Cobalt-based bulk glassy alloy with ultrahigh strength and soft magnetic properties, Nature Materials 2 (2003) 661. https://doi.org/10.1038/nmat982

[44]   L. Bie, Q. Li, D. Cao, H.X. Li, J. Zhang, C.T. Chang and Y.F. Sun, Preparation and properties of quaternary CoMoPB bulk metallic glasses, Intermetallics 71 (2016) 7. https://doi.org/10.1016/j.intermet.2015.12.008

[45]   H.J. Sun, Q.K. Man, Y.Q. Dong, B.L. Shen, H. Kimura, A. Makino and A. Inoue, Effect of Nb addition on the glass-forming ability, mechanical and soft-magnetic properties in $(Co_{0.942}Fe_{0.058})_{72-x}Nb_xB_{22.4}Si_{5.6}$ bulk glassy alloys, J. Alloys Comp., 504 (2010) S31. https://doi.org/10.1016/j.jallcom.2010.03.044

[46]   Q.K. Man, A. Inoue, Y.Q. Dong, J. Qiang, C.L. Zhao and B.L. Shen, A new CoFe-based bulk metallic glasses with high thermoplastic forming ability, Scr. Mater., 69 (2013) 553. https://doi.org/10.1016/j.scriptamat.2013.07.002

[47]   A. Zhukov, A. Talaat, J. M. Blanco, M. Ipatov, V. Zhukova, Features of amorphous microwires with spontaneous and induced magnetic bistability, IEEE Transactions on Magnetics 52 (2016) 2002504. https://doi.org/10.1109/TMAG.2015.2512907

[48]   A. Zhukov, A. Talaat, M. Churyukanova, S. Kaloshkin, V. Semenkova, M. Ipatov, J.M. Blanco and V. Zhukova, Engineering of magnetic properties and GMI effect in Co-rich amorphous microwires, Journal of Alloys and Compounds 664 (2016) 235. https://doi.org/10.1016/j.jallcom.2015.12.224

[49]   K.S. Nakayama, Y. Yokoyama, G. Xie, Q.S. Zhang, M. W. Chen, T. Sakurai and A. Inoue, Metallic glass nanowire, Nano Lett. 8 (2008) 516. https://doi.org/10.1021/nl0727462

[50] K.S. Nakayama, Controlled formation and mechanical characterization of metallic glassy nanowires, Adv. Mater. 22 (2010) 872. https://doi.org/10.1002/adma.200902295

[51] S.I. Iwasaki, Perpendicular magnetic recording—Its development and realization, Proc. Jp. Acad. Ser. B Phys. Biol. Sci. 85 (2009) 37. https://doi.org/10.2183/pjab.85.37

[52] K. Nielsch, R.B. Wehrspohn, S.F. Fischer, H. Kronmiller, J. Kirsehner and U. Gosele, Magnetic properties of 100 nm-period nickel nanowire arrays obtained from ordered porous-alumina templates, Mater. Res. Soc. Symp. Proc. 9 (2001) 636.

[53] A.P. Zhukov and B.L. Shtangeev, Cooling-induced phase transition in amorphous CoCrZr alloy, J. Appl. Phys. 73 (1993) 5716. https://doi.org/10.1063/1.353601

[54] A. Inoue, T. Zhang, W. Zhang and A. Takeuchi, Bulk Nd-Fe-Al amorphous alloys with hard magnetic properties, Materials Transactions JIM, 37 (1996) 99.

[55] A. Inoue and T. Zhang, Thermal stability and glass-forming ability of amorphous Nd-Al-TM (TM = Fe, Co, Ni or Cu) alloys, Mater. Sci. Eng. A, 226–228 (1997) 393. https://doi.org/10.1016/S0921-5093(97)80050-0

[56] R. Coehoorn, D.B. de Mooij, J.P.W.B. Duchateau and K.J.H. Buschow, Novel permanent magnetic materials made by rapid quenching, J. de Phys. C 8 (1988) 669. https://doi.org/10.1051/jphyscol:19888304

[57] Q.Y. Wu, D.H. Ping, B.S. Murty, H. Kanekiyo, S. Hirosawa and K. Hono, Influence of heating rate on the microstructure and magnetic properties of $Fe_3B/Nd_2Fe_{14}B$ nanocomposite magnets, Scripta Materialia 45 (2001) 355. https://doi.org/10.1016/S1359-6462(01)01042-9

[58] S. Hirosawa and H. Kanekiyo, Nanostructure and magnetic properties of chromium-doped $Fe_3B-Nd_2Fe_{14}B$ exchange-coupled permanent magnets, Mater. Sci. Eng. A 217/218 (1996) 367. https://doi.org/10.1016/S0921-5093(96)10284-7

[59] A. Inoue, A. Takeuchi, A. Makino and T. Masumoto, Hard magnetic properties of Fe-Nd-B alloys containing intergranular amorphous phase, IEEE Trans. Magn. 31 (1995) 3626. https://doi.org/10.1109/20.489590

[60] A. Manaf, R.A. Buckley and H.A. Davies, New nanocrystalline high-remanence Nd-Fe-B alloys by rapid solidification, J. Magn. Magn. Mater. 128 (1993) 302. https://doi.org/10.1016/0304-8853(93)90475-H

[61] A. Inoue, A. Takeuchi, A. Makino and T. Masumoto, Hard magnetic properties of nanocrystalline Fe-rich Fe-Nd-B alloys prepared by partial crystallization of amorphous phase, Mater. Trans. JIM 36 (1995) 962. https://doi.org/10.2320/matertrans1989.36.962

[62] K. Inomata, T. Sawa and S. Hashimoto, Effect of large boron additions to magnetically hard Fe - Pt alloys, J. Appl. Phys. 64 (1988) 2537. https://doi.org/10.1063/1.341638

[63] W. Zhang, D.V. Louzguine and A. Inoue, Synthesis and magnetic properties of Fe-Pt-B nanocomposite permanent magnets with low Pt concentrations, Applied Physics Letters, 85, (2004) 4998. https://doi.org/10.1063/1.1824172

[64] A.A. Kündig, N. Abe, M. Ohnuma, T. Ohkubo, H. Mamiya and K. Hono, Rapidly solidified $(FePt)_{70}P_{30}$ alloy with high coercivity, Appl. Phys. Lett. 85 (2004) 789. https://doi.org/10.1063/1.1776333

[65] A. Inoue, W. Zhang, T. Tsurui and D.V. Louzguine, Formation, crystallized structure and magnetic properties of Fe-Pt-B amorphous alloys, Materials Transactions 46 (2005) 891. https://doi.org/10.2320/matertrans.46.891

[66] D.V. Louzguine-Luzgin, W. Zhang and A. Inoue, Nanoscale precipitates and phase transformations in a rapidly-solidified Fe-Pt-B amorphous alloy, Journal of Alloys and Compounds, 402 (2005) 78. https://doi.org/10.1016/j.jallcom.2005.03.089

[67] J. Glanz, Making a bigger chill with magnets, Science 279 (1998) 2045. https://doi.org/10.1126/science.279.5359.2045

[68] V.K. Pecharsky and K.A. Jr Gschneidner, The giant magnetocaloric effect in $Gd_5(Si_2Ge_2)$, Phys. Rev. Lett. 78 (1997) 4494. https://doi.org/10.1103/PhysRevLett.78.4494

[69] V. Provenzano, A.J. Shapiro and R.D. Shull, Reduction of hysteresis losses in the magnetic refrigerant $Gd_5Ge_2Si_2$ by the addition of iron, Nature, 429 (2004) 853. https://doi.org/10.1038/nature02657

[70] Q. Luo, B. Zhang, D.Q. Zhao, M.X. Pan, R.J. Wang and W.H. Wang, Aging and stability of cerium-based bulk metallic glass, Appl. Phys. Lett., 88 (2006) 151915. https://doi.org/10.1063/1.2196231

[71] Q. Luo, D.Q. Zhao, M.X. Pan and W.H. Wang, Magnetocaloric effect in Gd-based bulk metallic glasses, Appl. Phys. Lett., 90 (2007) 211903. https://doi.org/10.1063/1.2741120

Chapter 7

Applications related to structural, functional, magnetic, chemical and biological properties

Bulk metallic glasses are suitable for different applications owing to their excellent structural, functional, magnetic, chemical and biological properties. This chapter is devoted to this subject.

Contents

7.1 Structural and functional materials

Bulk metallic glasses have important applications owing to high glass-forming ability, good casting ability, good formability in the supercooled liquid region as well as good mechanical and chemical properties. These materials are used in sportive goods [1,2], watches, electromagnetic wave shields, optical devices, power inductors, mini transformers, micro-geared motor parts, pressure sensors [3], Coriolis flow meters, coating materials, for medical instrument, etc... Metallic glasses are also useful as a strong light cord, strong light plate loaded in compression, elastic hinge, springs, diaphragms, knife blades, thermal shock resistant materials [4]. Various types of connection adapters have been tested for applications to advanced medical equipments, including endoscope and micro-pump, precision optics, and micro-machines.

Owing to large elastic limit of about 2 % the Zr–Cu–Al–Ni glassy alloy diaphragms produced by net-shape casting have been applied to pressure sensors with unique features such as high sensitivity, high-pressure endurance, and small size which cannot be achieved for conventional stainless steel diaphragms.

Even alloys with low glass-forming ability can be utilized when produced by using special techniques. A plate of the $Al_{86}Si_{0.5}Ni_{4.06}Co_{2.94}Y_6La_{0.5}$ amorphous matrix nanocomposite with a maximum thickness of 12 mm and high microhardness of 420 HV was prepared by spray-forming [5]. Nano-scale fcc-Al particles are homogeneously dispersed in the amorphous matrix.

Also, recent finding indicates that the long standing problem of the annealing-induced embrittlement in metallic glasses was solved using crystalline W coatings. The excess volume could be preserved in a $Zr_{50.7}Cu_{28}Ni_9Al_{12.3}$ bulk metallic glass upon annealing as it is not able to sink from BMG into crystalline W. Thus, the strength and ductility of the BMG are preserved [6].

Successful applications of metallic glasses in micromechanical devices was demonstrated by Saotome et al. [7,8]. The deformation characteristics of the $Zr_{65}Al_{7.5}Cu_{27.5}$, $Zr_{55}Al_{10}Cu_{30}Ni_5$ and $Pt_{48.75}Pd_{9.75}Cu_{19.5}P_{22}$ metallic glasses were obtained. The alloys exhibited the Newtonian viscous flow in the supercooled liquid state under low stresses and a superior micro-formability was achieved at the microscopic scale.

Microgeared motor parts were made of the Zr–Cu–Al–Ni and Ni-Zr-Ti-Fe-Co-Cu glassy alloys by a net-shape casting technique (Fig. 7.1). By using these parts, a small geared motor with a diameter of 1.5 mm and a length of 9.9 mm was produced and is applied present. A three-stage micro-geared motor made of the Zr-based glassy alloy gears has high torque which is about 20 times higher than that for a conventional vibration motor with a diameter of 4 mm used in the ordinary cell phones. In addition, the number of rotations to failure is about 34 times larger for the Zr-based glassy alloy gear and 313 times larger for the Ni-based glassy alloy gear than that for conventional tool steel gears [9]. Later, the diameter and length of the micro-geared motor have been reduced to 0.9 mm and 2.0 mm, respectively.

Fig. 7.1. A part of a microgeared motor.

Fig. 7.2. Wear rate as a function of the loads [(a) and (c)] at constant velocity and as a function of the velocity [(b) and (d)] at constant load for the as spun [(a) and (b)] and annealed [(c) and (d)] $Ni_{62}Nb_{38}$ metallic glass samples. Reproduced from [11] with permission of the American Institute of Physics.

When the size of the mechanical components is reduced to the micrometer and sub-micrometer level, the native surface oxide layer begins to play an important role in the contact mechanical behavior of metallic glasses. The nanoscale tribological properties of the $Ni_{62}Nb_{38}$ [10,11] and Pt-Cu-Ni-P [12] metallic glasses have recently been studied (Fig. 7.2). These materials were found to be suitable as wear resistant materials for micro-electro-mechanical applications.

Metallic glasses have high elastic strain limit and fracture toughness resulting from the amorphous structure without structural defects like dislocations and grain boundaries. These attractive functional properties are required for the micro-electro-mechanical systems. The Fe-base metallic glass with good soft magnetic properties is a promising material for mirror actuation [13]. The scanning mirror was fabricated using a Fe-based metallic glass (Fig. 7.3) in order to achieve high performance and mechanical rigidity. The optical tilting angle exceeding ±110 degrees was obtained by driving with Lorentz force and magnetostatic force at 100 mA. The mirror and the gimbal were actuated at different frequency in two axes (Fig. 7.4) and dual axial scanning was possible. The large scanning angle was possible due to good magnetic properties, large elastic strain limit and fracture toughness of the Fe-based metallic glass.

Fig. 7.3. SEM and optical microscope image for the micro-mirror structure. Reproduced from [13] with permission of the Japan Society of Applied Physics.

Fig. 7.4. Scanning patterns of the dual axial scanner. (a) Horizontal scanning pattern by mirror movement at 310 Hz. (b) Vertical scanning pattern by movement at 370 Hz. (c) dual axial scanning pattern by the mirror with dual AC at 300 and 360 Hz. Reproduced from [13] with permission of the Japan Society of Applied Physics.

Ti-based glassy alloy pipes of 2 mm in outer diameter, 0.2 mm in thickness, and 300 mm in length produced by net-shape casting technique exhibit high tensile strength and a large elastic strain of 0.02 and high corrosion resistance which are suitable for application as sensing element in a Coriolis flow meter to measure the mass flow of liquid or gas inside the glassy alloy pipe subjected to reinforced oscillation (Fig. 7.5). The sensitivity of the Coriolis flow meter using the Ti-based glassy alloy pipe was reported to be 28–53 times higher than that for conventional SUS316 pipe [14]. Such a Coriolis flow meter can be used in various industrial fields such as chemical, environmental, semiconductor, and medical science fields.

Fig. 7.5. A Ti-based glassy alloy pipe for the Coriolis flow meter.

Glassy alloys exhibit high hydrogen solubility and significant resistance to embrittlement. Such materials were found to be promising for future applications in separators for fuel cells owing to hydrogen permeation characteristics of Zr-Hf-Ni, Ni-Nb-Zr [15] and other glassy membranes. Hydrogen flow through $(Ni_{0.6}Nb_{0.4})_{100-x}Zr_x$ (x=0 to 40 at.%) metallic glassy membranes increased with temperature and the square-root of hydrogen pressures across the membrane. The hydrogen flow was proportional to the square-root of hydrogen pressure difference. The diffusion of hydrogen through the membrane was found to be a rate-controlling factor for hydrogen permeation. The permeability increased with increasing Zr content. The maximum hydrogen permeability for the $(Ni_{0.6}Nb_{0.4})_{70}Zr_{30}$ amorphous alloy was 1.3×10^{-8} $(mol/m \cdot s \cdot Pa^{1/2})$ at 673 K, higher than that of pure Pd. These permeation characteristics indicate the possibility of future practical use of the melt-spun amorphous alloys as a hydrogen permeable membrane.

Flexible, long metallic glassy nanowires were produced by the drawing process based on deformation above the glass transition temperature [16]. The free end wire with high elasticity was shown to oscillate forming a sine-wave pattern. Metallic glassy nano-wires may have analytical applications ranging from interconnects to sensors.

AMO-beads powders of the Fe–Ni–Cr–Mo–B–Si alloy were produced by water atomization [17] as shot peening balls and fine precise polishing medium. These glassy balls have much longer endurance times compared to the cast steel shots and high speed steel shots, in addition to the important advantage of extremely high resistance to powder explosion reaction in air. Good mechanical properties, such as high Vickers Hardness of HV900, high fracture strength of 3000 MPa and large elastic strain of 0.02 together with high corrosion resistance and a smooth outer surface make AMO-beads a good tool for the shot peening treatment. The AMO-beads are used to make the shot peening balls to high class of automobile high-alloy steel gears, for saving of steel gear weight by about 45 % by improving its fatigue strength by 50-80 % increasing the residual stress on the surface to above 2000 MPa.

The Fe–Cr–Mo–C–B system glassy alloy-coated layer exhibits a better corrosion resistance than SUS304 plates, higher Vickers hardness than hard chromium plates, and better wear resistance than SKD tool steels and FC cast iron. The glassy alloy-coating technique has been applied to surface glassy alloy layers inside solder-melting vessels in continuous solder casting machines. The Fe-Cr-W-C-B glassy alloy coatings are used as heat-resistant materials.

7.2 Magnetic applications

7.2.1 Soft magnetic applications

Fe-based BMGs exhibiting high GFA, low coercive force (H_c), high permeability (μ), high electrical resistivity, low core losses, high mechanical strength, large elastic strain and large supercooled liquid region leading to viscous flow deformation, and the saturation magnetization M_s approaching 1.5 T are utilized as magnetic materials [18]. Fe-based BMGs were commercialized under the trademarks of "Liqualloy", "SENNTIX", "GALOA" [19,20].

Liqualloy powder produced by water atomization can be deformed into flaky shape with a thickness of 2–3 μm to be mixed with a resin. The Liqualloy Fe–Cr–P–C–B–Si system magnetic cores produced by cold consolidation processes of the atomized glassy alloy powders with resin exhibit nearly constant relative permeability in a wide frequency range up to several megahertz, good linear relation between permeability and DC bias field. Low core loss conversion inductors are used in the transformers of notebook PCs. They are also used as power sensing devices with excellent temperature stability. The parts made of Liqualloy are also used as power inductors in CPU and graphics power supply circuits for notebook PC converters. The Liqualloy sheet between metal device part and loop antenna is able to increase transmission distance of magnetic flux line to the inside of loop antenna. A significant increase in the antenna sensitivity was obtained. The Liqualloy cores have better soft magnetic properties than those for the other soft magnetic cores, especially at high frequency. The resulting Liqualloy powder cores have been used as a power inductor in laptop-type personal computers because of higher efficiency and much smaller heat generation.

The Fe–Nb–B–Si, Fe–Nb–B–P and Fe–Nb–B–Si–P systems soft magnetic powder cores named SENNTIX were formed and exhibited low core losses [21]. A soft magnetic SENNTIX II alloy of the Fe–P–B–Nb–Cr system shows the high GFA and low H_c [22]. The use of SENNTIX-II powder cores in the Fe–Nb–Cr–B–P and Fe-Nb-Cr-B-P-Si systems reduced by more than 50% core loss compared to the existing metal powder and reduce thermal energy loss in personal computers, resulting in the extension of battery life time of notebook personal computers.

The high thermal stability of the structure and the high Curie temperature are required for high temperature application. A Co-based BMG can be applied as a material for the position sensor, antenna for radio-controlled watch, solenoid valve and magnetic sensor.

Co–based BMGs exhibit sharp signal, high output voltage and high mechanical strength. However, Co-based BMGs have higher materials cost compared to Fe-based glasses.

Fig. 7.6. Electric field control of ferromagnetism in the $Co_{28.6}Fe_{12.4}Ta_{4.3}B_{8.7}O_{46}$ magnetic semiconductor. (a) A schematic view of the experimental set-up for applying gate voltage on the thin film through a drop of ionic liquid. The thickness of an insulating HfO_2 layer is about 2 nm. (b) Illustration of electrically induced decrease of carrier concentration at a positive gate voltage. (c) Illustration of electrically induced increase of carrier concentration at a negative gate voltage. (d) Variation of M–H curves with different gate voltages measured at 300 K. The thickness of the a-CFTBO MS is 50 nm. Left inset shows the plot of the normalized M/Ms(0) versus magnetic field measured at different gate voltages for the $Co_{28.6}Fe_{12.4}Ta_{4.3}B_{8.7}O_{46}$ magnetic semiconductor with a thickness of 25 nm, Ms(0) denotes the saturation magnetization without any gate voltage. Right inset is an enlargement of the part indicated by a rectangle. Reproduced from [23] with permission of Nature Publishing Group.

301

A room temperature magnetic semiconductor was recently produced from a ferromagnetic metallic glass [23]. Introduction of oxygen into a ferromagnetic metallic glass allowed producing a room temperature p-type $Co_{28.6}Fe_{12.4}Ta_{4.3}B_{8.7}O_{46}$ magnetic semiconductor with a Curie temperature above 600 K with a unique combination of optical, electrical and ferromagnetic properties. p-n heterojunctions and electric field control of the room-temperature ferromagnetism obtained indicate its potential for various electronic devices applications (Fig. 7.6).

7.2.2 Hard magnetic applications

Hard magnetic alloys with large coercive force exceeding 10 kA/m can be used as permanent magnets. High magnetic induction is retained because of a strong resistance to demagnetization, for example, as a result of high anisotropy. Permanent magnets must have large coercive force, high saturation magnetization and the energy product BH_{max}. Ferromagnetic $Nd_{90-x}Fe_xAl_{10}$ bulk amorphous alloys exhibit high coercive force (of about 280 kA/m) [24]. The bulk amorphous $Nd_{70}Fe_{20}Al_{10}$ alloy shows ferromagnetism with the Curie temperature (T_c) of about 600 K. The remanence (B_r) and coercive force for the bulk $Nd_{60}Fe_{30}Al_{10}$ alloy are 0.122 T and 277 kA/m, respectively. The hard magnetic properties for the bulk amorphous alloys are presumably due to the development of homogeneous ferromagnetic clusters with large random magnetic anisotropy.

7.3   Corrosion resistant alloys

High corrosion resistance of some BMGs for example, Zr-based BMGs [25] and Zr-based BMGs containing Nb [26,27], makes them useful for chemical vessels and other applications. The $(Cu_{0.6}Hf_{0.25}Ti_{0.15})_{90}Nb_{10}$ bulk metallic glass composite prepared by copper mold casting containing ductile dendritic Nb-rich crystalline phase dispersed in the glassy matrix also exhibited much higher corrosion resistance in 1N HCl and 3 mass% NaCl solutions than the Nb-free alloy. Nb forms a highly protective surface film with higher chemical stability in HCl and NaCl solutions [28]. The compressive fracture strength and compressive plastic strain of the $(Cu_{0.6}Hf_{0.25}Ti_{0.15})_{90}Nb_{10}$ BMG composite are 2625 MPa and 12.5%, respectively. The ductile dendrites block the shear band propagation and promote generation of multiple shear bands.

The $(Ti_{0.45}Zr_{0.1}Pd_{0.1}Cu_{0.31}Sn_{0.4})_{100-x}M_x$ (M: Ta and Nb) BMG alloys with a large supercooled liquid region (ΔT_x) of 53~67 K do not contain toxic elements like Ni, Al and Be. These BMG alloys have good mechanical properties, remarkable GFA and large supercooled liquid region. The $Ti_{44.1}Zr_{9.8}Pd_{9.8}Cu_{30.38}Sn_{3.92}Nb_2$ bulk glassy alloy has a high potentiality to be applied in as dental biomaterial. Its compressive strength of 1990 MPa exceeds those values of crystalline Ti and conventional Ti6Al4V alloy. The

$Ti_{44.1}Zr_{9.8}Pd_{9.8}Cu_{30.38}Sn_{3.92}Nb_2$ BMG alloy shows high corrosion resistance (Fig. 7.7) with good passivity in a wide range and demonstrates lower passive current densities of approximately $10^{-2}A/m^2$ in 1 mass% lactic acid, $10^{-3}A/m^2$ in PBS (phosphate-buffered saline without calcium and magnesium salts solution containing 8 g/l NaCl, 0.2 g/l KCl, 1.15 g/l Na₂HPO₄, 0.2 g/l KH₂PO₄) and and $10^{-2}A/m^2$ in HBSS (Hank's balance without Ca and Mg or phenol red salts solution containing 8 g/l NaCl, 0.4 g/l KCl, 0.09 g/l Na₂HPO₄·7H2O, 0.06 g/l KH₂PO₄, 0.35 g/l NaHCO₃, 1.0 g/l glucose) aqueous solutions open to air at 310 K compared to those of Ti and the Ti-6Al-4V alloy [29].

Fig. 7.7. Anodic polarization curves of the $(Ti_{0.45}Zr_{0.1}Pd_{0.1}Cu_{0.31}Sn_{0.4})_{100-x}M_x(at.\%)$ alloys in 1 mass% lactic acid, PBS(-) and HBSS at 310K, as indicated. All samples were rods of 3 mm in diameter (the exposed area was 1 cm^2). Reproduced from Ref. [29] with permission of Elsevier.

The tribocorrosion degradation mechanisms of the $Zr_{55}Cu_{30}Ni_5Al_{10}$ and $Zr_{65}Cu_{18}Ni_7Al_{10}$ alloys applicable as potential load-bearing implant materials were studied. The $Zr_{65}Cu_{18}Ni_7Al_{10}$ alloy exhibited higher plasticity but relatively reduced wear resistance during the ball-on-disc tests owing to its lower hardness of HV=5.7±1.8 GPa compared to that (HV=6.8±1.6 GPa) of the $Zr_{55}Cu_{30}Ni_5Al_{10}$ alloy. Vickers hardness in the worn of the wear tracks in dry condition was higher than in the unworn areas indicating possible formation of the intermetallic compounds. The spontaneous passivation of these BMGs in Phosphate Buffer Saline solution was mainly attributed to the formation of a passive film. The galvanic coupling was established between the depassivated wear track and the surrounding passive area as a degradation mechanism for the $Zr_{65}Cu_{18}Ni_7Al_{10}$ BMG in the tribological corrosion environment. The $Zr_{55}Cu_{30}Ni_5Al_{10}$ alloy showed a higher stability against tribological corrosion [30].

7.4 Surface oxides for electronic devices

Oxidation is usually not desired as it reduces surface quality and may induce surface crystallization of the glassy matrix. Fine-grained microstructure consisting of ZrO_2 and Cu particles was found on the surface of $Zr_{70}Cu_{30}$ ribbons stored for a few years at room temperature owing to surface oxidation [31]. An amorphous oxide growing on the metal substrate can be stable up to a certain critical thickness [32]. Usually surface oxides on BMGs are a few nanometers thick [33]. However, a native surface oxide on a Cu-Zr-Al alloy studied recently was found to reach 30 nm [34].

The local current-voltage characteristics (CVCs) of the Ni-Nb/NiNb oxide/Pt system, carried out in contact mode of an atomic force microscope (AFM) showed n-type conductivity in the case of native oxides, while with the artificial one it exhibited p-type conductivity. X-ray photoelectron spectroscopy (XPS) measurements revealed that the structure of natural Ni-Nb oxide film consists of Ni-NbO$_x$ top layer and nickel enriched bottom layer which provides n-type conductivity. In the artificial oxide film Nb is oxidized completely to Nb_2O_5, Ni atoms migrate into bulk Ni-Nb matrix and the electron depletion layer is formed at the Ni-Nb / Nb_2O_5 interface providing p-type conductivity (Fig. 7.8) [35]. An electronic conducting device can be made using this principle.

Fig. 7.8. A family of CVCs recorded for different cantilever loads for native (a) and artificial (b) Ni-Nb oxide films. Load ranges were from 39 to 87 nN for (a) and from 110 to 160 nN for (b). Insets show dependence of maximum current in conductive CVC region on applied cantilever load. Reproduced from Ref. [35] with permission of the American Institute of Physics.

7.5 Cleaning water pollutions

The outstanding efficiency of the $Fe_{73}Si_7B_{17}Nb_3$ metallic glass powders in degrading organic water contaminants was reported. The powders completely decompose the $C_{32}H_{20}N_6Na_4O_{14}S_4$ azo dye in an aqueous solution in short time, about 200 times faster than the conventional Fe powders [36]. MgZn-based metallic glass powders also have excellent functional ability in degrading azo dyes which are typical organic water pollutants. Their azo dye degradation efficiency is about 1000 times higher than that of commercial crystalline Fe powders, and 20 times higher than the Mg-Zn alloy crystalline counterparts. The high Zn content in the amorphous Mg-based alloy enables a greater corrosion resistance in water and higher reaction efficiency with azo dye compared to crystalline Mg. Even under complex environmental conditions, the MgZn-based metallic glass powders retain high reaction efficiency (Fig. 7.9) [37].

Fig. 7.9. (a) The UV absorption spectrum of the azo dye solutions decolored by G-MgZn powders for different time at 30 °C. The absorption peak disappears in 10 minutes. (b) The degradation process by 4 different powders. (c) The comparison of reaction efficiency of the 4 metallic powders. (d) The high-resolution SEM image of the surface morphology of the reacted $Mg_{73}Zn_{21.5}Ca_{5.5}$ glassy powders. The nano-size reaction products distribute homogeneously on the surface of powder particle indicating high reaction activity. Reproduced from Ref. [37] with permission of Nature Publishing Group.

7.6   Materials for medicine

Owing to higher strength, larger elasticity and better corrosion resistance metallic glasses hold promise as a new generation of biomaterials. Both *in vitro* and *in vivo* testing had been done to evaluate their performance as biomaterials. They can be separated to non-biodegradable, typically: Ti-based BMGs, Zr-based BMGs, Fe-based BMGs and biodegradable, typically: Mg-based BMGs, Ca – based BMGs, Zn-based BMGs, Sr-based BMGs [38]. For example, non-biodegradable BMGs belong to Ni-free Ti-Zr-Cu-Pd [39,40] and similar alloys [41,42], Zr-Cu-Fe-Al BMGs [43], Zr-Ti-Cu-Al BMGs [44], these alloys with Nb [45], Cu and Ni-free Zr-based alloys [46]. The $(Ti_{0.45}Zr_{0.1}Pd_{0.1}Cu_{0.31}Sn_{0.4})_{98}Nb_2$ glassy alloy is also a good candidate [29]. The $Fe_{41}Co_7Cr_{15}Mo_{14}C_{15}B_6Y_2$ BMG [47] and similar alloys [48] showed good corrosion resistance in both simulated body fluids.

The $Zr_{56}Al_{16}Co_{28}$ bulk metallic glass suitable for biomedical application exhibited excellent stress corrosion cracking resistance in various simulated body fluids including Hanks', PBS and 0.9% NaCl solutions open to air at 37 °C at a strain rate ranged from 5×10^{-7} to 5×10^{-4} s^{-1} [49].

Porous bulk metallic glassy samples can also be used as bio-implants or damping materials owing to their reduced Young's Modulus compared to monolithic BMGs.

Multi-component BMGs usually contain toxic elements such as Be, Al, Ni et al., thereby degrading such bioengineering applicability. However, the studies on application of BMGs as biomaterials showed that Be-bearing BMGs are, in general, biocompatible with tissue cells [50].

Biodegradable BMGs were developed in the Mg–Zn–Ca [51,52] and Ca-Mg-Zn $(Ca_{65}Mg_{15}Zn_{20})$ [53] systems. Alloying with Yb improved the ductility of Mg-Zn-Ca BMGs [54]. Also in comparison with that of Yb-free control, an *in vitro* cell culture study confirms an improved biocompatibility of these Mg-based BMGs alloyed with Yb. The $Ca_{48}Zn_{30}Mg_{14}Yb_8$ BMG exhibited a low degradation rate without observable hydrogen evolution in Hank's solution [55]. The Ca-Li-Mg-Zn glasses are very light and have good thermoplastic formability close to room temperature but degrade rapidly [56]. Zn-based metallic glasses like $Zn_{38}Ca_{32}Mg_{12}Yb_{18}$ can also be applied [57].

The cellular responses to surfaces of biomaterials are significantly influenced by the surface roughness, topography and chemical composition. The ability to tailor surfaces to control cellular growth may pave the way for the design of novel biomaterials that can be integrated in the human body more effectively. A coating of submicro-nano hierarchically structured $Ti_{30.9}Zr_{15.0}Cu_{21.6}Pd_{30.2}$ metallic glass was created [58]. A submicro-nanostructured morphology on the surface of glassy samples was produced without any electrochemical treatment. This submicro-nano hierarchically structured glassy layer promoted cell attachment and proliferation. This indicates that introduction of a thin layer of submicro-nano structured metallic glass composite could be optimal to modify Ti-based BMGs. The $Ti_{30.9}Zr_{15.0}Cu_{21.6}Pd_{30.2}$ glassy alloy shows much finer submicro-scale surface morphology. Spherical particles ranging from 100 nm to 300 nm in size are dispersed in the relatively flat matrix. At higher magnification, a more detailed structure can be detected. As shown in Fig. 7.10, the spherical particles are further composed of much smaller particles of about 10 nm in size. The matrix also consists of similar particles of 10 nm in size indicating formation of a hierarchical $Ti_{30.9}Zr_{15.0}Cu_{21.6}Pd_{30.2}$ glassy structure.

Fig. 7.10. SEM images of the $Ti_{30.9}Zr_{15.0}Cu_{21.6}Pd_{30.2}$ glassy alloy surface morphologies at different magnification. Reproduced from Ref. [58] with permission of the Royal Society of Chemistry.

According to the binding energy states of the elements involved in the $Ti_{30.9}Zr_{15.0}Cu_{21.6}Pd_{30.2}$ Ti2p spectrum was assigned to be TiO_2 and Ti metallic states. Ti is mainly combined in oxide on the surface layer. Zr3d spectrum indicated that Zr is mainly combined in ZrO_2 oxide. Cu and Pd sustain their metallic states. The XRD pattern obtained for the top thin layer indicated that the TiO_2 and ZrO_2 oxides are amorphous. They protect the underlying metallic alloys and enhance the bioactivity.

Biocompatibility of the $Ti_{30.9}Zr_{15.0}Cu_{21.6}Pd_{30.2}$ glassy alloy NGC-1 with a submicron-scale morphology (Fig. 7.11) was compared with another $Ti_{30.9}Zr_{15.0}Cu_{21.6}Pd_{30.2}$ glassy alloy NGC-2 sample consisting of clusters with a size of about 30 nm that are composed of nanometer-sized particles with a diameter of around 5 nm produced at different sputtering conditions. The NGC-2 shows a lower cell activity compared to that of NGC-1. However, the cell bioactivity of NGC-2 is still comparable with that of the Ti controls (Fig. 7.11).

Fig. 7.11. Cell proliferation on NGC-1, NGC-2, pure Ti and Ti alloy surfaces: (a) live/ dead staining; (b) cell number, magnification: 20x. Asterisk (*) indicates statistical significance when compared with NGC-2 and pure Ti, and Ti alloy (n=6 for all the conditions). Significant differences (unpaired Student's t-test): P < 0.05. Reproduced from Ref. [58] with permission of the Royal Society of Chemistry.

Similar results were obtained for the Zr-Pd metallic nanoglasses [59]. Fine nano-structured $Zr_{62.5}Pd_{37.5}$ metallic glassy alloy exhibit a good thermal stability versus crystallization and good resistance to oxidation in dry air up to 573-673 K. The samples also show a very high corrosion resistance and spontaneous passivation in a simulated body fluid. Spontaneous passivation is achieved and low current density of about 3.5 mA/m^2 maintains up to a very high potential of 1 V. Osteoblast cells cultivation on the nanoglass show its good biocompatibility. The material has potential in biochemistry as a basis for biosensors and artificial tissue engineering.

A combination of magnetostriction and soft magnetic properties makes BMGs ideal candidates for developing high performance biosensors. The magnetoelastic biosensors are based on Fe–B amorphous metallic glasses. The sensors based on amorphous alloys show a very high elastic–magnetic energy conversion efficiency. The detection of Salmonella in liquid using these phage-based ME biosensors was demonstrated. The ME biosensor exhibited high sensitivity and a detection limit higher than 50 CFU/mL [60].

7.7 Catalysts

Similar to the nano-structured metallic glasses discussed above Au-based metallic nano-glasses (Fig. 7.12) with large surface area produced from a BMG forming alloy using magnetron sputtering (Fig. 7.13) [61] open up a new application area of such a material as a catalyst. This material showed a good catalytic activity for the following reaction of dimethylphenylsilane (PhMe$_2$Si-H) with water in the presence of Au-based metallic nanoglass. The reaction proceeded at room temperature for 24 h and the desired dimethylphenylsilanol was obtained (Table 7.1).

Fig. 7.12. SEM image of the as-deposited Au-based nanostructured metallic glass.

Fig. 7.13. (a) Schematic diagram for synthesis of the Au-based nanostructured metallic glassy samples. (b) The steps for preparation of powder targets. The right inset of (b) is the surface morphology of the powder targets. Reproduced from [61] with permission of Elsevier.

Table 7.1. Material-catalyzed oxidation of dimethylphenylsilane[a]. Reproduced from [61] with permission of Elsevier.

Entry	Route	Yield[b]
1	NA3	93
2	Reuse 1	94
3	Reuse 2	95
4	Reuse 3	100
5	Reuse 4	98

[a] Reactions were carried out using 1 (1.0 mmol), H_2O (0.1 mL), and NA3 (0.87 mol%) in 1.5 mL of acetone at RT for 24 h. [b] Isolated yield.

After electrolytic polishing of native surface oxide the Zr-Pd nanoglassy sample with an average grain size of as small as 6 nm (see Fig. 3.10) also exhibit catalytic activity in Suzuki-coupling reaction [59].

$Pt_{57.5}Cu_{14.7}Ni_{5.3}P_{22.5}$ bulk metallic glass nanowires, are applicable in electrochemical devices. High surface area nanowires exhibited good catalysts with high conductivity and activity for methanol and ethanol oxidation. After 1000 cycles, these nanowires maintain 96% of their performance exceeding that of a conventional Pt/C catalyst [62].

7.8 Final remarks and future prospects

Due to high cost of the main component palladium for a long time the initially-discovered BMGs were not of much interest to the scientists and engineers until they have been produced in alloys based on the industrial metals such as iron, copper, magnesium and titanium, which has opened many opportunities for their application. Bulk metallic glasses having high strength, hardness, wear resistance, large elastic deformation and high resistance to corrosion are readily applicable as structural and functional materials. At the same time the fundamental not fully solved scientific problems are: description of the structure of metallic glasses, processes of their plastic deformation and glass-transition on cooling, while the technical challenge is related to further increase in the tensile plasticity and impact fracture toughness of BMGs and their composites. Based on the full set of the results obtained up to date one can expect further development of the field and significant progress in it in the near future.

References

[1]   H. Kakiuchi, A. Inoue, M. Onuki, Y. Takano and T. Yamaguchi, Application of Zr-based bulk glassy alloys to golf clubs, Mater. Trans., 42 (2001) 678. https://doi.org/10.2320/matertrans.42.678

[2]   W.L. Johnson, Bulk amorphous metal–An emerging engineering material, JOM, 54 (2002), 40. https://doi.org/10.1007/BF02822619

[3]   A. Inoue and N. Nishiyama, New bulk metallic glasses for applications as magnetic-sensing, chemical, and structural materials, MRS Bull., 32 (2007) 651. https://doi.org/10.1557/mrs2007.128

[4]   A.I. Salimon, M.F. Ashby, Y. Bréchet and A.L. Greer, Bulk metallic glasses: what are they good for? Materials Science and Engineering: A 375–377 (2004) 385. https://doi.org/10.1016/j.msea.2003.10.167

[5]   L. Zhuo, B. Yang, H. Wang and T. Zhang, Spray formed Al-based amorphous matrix nanocomposite plate, J. Alloys Compd., 509 (2011) L169. https://doi.org/10.1016/j.jallcom.2011.02.125

[6]   Z.Q. Chen, L. Huang, F. Wang, P. Huang, T.J. Lu and K.W. Xu, Suppression of annealing-induced embrittlement in bulk metallic glass by surface crystalline coating, Materials and Design 109, (2016), 179. https://doi.org/10.1016/j.matdes.2016.07.069

[7]   Y. Saotome, S. Miwa, T. Zhang and A. Inoue, The micro-formability of Zr-based amorphous alloys in the supercooled liquid state and their application to micro-dies, Journal of Materials Processing Technology, 113, (2001) 64. https://doi.org/10.1016/S0924-0136(01)00605-7

[8]   Y. Saotome, K. Imai, S. Shioda, S. Shimizu, T. Zhang and A. Inoue, The micro-nanoformability of Pt-based metallic glass and the nanoforming of three-dimensional structures, Intermetallics 10, (2002) 1241. https://doi.org/10.1016/S0966-9795(02)00135-8

[9]   A. Inoue, B.L. Shen and A. Takeuchi, Developments and applications of bulk glassy alloys in late transition metal base system, Mater. Trans. 47 (2006) 1275. https://doi.org/10.2320/matertrans.47.1275

[10]   A. Caron, C.L. Qin, L. Gu, S. Gonzalez, A. Shluger, H.-J. Fecht, D.V. Louzguine-Luzgin and A. Inoue, Structure and nano-mechanical characteristics of surface oxide layers on a metallic glass, Nanotechnology, 22 (2011) 095704. https://doi.org/10.1088/0957-4484/22/9/095704

[11]   A. Caron, P. Sharma, A. Shluger, H.-J. Fecht, D.V. Louzguine-Luzgin and A. Inoue, Effect of surface oxidation on the nm-scale wear behavior of a metallic glass, Journal of Applied Physics, 109 (2011) 083515. https://doi.org/10.1063/1.3573778

[12]   A. Caron, D.V. Louzguine-Luzgin and R. Bennewitz, Structure vs chemistry: friction and wear of Pt-based metallic surfaces, ACS Appl. Mater. Interfaces, 5 (2013) 11341. https://doi.org/10.1021/am403564a

[13]   J.-W. Lee, Y.-C. Lin, N. Chen, D.V. Louzguine, M. Esashi and T. Gessner, Development of the large scanning mirror using Fe-based metallic glass ribbon, Japanese Journal of Applied Physics, 50 (2011) 087301. https://doi.org/10.7567/JJAP.50.087301

[14]   C. Ma, N. Nishiyama and A. Inoue, Fabrication and high performance characteristics of metallic glassy alloy tubes, J. Metastab. Nanocryst., 24-25 (2005) 515. https://doi.org/10.4028/www.scientific.net/JMNM.24-25.515

[15]   S. Yamaura, Y. Shimpo, H. Okouchi, M. Nishida, O. Kajita, H.M. Kimura and A. Inoue, Hydrogen permeation characteristics of melt-spun Ni-Nb-Zr amorphous alloy membranes, Materials Transactions, 44 (2003) 1885. https://doi.org/10.2320/matertrans.44.1885

[16]   K.S. Nakayama, Y. Yokoyama, T. Ono, M. W. Chen, K. Akiyama, T. Sakurai and A. Inoue, Controlled formation and mechanical characterization of metallic glassy nanowires, Advanced Materials, 22 (2010) 863. https://doi.org/10.1002/adma.200902295

[17]   A. Inoue, I. Yoshii, H. Kimura, K. Okumura and J. Kurosaki. Enhanced shot peening effect for steels by using Fe-based glassy alloy shots, Mater. Trans. 44 (2003) 2391. https://doi.org/10.2320/matertrans.44.2391

[18]   A. Inoue, F.L. Kong, Q.K. Man, B.L. Shen, R.W. Li and F. Al-Marzouki, Development and applications of Fe- and Co-based bulk glassy alloys and their prospects, Journal of Alloys and Compounds 615 (2014) S2. https://doi.org/10.1016/j.jallcom.2013.11.122

[19]   H. Koshiba, Y. Naito, T. Mizushima and A. Inoue, Development of the Fe-based glassy alloy "LiqualloyTM" and its application to powder core, Materia Japan, 47 (2008) 39 (in Japanese). https://doi.org/10.2320/materia.47.39

[20]   H. Matsumoto, A. Urata, Y. Yamada and A. Inoue, Novel $Fe_{(97-x-y)}P_xB_yNb_2Cr_1$ glassy alloys with high magnetization and low loss characteristics for inductor core materials, IEEE Trans. Magn., 46 (2010) 373. https://doi.org/10.1109/TMAG.2009.2033708

[21]   H. Matsumoto, A. Urata, Y. Yamada and S. Yoshida, "MPCG" for large-current using ultra low loss dust material "SENNTIX" formed by one-piece construction, Nec Technical Journal, 2 (2007) 66 (in Japanese).

[22]   H. Matsumoto, A. Urata, Y. Yamada and A. Inoue, FePBNbCr soft-magnetic glassy alloys with low loss characteristics for inductor cores, J. Alloys Comp., 504 (2010) S139. https://doi.org/10.1016/j.jallcom.2010.03.029

[23]   W.J. Liu, H.X. Zhang, J.A. Shi, Z.C. Wang, C. Song, X.G. Wang, S. Lu, X.G. Zhou, L. Gu, D.V. Louzguine-Luzgin, M.W. Chen, K.F. Yao and N. Chen, A

room-temperature magnetic semiconductor from a ferromagnetic metallic glass, Nature Communications 7 (2016) 13497. https://doi.org/10.1038/ncomms13497

[24] A. Inoue and T. Zhang, Thermal stability and glass-forming ability of amorphous Nd-Al-TM (TM = Fe, Co, Ni or Cu) alloys, Mater. Sci. Eng. A, 226–228 (1997) 393. https://doi.org/10.1016/S0921-5093(97)80050-0

[25] S. Hiromoto, K. Asami, A.P. Tsai and T. Hanawa, Surface characterization of amorphous Zr-Al-(Ni, Cu) alloys immersed in cell-culture medium, Mater. Trans. JIM 43 (2002) 261-266. https://doi.org/10.2320/matertrans.43.261

[26] S. Pang, T. Zhang, H. Kimura and K. Asami, A. Inoue, Corrosion behavior of Zr-(Nb-)Al-Ni-Cu glassy alloys, Mater. Trans. JIM 41 (2000) 1490. https://doi.org/10.2320/matertrans1989.41.1490

[27] V.R. Raju, U. Kühn, U. Wolff, F. Schneider, J. Eckert, R. Reiche and A. Gebert, Corrosion behaviour of Zr-based bulk glass-forming alloys containing Nb or Ti, Mater. Lett., 57 (2002) 173. https://doi.org/10.1016/S0167-577X(02)00725-5

[28] C.–L. Qin, W. Zhang, K. Amiya, K. Asami and A. Inoue, Mechanical properties and corrosion behavior of $(Cu_{0.6}Hf_{0.25}Ti_{0.15})_{90}Nb_{10}$ bulk metallic glass composites, Materials Science and Engineering: A 449–451 (2007) 230. https://doi.org/10.1016/j.msea.2006.02.244

[29] J.J. Oak, D.V. Louzguine-Luzgin and A. Inoue, Investigation of glass-forming ability, deformation and corrosion behavior of Ni-free Ti-based BMG alloys designed for application as dental implants, Materials Science and Engineering: C 29 (2009) 322. https://doi.org/10.1016/j.msec.2008.07.009

[30] G.-H. Zhao, R.E. Aune, H. Mao and N. Espallargas, Degradation of Zr-based bulk metallic glasses used in load-bearing implants: A tribocorrosion appraisal, Journal of the Mechanical Behavior of Biomedical Materials 60 (2016) 56. https://doi.org/10.1016/j.jmbbm.2015.12.024

[31] U. Koster and L. Jastrow, Oxidation of Zr-based metallic glasses and nanocrystalline alloys, Mater. Sci. Eng. A 449 (2007) 57. https://doi.org/10.1016/j.msea.2006.02.316

[32] F. Reichel, L.P.H. Jeurgens and E.J. Mittemeijer, The thermodynamic stability of amorphous oxide overgrowths on metals, Acta Materialia 56 (2008) 659. https://doi.org/10.1016/j.actamat.2007.10.023

[33] S.K. Sharma, T. Strunskus, H. Ladebusch, V. Zaporojtchenko and F. Faupel, XPS study of the initial oxidation of the bulk metallic glass $Zr_{46.75}Ti_{8.25}Cu_{7.5}Ni_{10}Be_{27.5}$, J. Mater. Sci. 43 (2008) 5495. https://doi.org/10.1007/s10853-008-2834-4

[34] D.V. Louzguine-Luzgin, C.L. Chen, L.Y. Lin, Z.C. Wang, S.V. Ketov, M.J. Miyama, A.S. Trifonov, A.V. Lubenchenko and Y. Ikuhara, Bulk metallic glassy surface native oxide: Its atomic structure, growth rate and electrical properties, Acta Materialia, 97 (2015) 282. https://doi.org/10.1016/j.actamat.2015.06.039

[35] A.S. Trifonov, A.V. Lubenchenko, V.I. Polkin, A.B. Pavolotsky, S.V. Ketov and D.V. Louzguine-Luzgin, Difference in charge transport properties of Ni-Nb thin films with native and artificial oxide, Journal of Applied Physics, 117 (2015) 125704. https://doi.org/10.1063/1.4915935

[36] J.-Q. Wang, Y.-H. Liu, M.-W. Chen, G.-Q. Xie, D.V. Louzguine-Luzgin, A. Inoue and J. H. Perepezko, Rapid degradation of azo dye by Fe-based metallic glass powder, Advanced Functional Materials, 22 (2012) 2567. https://doi.org/10.1002/adfm.201103015

[37] J.-Q. Wang, Y.-H. Liu, M.-W. Chen, D.V. Louzguine-Luzgin, A. Inoue and J.H. Perepezko, Excellent capability in degrading azo dyes by MgZn-based metallic glass powders, Scientific Reports 2 (2012) 418. https://doi.org/10.1038/srep00418

[38] H.F. Li and Y.F. Zheng, Recent advances in bulk metallic glasses for biomedical applications, Acta Biomaterialia 36 (2016) 1. https://doi.org/10.1016/j.actbio.2016.03.047

[39] S.L. Zhu, X.M. Wang, F.X. Qin and A. Inoue, A new Ti-based bulk glassy alloy with potential for biomedical application, Mater. Sci. Eng. A, 459 (2007) 233. https://doi.org/10.1016/j.msea.2007.01.044

[40] J.-J. Oak, D.V. Louzguine-Luzgin and A. Inoue, Fabrication of Ni-free Ti-based bulk-metallic glassy alloy having potential for application as biomaterial, and investigation of its mechanical properties, corrosion, and crystallization behavior, J. Mater. Res., 22 (2007) 1346. https://doi.org/10.1557/jmr.2007.0154

[41] Y.B. Wang, H.F. Li, Y. Cheng, Y.F. Zheng and L.Q. Ruan, In vitro and in vivo studies on Ti-based bulk metallic glass as potential dental implant material, Mater. Sci. Eng. C, 33 (2013) 3489. https://doi.org/10.1016/j.msec.2013.04.038

[42] S.J. Pang, Y. Liu, H.F. Li, L.L. Sun, Y. Li and T. Zhang, New Ti-based Ti-Cu-Zr-Fe-Sn-Si-Ag bulk metallic glass for biomedical applications, J. Alloys Compd., 625 (2015) 323. https://doi.org/10.1016/j.jallcom.2014.07.021

[43] L. Huang, C. Pu, R.K. Fisher, D.J.H. Mountain, Y. Gao and P.K. Liaw, A Zr-based bulk metallic glass for future stent applications: materials properties, finite element modeling, and in vitro human vascular cell response, Acta Biomater., 25 (2015) 356. https://doi.org/10.1016/j.actbio.2015.07.012

[44] J. Li, L.L. Shi, Z.D. Zhu, Q. He, H.J. Ai and J. Xu, $Zr_{61}Ti_2Cu_{25}Al_{12}$ metallic glass for potential use in dental implants: biocompatibility assessment by in vitro cellular responses, Mater. Sci. Eng. C, 33 (2013) 2113. https://doi.org/10.1016/j.msec.2013.01.033

[45] L. Liu, C.L. Qiu, C.Y. Huang, Y. Yu, H. Huang and S.M. Zhang, Biocompatibility of Ni-free Zr-based bulk metallic glasses, Intermetallics, 17 (2009) 235. https://doi.org/10.1016/j.intermet.2008.07.022

[46] Z. Liu, K.C. Chan and L. Liu, Development of Ni- and Cu-Free Zr-based bulk metallic glasses for biomedical applications, Mater. Trans., 52 (2011) 61. https://doi.org/10.2320/matertrans.M2010068

[47] Y.B. Wang, H.F. Li, Y. Cheng, S.C. Wei and Y.F. Zheng, Corrosion performances of a Nickel-free Fe-based bulk metallic glass in simulated body fluids, Electrochem. Commun., 11 (2009) 2187. https://doi.org/10.1016/j.elecom.2009.09.027

[48] Y.B. Wang, H.F. Li, Y.F. Zheng, M. Li, Corrosion performances in simulated body fluids and cytotoxicity evaluation of Fe-based bulk metallic glasses, Mater. Sci. Eng. C, 32 (2012) 599. https://doi.org/10.1016/j.msec.2011.12.018

[49] A. Kawashima, T. Wada, K. Ohmura, G. Xie and A. Inoue, A Ni- and Cu-free Zr-based bulk metallic glass with excellent resistance to stress corrosion cracking in simulated body fluids, Mater. Sci. Eng. A, 542 (2012) 140. https://doi.org/10.1016/j.msea.2012.02.047

[50] Y.B. Wang, Y.F. Zheng, S.C. Wei and M. Li, In vitro study on Zr-based bulk metallic glasses as potential biomaterials, J. Biomed. Mater. Res. B, 96B (2011) 34. https://doi.org/10.1002/jbm.b.31725

[51] B. Zberg, P.J. Uggowitzer and J.F. Loffler, MgZnCa glasses without clinically observable hydrogen evolution for biodegradable implants, Nature Materials, 8 (2009) 887. https://doi.org/10.1038/nmat2542

[52] X.N. Gu, Y.F. Zheng, S.P. Zhong, T.F. Xi, J.Q. Wang, W.H. Wang, Corrosion of, and cellular responses to Mg-Zn-Ca bulk metallic glasses, Biomaterials, 31 (2010) 1093. https://doi.org/10.1038/nmat2542

[53] Y.B. Wang, X.H. Xie, H.F. Li, X.L. Wang, M.Z. Zhao, E.W. Zhang, Y.J. Bai, Y.F. Zheng and L. Qin, Biodegradable CaMgZn bulk metallic glass for potential skeletal application, Acta Biomaterialia, 7 (2011) 3196. https://doi.org/10.1016/j.actbio.2011.04.027

[54] H.-J. Yu, J.-Q. Wang, X.-T. Shi, D.V. Louzguine-Luzgin, H.-K. Wu and J.H. Perepezko, Ductile biodegradable Mg-based metallic glasses with excellent biocompatibility, Advanced Functional Materials, 23 (2013) 4793. https://doi.org/10.1002/adfm.201203738

[55] W. Jiao, H.F. Li, K. Zhao, H.Y. Bai, Y.B. Wang, Y.F. Zheng and W.H. Wang, Development of CaZn based glassy alloys as potential biodegradable bone graft substitute, Journal of Non-Crystalline Solids, 357 (2011) 3830. https://doi.org/10.1016/j.jnoncrysol.2011.08.003

[56] J.Fu, L.J.Qiang, W.X.Feng, L.Kun, Z.Bo, Z.H.Yang, B.M.Xiang, M.X. Pan and W.H. Wang, Glassy metallic plastics, Science China Physics, Mechanics and Astronomy, 53, (2010), 409. https://doi.org/10.1007/s11433-010-0138-6

[57] W. Jiao, K. Zhao, X.K. Xi, D.Q. Zhao, M.X. Pan and W.H. Wang, Zinc-based bulk metallic glasses, Journal of Non-Crystalline Solids, 356 (2010) 1867. https://doi.org/10.1016/j.jnoncrysol.2010.07.017

[58] N. Chen, X. Shi, R. Witte, K.S. Nakayama, K. Ohmura, H.K. Wu, A. Takeuchi, H. Hahn, M. Esashi, H. Gleiter, A. Inoue and D.V. Louzguine, A novel Ti-based nanoglass composite with submicron–nanometer-sized hierarchical structures to modulate osteoblast behaviors, J. Mater. Chem. B, 1 (2013) 2568. https://doi.org/10.1039/c3tb20153h

[59] S.V. Ketov, X.T. Shi, G.Q. Xie, R. Kumashiro, A.Yu. Churyumov, A.I. Bazlov, N. Chen, Y. Ishikawa, N. Asao, H.K. Wu and D.V. Louzguine-Luzgin, Nanostructured Zr-Pd metallic glass thin film for biochemical applications, Scientific Reports, 5 (2015) 7799. https://doi.org/10.1038/srep07799

[60] S. Li, S. Horikawa, M.K, Park, Y. Chai, V.J. Vodyanoy and B.A. Chin, Amorphous metallic glass biosensors, Intermetallics 30 (2012) 80. https://doi.org/10.1016/j.intermet.2012.03.030

[61] N. Chen, R. Frank, N. Asao, D.V. Louzguine-Luzgin, P. Sharma, J.Q. Wang, G.Q. Xie, Y. Ishikawa, N. Hatakeyama, Y.C. Lin, M. Esashi, Y. Yamamoto and A. Inoue, Formation and properties of Au-based nanograined metallic glasses, Acta Materialia 59, (2011) 6433. https://doi.org/10.1016/j.actamat.2011.07.007

[62]   M. Carmo, R.C. Sekol, S. Ding, G. Kumar, J. Schroers and A.D. Taylor, Bulk metallic glass nanowire architecture for electrochemical applications, ACS Nano, 5 (2011) 2979. https://doi.org/10.1021/nn200033c

Keywords

About the Author

Dmitri V. LOUZGUINE

Address: WPI Advanced Institute for Materials Research, Tohoku University, 2-1-1 Katahira, Aoba-Ku, Sendai, 980-8577, Japan.

Since 2007 he is the Professor and Principal Investigator at WPI Advanced Institute for Materials Research, Tohoku University, Japan.

RESEARCH ACTIVITIES
Investigation of the formation mechanisms, structure, phase transformations, deformation behavior, mechanical and physical properties of metallic glasses, nanostructured glassy-crystalline/quasicrystalline materials, composite materials having a mixed structure containing nanoparticles embedded in the glassy matrix and high-strength crystalline alloys. Long-term experience in various types of structural analysis techniques, in particular, Transmission Electron Microscopy (TEM) of nanostructured and glassy materials including nanobeam diffraction and high-resolution TEM imaging. X-ray diffraction analysis including creation of the radial distribution functions from Synchrotron-radiation XRD data.
Computer-simulation to materials science (phase transformations, crystallography).
In general, the area of expertise is Physical Metallurgy of metals and alloys which establishes the relationships between the structure of materials and their properties. This subject includes the following topics: solidification, micro- and nano-structure, crystallography, phase transformations, processing, mechanical properties and deformation behavior of metallic materials.

www.ingramcontent.com/pod-product-compliance
Lightning Source LLC
Chambersburg PA
CBHW071323210326
41597CB00015B/1333